Toward a Global History of Soil

Agriculture and the Making of Sciences 1100–1700

TEXTS, PRACTICES, AND KNOWLEDGE TRANSMISSION IN ASIA

VOLUME 1

Series editors

Dagmar Schäfer, XU Chun, Bethany J. Walker
and Aleksandar Shopov

The titles published in this series are listed at *brill.com/amos*

Toward a Global History of Soil

Sciences, Practices, and Materialities, 1300–1750

Edited by

Justin Niermeier-Dohoney
Aleksandar Shopov

BRILL

LEIDEN | BOSTON

Cover illustration: The cover image refers to a story that stresses the importance of the quest (talab). The painting is also a portrayal of activities related to agriculture and soil, from the plowing of a field to the weighing of harvested melons.

Farid al-Din 'Attar, *Shaikh Mahneh and the Villager*, fol. 49r from *Mantiq al-Tayr*, 1487, painting, 7 3/4 × 5 3/4 (19.7 × 14.6 cm), The Metropolitan Museum of Art, New York, https://www.metmuseum.org/art/collection/search/451733.

The Library of Congress Cataloging-in-Publication Data is available online at https://catalog.loc.gov
LC record available at https://lccn.loc.gov/2025034310

Typeface for the Latin, Greek, and Cyrillic scripts: "Brill". See and download: brill.com/brill-typeface.

ISSN 2772-9834
ISBN 978-90-04-72748-9 (hardback)
ISBN 978-90-04-74848-4 (e-book)
DOI 10.1163/9789004748484

This book is printed on acid-free paper and produced in a sustainable manner.

Contents

PART 3
Governing the Soil: Taxonomy, Expertise, and the State

PART 4
Soil, Specialization, and Experimental Culture

Acknowledgments

This book came out of the highly stimulating discussions in the working group "Agriculture and the Making of Sciences (1100–1700)," in the Department on Artifacts, Action, Knowledge (AAK) of the Max Planck Institute for the History of Science (MPIWG). We are deeply indebted to Dagmar Schäfer and her generous dedication of funds from the Gottfried Wilhelm Leibniz Prize she was awarded in 2020, which enabled numerous predoctoral, postdoctoral, and visiting fellowships. Without Dagmar's initial suggestions, prompts, and ongoing commitment, this volume would never have come to light.

Through our affiliation with Department AAK, we had the opportunity of working with Chun Xu, leader of the working group "Agriculture and the Making of Sciences (1100–1700)," and all of the participants: Bethany Walker, Himmet Taşkömür, Heba Mahmoud Saad Abdelnaby, Deniz Karakaş, Tarek Sabraa, Riaz Howey, Anthony Quickel, Mengxi Zhao, Sarah Chen Huang, Anil Paralkar, Gianamar Giovannetti-Singh, Christoph Hess, Kaveh Yazdani, Xinyi Wen, and Xinchang Li. They each played a formative role in shaping ideas and approaches, through their active participation in our meetings, and several contributed chapters to this volume. Many thanks also to the moderators who facilitated at the 2021 conference during which the chapters in this volume were first presented as working papers: Antonio Clericuzio, Shirine Hamadeh, Annapurna Mamidipudi, and Mackenzie Cooley provided excellent commentary and thought-provoking discussions in the early stages of the project.

One of the advantages offered by the MPIWG is their excellent academic support infrastructure. We would like to express our special gratitude to the publications managers in Department AAK, Gina Partridge-Grzimek and Melanie Luise Glienke, supported by Benjamin Carter, for their tireless efforts in editing the entire volume, and to the graduate and student assistants in the department, Fatma Nur Özdemir, Spencer Forbes, Lennart Holst, and Rebecca Schmitt. In addition, we would like to acknowledge the vital assistance in acquiring books and other library materials provided by Esther Chen, head of the MPIWG Library, and Cathleen Paethe, subject librarian for Chinese studies.

Thanks are also due to Paola La Battaglia and Chunyan Shu at De Gruyter Brill, and the two anonymous reviewers for their invaluable comments and suggestions.

Finally, we would like to extend our heartfelt thanks to the members of our families: Marisa Mandabach and Kostadin Shopov, and Carly, Wyatt, and Emerson Niermeier-Dohoney, for their love, patience, and support along the way.

Illustrations and Tables

Figures

Maps

Tables

Notes on Contributors

Heba Mahmoud Saad Abdelnaby
is professor of Islamic archeology and civilization at Alexandria University and the international representative of the Historians of Islamic Art Association. She received her PhD from Alexandria University, was a Fulbright Scholar at Mary Baldwin University, VA, and a visiting scholar at George Washington University, DC, and the Max Planck Institute for the History of Science, Berlin. Her research focuses on the Islamic art, architecture, civilization, and heritage of Egypt, especially during the Mamluk period. During the past 5 years, she has been working on the environmental history of Egypt. Her most recent book is titled *Birds in the Mamluk Period* (*Al-ṭuyūr fī al-ʿaṣr al-Mamlūkī*, 2021). Throughout 2023–2025, she was the cultural attaché and director of the Egyptian Cultural Center in Washington, DC, and Rabat.

Dulce Freire
is professor at the Faculty of Economics and researcher at the Centre for Interdisciplinary Studies, both at the University of Coimbra. She has been researching rural and agrarian history within Portuguese and Iberian contexts since the 1990s and received her PhD from Universidade NOVA de Lisboa. In recent years she has been coordinating various scientific projects related to changes in agriculture, food, agrobiodiversity, society, and public policy since the sixteenth century. She is currently the principal investigator of the ERC StG project ReSEED—Rescuing Seeds' Heritage: Engaging in a New Framework of Agriculture and Innovation since the 18th Century.

Alberto González Remuiñán
holds a PhD (2019) in contemporary history from the University of Santiago de Compostela. His research interests include the rural, social, economic, and agricultural histories of the eighteenth and nineteenth centuries of the Iberian Peninsula, especially the dynamics of the local peasant communities. He is currently a postdoctoral researcher in the ERC StG project project ReSEED—Rescuing Seeds' Heritage: Engaging in a New Framework of Agriculture and Innovation since the 18th Century, University of Coimbra, and member of the Centre of Interdisciplinary Studies at the University of Coimbra.

Jörg Henning Hüsemann
is a lecturer for Chinese culture and history at the Institute of East Asian Studies (Chinese Studies) at Leipzig University. In 2012 he received his PhD from Hamburg University for a study on a medieval geographical treatise,

published as a monograph in 2017. Currently he is working on his second book on the history of fertilizers in early modern China.

Deniz Karakaş
is an art and architectural historian and holds a PhD from Binghamton University (SUNY). Between 2013–2021, Deniz worked as a postdoctoral fellow and a visiting lecturer at several liberal arts colleges and research universities in the US and Turkey. Currently, she is working on her first book manuscript about water resources management and use in Istanbul during the seventeenth and eighteenth centuries, which proposes a new way of writing architectural history, with water as its core site. She is also in the early stages of integrating geospatial data to analyze the early modern visual representations of urban waters and subterranean settings in the Ottoman Empire.

Zubair Khalid
teaches at the Shaikh-Ul-Aalam Centre for Multidisciplinary Studies, University of Kashmir, Srinagar. His PhD from the Centre for Historical Studies, Jawaharlal Nehru University, titled "State and Polity in the Textual and Material Culture of Medieval Kashmir (1339–1586)," focused on the region's textual and material cultures under the rule of Kashmiri sultans between the fourteenth and sixteenth centuries. His interests include the history of Islam and Sufism, and the intellectual and material networks of Kashmir.

Monika Kozłowska-Szyc
is a historian and PhD candidate at the Faculty of History, University of Bialystok. Her research interests include socio-economic history and environmental history. Currently she is preparing a dissertation that explores the impact of climate change on the rural economy of pre-industrial Poland.

Sarah Newman
is an anthropological archaeologist and assistant professor of anthropology and social sciences at the University of Chicago. She holds a PhD from Brown University. Her research explores the deep histories of environmental issues, including waste and reuse, landscape transformations, and human-animal relationships. Her most recent book is *Unmaking Waste: New Histories of Old Things* (University of Chicago Press, 2023).

Justin Niermeier-Dohoney
is assistant professor of the history of science and medicine at the Florida Institute of Technology. He received his PhD from the University of Chicago in 2018, and has held postdoctoral fellowships in the Social Sciences Division,

the History Department, and the College at the University of Chicago and at the Max Planck Institute for the History of Science, Berlin. He has also taught as an adjunct professor of humanities and social sciences in the Honors Program at Indiana University Southeast. His research focuses on the history of early modern science in Britain and the Atlantic world with concentration on alchemy, agriculture, climate, and the environment.

Nicolas Roth
holds a PhD in South Asian studies and most recently served as the Visual Resources Librarian for Islamic Art and Architecture at the Fine Arts Library of Harvard University. His research focuses on the history of horticulture and horticultural practices and their reflections in literature and the visual arts, and in his free time he can usually be found tending to his garden.

Aleksandar Shopov
is assistant professor of early modern Ottoman history at Binghamton University (SUNY). His work focuses on the history of science and technology, and environmental history between 1300 and 1700. He received his PhD from Harvard University in 2016, and has held fellowships at the Max Planck Institute for the History of Science, Berlin; Dumbarton Oaks Library; and the Rachel Carson Center for Environment and Society. His publications investigate early modern Ottoman knowledge and practices related to plants, including topics such as flower breeding, grafting, urban farming, rice growing, balsam oil extraction, and agricultural manuscripts. He is currently writing a monograph on urban agriculture in early modern Istanbul.

Himmet Taşkömür
is a senior faculty member in the Department of Near Eastern Languages and Civilizations at Harvard University. He works on Ottoman legal and intellectual history; the history of Ottoman institutions; Ottoman economic and social history; book history; and comparative Ottoman, Arabic, and Persian literatures and philology. His recent publications include "Books on Islamic Jurisprudence, Schools of Law, and Biographies of Imams from the Hanafi School," and the transliteration of MS Török F. 59, prepared with Hesna Ergün Taşkömür, both in *Treasures of Knowledge: An Inventory of the Ottoman Palace Library* (1502/3–1503/4), ed. Gülru Necipoğlu, Cemal Kafadar, and Cornell H. Fleischer, 2 vols. (Leiden: Brill, 2019).

Introduction: Toward a Global History of Soil

Sciences, Practices, and Materialities, 1300–1750

Justin Niermeier-Dohoney and Aleksandar Shopov

1 The Grounded Cosmogonies of Soil

Shared by the Haudenosaunee people living in what European settlers renamed the state of New York, the Skywoman story centers on the making of soil. It relates that, when Skywoman fell from the Skyworld into the Dark Ocean, the muskrat gave his life to carry mud from the depths of the ocean for Skywoman to live on.

> Skywoman bent and spread the mud with her hands across the shell of the turtle. Moved by the extraordinary gifts of the animals, she sang in thanksgiving and then began to dance, her feet caressing the earth. The land grew and grew as she danced her thanks, from the dab of mud on Turtle's back until the whole earth was made. Not by Skywoman alone, but from the alchemy of all the animals' gifts coupled with her deep gratitude. Together they formed what we know today as Turtle Island, our home.[1]

The story conveys an understanding of soil not only as a material that mediates contact between vegetal, animal, human, and spiritual worlds, but also as matter that can be multiplied or increased. This generative and regenerative power is reflected in the Haudenosaunee practice of the Three Sisters Garden, in which corn and squash are planted together with nitrogen-fixing beans on soil mounds resembling turtle shells.[2]

---

1 Robin Wall Kimmerer, *Braiding Sweetgrass: Indigenous Wisdom, Scientific Knowledge, and the Teachings of Plants* (Minneapolis: Milkweed Publishing, 2013), 4.

2 Kimmerer, *Braiding Sweetgrass*, 4. Notably, a number of traditional cosmogonies from cultures around the world—including the Yoruba, Māori, and ancient Xia dynasty China, along with the Abrahamic religious traditions—contain stories in which soil is generated and multiplied by a particular divine figure, or in which soil is the substance out of which humanity is formed. See, e.g., Harold Scheub, *A Dictionary of African Mythology* (Oxford: Oxford University Press, 2000), 196; Te Ahukaramū Charles Royal, "First Peoples in Māori Traditions: Tāne, Hineahuone, and Hina," *Te Ara Encyclopedia of New Zealand*, published February 8, 2005, https://teara.govt.nz/en/first-peoples-in-maori-tradition; Anthony Christie, *Chinese*

 | DOI:10.1163/9789004748484_002

Ideas about soil as a substance that can be amended or increased are part of the historical experiences of cultures across the world, yet histories of science have overwhelmingly emphasized such ideas as a unique feature of Western European epistemology and technology.[3] This view is related to a perception of Europe as the epicenter of transformations of land regimes that led to capitalist development and modernity. It is perhaps also a legacy of European colonial regimes, which made soil a central material in the conquest and subjugation of Indigenous peoples. Indeed, since the early sixteenth century, European natural philosophers had argued that settler colonialists had to improve and reform the "virgin soils" of the American continents if they were to create livable conditions for Europeans.[4] In 1783, immediately following the American Revolution and the conquest of the already agriculturally prosperous Haudenosaunee Confederacy located on lands in present-day New York State, George Washington proclaimed that the newly independent American citizens were "placed in the most enviable condition, as the sole Lords and Proprietors of a vast Tract of Continent, comprehending all the various soils and climates of the World."[5] It is George Washington who is celebrated in the national historiography as an expert in his time in the improvement of soil on his properties, while the history of Haudenosaunee soil expertise is ignored and marginalized.

Toward a Global History of Soil is an invitation to reconsider this framing of the histories of knowing, producing, and amending soil within a teleology that connects the intensification of agriculture in late medieval and early modern Europe to the rise of agrarian capitalism, the Industrial Revolution, and modernity itself. In that telling, modern "soil science" was born as a subplot within this narrative. This volume offers fresh perspectives on the history of

---

Mythology, 3rd ed. (London: Hamlyn, 1975), 87–88; Genesis 2:7 and 3:23; Qur'an 7:12 and 15:28–29.

3 See, e.g., Benno P. Warkentin, ed., *Footprints in the Soil: People and Ideas in Soil History* (Amsterdam: Elsevier, 2006).

4 E.g., Rebecca Earle, *The Body of the Conquistador: Food, Race, and the Colonial Experience in Spanish America, 1492–1700* (Cambridge: Cambridge University Press, 2012); Kate Luce Mulry, *An Empire Transformed: Remolding Bodies and Landscapes in the Restoration Atlantic* (New York: New York University Press, 2021); Joyce E. Chaplin, *Subject Matter: Technology, the Body, and Science on the Anglo-American Frontier, 1500–1676* (Cambridge, MA: Harvard University Press, 2001). On the creation of "Neo-Europes" as the goal of settler colonialism, at least in temperate regions around the globe, the foundational text is Alfred W. Crosby, *Ecological Imperialism: The Biological Expansion of Europe, 900–1900* (Cambridge: Cambridge University Press, 1986).

5 George Washington, "From George Washington to The States, 8 June 1783," Founders Online, National Archives, accessed February 28, 2024, https://founders.archives.gov/documents/Washington/99-01-02-11404.

knowledge and practices around soil before the nineteenth century. The chapters demonstrate that in the period from roughly 1300 to 1750, preceding the elevation of soil into a European science, societies across the Americas and Afro-Eurasia were examining, questioning, and testing soils, and classifying them in changing ways. This knowledge about soil was closely linked with agricultural practices that were themselves related to social and economic transformations; as such, it often moved into or interacted with other bodies of knowledge. Soil was tied to material histories that were far from static. Agriculture was the most common form of labor across many cultures in America and Afro-Eurasia, forming both their primary mode of subsistence and their economic foundations. However, the period between 1300 and 1750 also saw a global intensification of agricultural production and the increased systematization of soil knowledge and practices.

By studying the material culture of soil knowledge, we demonstrate that in societies across the late medieval and early modern world, individuals belonging to various social groups began to investigate and manipulate soils in systematic and methodologically sophisticated ways that both benefited from contemporary knowledge-making practices and propelled them forward. This knowledge often traveled along multidirectional and circuitous routes. By subjecting soils to rigorous empirical and experimental analysis and recording in agriculture, soil experts from late medieval Ilkhanid Tabriz, Mamluk Cairo, and Kashmir to early modern England, the Ottoman Empire, and Ming dynasty China created and disseminated new types of knowledge about soil in spaces *outside* of the agrarian sphere—in areas as diverse as alchemy and veterinary medicine, hydraulic engineering and poetry, flower breeding and mysticism—even as soil knowledge produced in these novel contexts often led back to farming practices. The chapters in this volume ask what new knowledge people across the late medieval and early modern world unleashed when they worked with soil, and how did they change history with it? To answer these questions, this book aims to be the first to examine how the emergence of systematic attempts to understand the nature of soil informed the development of early modern sciences.

2 Histories and Historiographies of Soil: Knowledge-Making Practices before Soil Science

Historians have not neglected the study of soil. Economic and agricultural historians have long argued that the productive capacities of soils and the maintenance of their nutrients have been essential factors in understanding the wealth and stability of cultures dependent on farming. Environmental historians have long pleaded with us to pay particular attention to the treatment, maintenance, and abuse of soils as a key to understanding the relationship a society has with nature. Historians from many subfields have posited that understanding how soil is understood is essential for understanding the politics of land use, the financial situation of agrarian states, and long-term demographic changes.[6] Yet, in most histories, soil often appears as a passive recipient of human action—or simply a fixed, natural feature and a stage on which more important actors perform—rather than an object (or agent) of change.

Among the first to seriously engage with soils as an essential historical actor, at least in the Western European tradition, were scholars from the French *Annales* historiographical school.[7] To understand a soil, they argued, was to understand a place and the people that inhabited it. In the early twentieth century, the French geographer Paul Vidal de La Blache, mentor to Lucien Febvre, one of the founders of the *Annales* school, referred to soil as its own "historical personage" which "acts by means of the pressure that it exerts on habits," and which "governs the oscillations of history."[8] Inspired in part by Georges Lefebvre's *Paysans du Nord pendant la Révolution française*, which dealt with the internecine disputes between the peasantry and the Ancien Régime leading up to the French Revolution, later *Annalistes* became motivated by questions about social and political control over natural resources, including soil.[9] In *The Historian's Craft*, Marc Bloch discussed the silting of the harbor of Bruges that started in the eleventh century because of a confluence of geological and social factors, including the construction of dykes that changed the direction of the channel. Bloch notes that "the act of a society remodeling the soil upon which it lives in accordance with its needs is, as anyone recognizes instinctively, an eminently 'historical' event."[10] Fernand Braudel went further in the succeeding generation, claiming that all history was "tied to the soil,"

6 See, e.g., J. R. McNeill and Verena Winiwarter, eds., *Soils and Societies: Perspectives from Environmental History* (Isle of Harris, UK: White Horse Press, 2006); Hillary Eklund, ed., *Ground-Work: English Renaissance Literature and Soil Science* (Pittsburgh: Duquesne University Press, 2017); Edward R. Landa and Christian Feller, eds., *Soil and Culture* (Dordrecht: Springer, 2010); Warkentin, *Footprints in the Soil*.

7 For a brief analysis, and criticism, of the role of soils in *Annales* school history, see Samuel Kinser, "*Annaliste* Paradigm? The Geohistorical Structuralism of Fernand Braudel," *American Historical Review* 86, no. 1 (1981): 63–105, esp. 68–77.

8 Paul Vidal de La Blache, *Tableau de la géographie de la France* (Paris: Hachette, 1908), 384, authors' translation.

9 Georges Lefebvre, *Les paysans du Nord pendant la Révolution française* (Bari: Laterza, 1959).

10 Marc Bloch, *The Historian's Craft*, trans. Peter Putnam (New York: Vintage, 1953), 25. On soil and land reclamation in the Dutch Republic, see, e.g., Robert J. Hoeksema, "Three

but that this "geohistory" was largely "immobile," meaning that its characteristic patterns and the human activities that existed within it lasted for periods of time far exceeding that of an individual lifetime and sometimes that of an entire civilization.[11] It was precisely Braudel's focus on long-term environmental history, including the history of soils, that led him to eschew *l'histoire événementielle* (the history of events) and *l'histoire conjoncturelle* (conjunctural history) in favor of an emphasis on the *longue durée*.

Although sometimes charged with environmental determinism for these assertions, more recent criticism of *Annales* geohistory has revolved around the geological agency of humans in the Anthropocene: *longue durée* environmental structures, like the soil of a nation, are not the primary drivers of historical change; rather, anthropogenic disruptions of seemingly glacial natural processes are.[12] For the *Annalistes*, it was the very *un*changing nature of the environment within a short timescale that makes it important for structuring history; for historians in the Anthropocene, on the contrary, the importance lies in its abrupt changeability. For example, geomorphologist David R. Montgomery has marshaled significant geological and archaeological evidence to argue that it is precisely because processes like rock weathering, soil exhaustion, and soil erosion are often too slow to be observed in an individual lifetime but fast enough on the scale of centuries to significantly shorten the duration of civilizations that the careless treatment of soil has been many cultures' Achilles' heel.[13] In this interpretation, studying how people have manipulated, exploited, and learned from working with soils is a crucial factor both in linking time scales and providing a linchpin for understanding even long-term changes.

Historians of science have been slightly more attentive to soil as an object of analysis and as a physical and chemical substance subject to experimentation. However, many have typically begun their narratives in the mid-nineteenth

---

Stages in the History of Land Reclamation in the Netherlands," *Irrigation and Drainage* 56, no. S1 (2007): 113–126.

11 Fernand Braudel, *La Méditerranée et le monde méditerranéen à l'époque de Philippe II*, 2nd ed., vol. 1 (Paris: Armand Colin, 1967), xiii, 294–304; Fernand Braudel, "History and the Social Sciences: The *Longue Durée*," in *On History*, trans. Sarah Matthews (Chicago: University of Chicago Press, 1982), 25–54, on 27.

12 Kinser, "*Annaliste* Paradigm," 69. On humanity as a "geological agent," see Dipesh Chakrabarty, *The Climate of History in a Planetary Age* (Chicago: University of Chicago Press, 2021). Bloch, to his credit, seemed to recognize this. In his discourse on society's uses of soil, he wrote: "What is it that seems to dictate the intervention of history? It is the presence of the human element." Bloch, *Historian's Craft*, 25.

13 David R. Montgomery, *Dirt: The Erosion of Civilizations*, 2nd ed. (Berkeley: University of California Press, 2012).

century, when something approaching modern soil science began to emerge. German amateur naturalist Friedrich Albert Fallou's separation of soil knowledge from geology and agriculture, and the Russian geologist and geographer Vasily Dokuchaev's mapping and genetic analyses of soil are taken as typical starting points.[14] "Soil," it seems, is a topic for environmental historians, economic historians, and historians of agriculture studying many millennia of history, but "soil science" is the purview of historians of science studying the last two centuries of the modern era.

Yet, as historians of science themselves have long noted, the development of systematic knowledge-making practices often significantly predates the specialized scientific fields that arose from earlier experience, methods, traditional bodies of knowledge, and discrete subjects of inquiry.[15] In the process, scholars have transformed areas that were once narrowly and anachronistically categorized into broader, more inclusive areas of study united by shared practices as well as shared subject matter. For instance, historians now regularly refer to "astronomy" and "astrology" in the premodern world as "the astral sciences," or the even broader "astral knowledge," to avoid excluding practices that were once a part of these sciences, even if they no longer are. Although "soil science" as such did not exist during the periods addressed in our chapters, many of the *practices* that are now included in it did. The "artisanal epistemology" of skilled craftspeople, to use Pamela H. Smith's term, demonstrates that knowledge was created by making and doing, and those who worked with their hands—and in our cases, those whose hands got dirty working in the soil—better emblematize the values of experiment, observation, and manipulation than the elite natural philosophers who have often been recognized as the pioneers of scientific practice.[16] In this volume, we do not shy away from the word "science" when discussing these practices or subjects of inquiry, but we do consciously

14 See, for example, Eric C. Brevik and Alfred E. Hartemink, "Early Soil Knowledge and the Birth and Development of Soil Science," *Catena* 83, no. 1 (2010): 23–33. Of course, this is not universally true. One of the earliest comprehensive book-length histories of soil science devoted nearly one-third of its over 300 pages to the premodern era. See I. A. Krupenikov, *History of Soil Science: From Its Inception to the Present*, trans. A. K. Dhote (New Delhi: Oxonian Press, 1991). This book was originally published in Russian in 1981.

15 The locus classicus for this argument is Edgar Zilsel, "The Sociological Roots of Science," *American Journal of Sociology* 47, no. 4 (1944): 544–562. On more recent critical histories, see, e.g., Pamela O. Long, *Artisans/Practitioners and the Rise of the New Sciences, 1400–1600* (Corvallis: Oregon State University Press, 2011); Pamela H. Smith, Amy Meyers, and Harold J. Cook, eds., *Ways of Making and Knowing: The Material Culture of Empirical Knowledge* (Chicago: University of Chicago Press, 2014).

16 See, most recently, Pamela H. Smith, *From Lived Experience to the Written Word: Reconstructing Practical Knowledge in the Early Modern World* (Chicago: University of Chicago Press, 2022).

use the word in the broader sense, echoing its Latin root, *scientia*, meaning "knowledge." This encompasses knowledge of nature and the material world that has been subjected to a systematized and organized rigor, as was increasingly the case with soil across the late medieval and early modern globe.

To their credit, *Annales* historians recognized the importance of soil in their grand narratives, and historians of science have illuminated the practices that accompanied direct material engagement with soils, but neither group has treated soil in the context of global histories. Focusing on discrete geographical spaces, some have argued for specific "revolutionary" frameworks to describe agricultural change. Thus, there was a "Medieval Agricultural Revolution," an "Arab Agricultural Revolution," a "British Agricultural Revolution," and a "Columbian Exchange," each of which describes a series of processes by which the management of land, the deployment of new technologies, and the widespread movement of plants and animals irrevocably altered the economies, cultures, and ecologies of the globe.[17] From the *sāqiyah* of Andalusia to the heavy plow of northwestern Europe to the commercial tea and mulberry plantations of late Song China, innovations in agricultural practices between roughly the ninth and fourteenth centuries meant that direct, substantial material engagement with soils became the norm not only for farmers but for individuals from all walks of life. This was a world of growing interconnectivity in Afro-Eurasia in which intensified Abbasid-Tang trade networks, an explosion of wealth from West African gold mining in the Mali and Songhai Empires, and the forging of Indian Ocean circuits of exchange—possibly buttressed by the favorable climate of the Medieval Warm Period—facilitated a profusion of knowledge in soils.[18]

---

17 These were—and are—highly contested categories. See, e.g., Georges Duby, "La Révolution agricole médievale," *Revue de géographie de Lyon* 29, no. 4 (1954): 361–366; Lynn White Jr., *Medieval Technology and Social Change* (Oxford: Oxford University Press, 1962), esp. 39–78; Andrew M. Watson, "The Arab Agricultural Revolution and Its Diffusion, 700–1100," *Journal of Economic History* 34, no. 1 (1974): 8–35; Mark Overton, *Agricultural Revolution in England: The Transformation of the Agrarian Economy, 1500–1850* (Cambridge: Cambridge University Press, 1996); Alfred W. Crosby Jr., *The Columbian Exchange: Biological and Cultural Consequences of 1492* (Westport, CT: Greenwood Press, 1972).

18 Expiración García Sánchez, "Agriculture in Muslim Spain," in *The Legacy of Muslim Spain*, ed. Salma Khadra Jayyusi (Leiden: Brill, 1992), 987–999; White, *Medieval Technology and Social Change*, 41–57; Qi Xia 漆侠, *Zhongguo jingji tongshi: Songdai jingji juan* 中国经济通史:宋代经济卷 [General economic history of China: Song Dynasty] (Beijing: Jingji ribao chubanshe, 1999), 1:856. On the Medieval Warm Period of c. 950–1250, see Michael E. Mann et al., "Global Signatures and Dynamical Origins of the Little Ice Age and Medieval Climate Anomaly," *Science* 326 (2009): 1256–1260. Notably, this did not have the same effect across the globe. While climatic conditions were favorable for agriculture in Europe and southwest Asia, for instance, the same phenomenon may have caused

Despite their global scope, these histories rely on specific practices originating in specific places to explain later historical developments. They are, paradoxically, regionally specific global histories. In this volume, we consciously strive for what Sanjay Subrahmanyam has called "connected histories," meaning not just histories of cross-cultural relations across regions but of the purposeful combination of subjects—and in our case practices—that have been conventionally treated as separate.[19] Just as we challenge the teleological framework that draws a line from the European agricultural revolution directly to industrial capitalism and modernity, we also challenge the oftentimes reflexive equation of "global" with "globalization," usually beginning with fifteenth-century European imperial expansion and the compulsory networks it forged. Rather, in our attempt to draw connections across both time and space, we define the "global" as a culturally determined yet existentially coherent and ontologically whole geographic space. These "globalizing cosmologies," to use Caroline Dodds Pennock and Amanda Power's term, enabled societies to understand their relationship to other places around the world and structured how they engaged with other societies.[20] Though the chapters in this volume do not cover the entire globe, each one does investigate peoples and societies that, in one way or another, consciously sought to understand their own soils in the context of knowledge and practice from around the world.

New efforts to connect the histories of different places under overarching themes have been instructive. For example, in their attempts to define a "global Middle Ages," uniting that period using similar motifs that can be applied across cultures and civilizations, historians Catherine Holmes and Naomi Standen have argued for a "great intensification"—particularly during the aforementioned agriculturally "revolutionary" time frames. Intensified economic activity, urbanization, multicentric politics (rather than hegemonic ones), and sometimes radical social reorganizations generated a "great diversification" in ideas and practices that cannot be fully understood when viewed as a discrete phenomenon within any one culture or civilization.[21] Rather,

drought in the Yucatán and what is now the southwestern United States, leading, in part, to the agricultural decline of the Mayan, Hohokam, and Ancestral Puebloan civilizations.

19 Sanjay Subrahmanyam, *Connected Histories: Essays and Arguments* (London: Verso, 2022).

20 Caroline Dodds Pennock and Amanda Power, "Globalizing Cosmologies," *Past and Present* 238, Issue Supplement 13 (2018): 88–115, esp. 91–94. See also Denis Cosgrove, *Geography and Vision: Seeing, Imagining and Representing the World* (London: I. B. Taurus, 2008).

21 Catherine Holmes and Naomi Standen, "Introduction: Towards a Global Middle Ages," *Past and Present* 238, Issue Supplement 13 (2018): 1–44, on 36–37. On the "intensification" and "diversification" theses, Holmes and Standen cite R. I. Moore, "The Eleventh Century

they suggest that the unifying theme of "experimentation," which emphasizes the role of human agency in generating material circumstances, provides one such link in the chain connecting practices, texts, goods, and knowledge across the late medieval and early modern world. Given how essential they were to the agriculture underpinning these developments, soils make an excellent case study in comparative experimental practices.

3 Sciences, Practices, and Materialities: The Mobility of Soil Knowledge, c. 1300–1750

Soil knowledge and the experimental techniques informing it began to move rapidly across the globe in the late medieval and early modern period. The Mongol cultural expansion, following the conquests in the first half of the thirteenth century, had established conditions allowing for the swift movement of goods, practices, texts, knowledge, technologies, plants, animals, and people across Eurasia along the well-trodden and long-established Silk Roads.[22] Just as important as famous overland travelers like the Chinese Buddhist monk Xuanzang 玄奘 (602–664), the Maghrebi scholar Ibn Battuta (1304–c. 1369), or the Venetian trader Marco Polo (1254–1324) were the thousands of anonymous physicians, miners, carpenters, merchants, cooks, and others who moved to China, or from there to western Asia and eastern Europe, and may have brought with them information about which crops grew best in which soils and whether best agricultural practices in one climate might be applicable in another.[23] At the beginning of the fifteenth century, the treasure fleets of Zheng He 鄭和 (1371–1435) crisscrossed the waves of the Indian Ocean from the Swahili coast to Java, while, at the end of the same century, Portuguese and Spanish traders enlarged their own economic and cultural sphere through

---

in Eurasian History: A Comparative Approach to the Convergence and Divergence of Medieval Civilizations," *Journal of Medieval and Early Modern Studies* 31, no. 1 (2003): 1–21; R. I. Moore, "The Transformations of Europe as a Eurasian Phenomenon," in *Eurasian Transformations, Tenth to Thirteenth Centuries: Crystallizations, Divergences, Renaissances*, ed. Johann Arnason and Björn Wittrock (Leiden: Brill, 2004), 77–98.

22 Pamela H. Smith, ed., *Entangled Itineraries: Materials, Practices, and Knowledges across Eurasia* (Pittsburgh: University of Pittsburgh Press, 2019). On the so-called cultural "Pax Mongolica" ushered in after these conquests, see, e.g., Thomas T. Allsen, *Culture and Conquest in Mongol Eurasia* (Cambridge: Cambridge University Press, 2001).

23 See, e.g., Susan Whitfield, *Silk, Slaves, and Stupas: Material Culture of the Silk Roads* (Berkeley: University of California Press, 2018). On the types of craftsmen moving between eastern and western parts of the Mongol Empire, see Allsen, *Culture and Conquest in Mongol Eurasia*.

voyages across those same waves along with the first sustained transatlantic contact with the Americas.

Meanwhile, regionally specific climate change transformed agricultural circumstances in different parts of the globe. In the North Atlantic, the fortuitous climatic conditions of the Medieval Warm Period ended, leading to a climate in both Europe and North America that made farming much more arduous and often much less productive. The so-called Little Ice Age posed major challenges to agriculture. This widespread cooling event occurred between roughly 1300 and 1850, during which temperatures averaged around one degree Celsius lower than either the relatively stable six-millennia interglacial Holocene climate that had preceded it or the twentieth-century temperature rebound that followed it.[24] Though its causes remain up for debate, many historians have pointed to these cooler temperatures and their impact on agricultural output as one reason for the increase in poor harvests, food dearth, and famines, as well as the precipitous rise in agricultural experimentation and technological innovation as a response. Some, like Geoffrey Parker and Sam White, have gone so far as to argue that this climatological event may be partly responsible for what was once called the "general crisis" of the long seventeenth century, in which agricultural decline contributed to the escalation of violent conflict, including extraordinarily destructive events like the Thirty Years War in Europe, the Manchu conquest in China, and civil wars or uprisings in Stuart England, late Sengoku Japan, Mughal India, and the Kingdom of Kongo in Central Africa.[25] Nevertheless, the Little Ice Age remains a controversial topic, with some historians and climatologists arguing for a much more limited or regionally specific influence,[26] and others, like Morgan Kelly and Cormac Ó Gráda, arguing against its relevance for historical change in human affairs.[27] Data from

24 On definitions and dating of the Little Ice Age, see, e.g., Michael Mann, "The Little Ice Age," in *The Encyclopedia of Global Environmental Change*, ed., Michael McCracken and John S. Parry (Chichester: John Wiley and Sons, 2002), 504–509.

25 Geoffrey Parker, *Global Crisis: War, Climate Change, and Catastrophe in the Seventeenth Century* (New Haven: Yale University Press, 2013); Sam White, *A Cold Welcome: The Little Ice Age and Europe's Encounter with North America* (Cambridge, MA: Harvard University Press, 2017).

26 On its regional variation, see, e.g., Liu Yang, Zheng Jingyun, Hao Zhixin, and Zhang Xuezheng, "Regional Differences for Temperature Changes in the Medieval Warm Period and Little Ice Age over Europe and Asia," *Quaternary Sciences* 41, no. 2 (2021): 462–473.

27 On this debate, see Morgan Kelly and Cormac Ó Gráda, "The Waning of the Little Ice Age," *Journal of Interdisciplinary History* 44, no. 3 (2014): 301–325; Sam White, "The Real Little Ice Age," *Journal of Interdisciplinary History* 44, no. 3 (2014): 327–352; Morgan Kelly and Cormac Ó Gráda, "Debating the Little Ice Age," *Journal of Interdisciplinary History* 45, no. 1 (2015): 57–68.

climate science strongly suggests that effects were much more pronounced in the northern hemisphere, with little to no impact on South America, southern Africa, and Australasia, while some places, like early Tokugawa Japan and the Dutch Republic, actually experienced a robust increase in population, agricultural output, and, in the case of the Dutch, expansive trade networks.[28] Other cases are more ambiguous. In England, a place that saw both colder temperatures and shorter growing seasons, farmers and agrarian reformers responded by adopting many new technologies to improve yields. Alternative cultivation techniques, experimental botany, crop acclimatization, chemical soil additives, and new technological farming equipment, all emerged as practical solutions in the wake of these geohistorical changes, often forged and spread through new transcultural connections.

As Lorraine Daston and Glenn W. Most have pointed out, the study of practices can often reveal relationships that are not evident when one studies only content or context. "The focus on practices is essential ... [because] it directs attention to what practitioners actually do," they write. "The origins of practices often connect different disciplines with a common context [because] practices endure while classifications of knowledge vary over epochs and among cultures."[29] In their example, nineteenth-century astronomy and philology—two fields with virtually no overlapping content—possessed strikingly similar practices for seeking to define true values in both stellar positions and cross-linguistic etymologies by reducing variability based on systematic and random errors. As we show, from the fourteenth through eighteenth centuries, efforts were underway across the globe to subject knowledge about soil to more systematic and methodological rigor in areas with little related content but highly analogous practices. Evidence for practical soil knowledge can be found not just in agricultural manuals and farming records but also in very different kinds of texts, ranging from Mamluk Egyptian medical treatises and Western European alchemical recipes to Sufi mystical poetry and Ottoman cadastral documents. Examining practices from these seemingly disparate disciplines shows just how much agrarian knowledge of soils informed other practices, while physicians, alchemists, poets, and surveyors contributed their own distinct knowledge of soil that often filtered back into the world of agriculture.

---

28 Parker, *Global Crisis*, 484–506; Dagomar Degroot, *The Frigid Golden Age: Climate Change, the Little Ice Age, and the Dutch Republic, 1560–1720* (Cambridge: Cambridge University Press, 2018).

29 Lorraine Daston and Glenn W. Most, "History of Science and the History of Philology," *Isis* 106, no. 2 (2015): 378–390, on 390. On the study of scientific practices as a subset of the broader study of knowledge, see Lorraine Daston, "History of Science and History of Knowledge," *KNOW: A Journal on the Formation of Knowledge* 1, no. 1 (2017): 231–254.

Such knowledge closely tracked with practices around soil, and often moved into or interacted with other bodies of knowledge.

Soil was a topic of a global practical engagement tied to material histories that were far from static. Even as greater scholarly attention has been paid to the growing global circulation of goods, peoples, and ideas in the early modern world across both overland Afro-Eurasian trading networks and their oceanic counterparts across the Indian, Atlantic, and Pacific, soils have remained noticeably motionless. Most histories of things *grown* in soils are commodities histories that are characterized by narratives of the movements of goods from place to place and the social, cultural, economic, and ecological consequences of that mobility.[30] Yet soils are a part of the landscape of particular locations, quite literally embedded in the cultures and polities in which they are found. Consequently, practices related to soil have, to a large extent, been depicted as static and inherent, fixed to particular spaces waiting to be "discovered" in the age of European colonial expansion. But we argue that practices related to soil were neither static nor confined to particular regions.

In some cases, similarities in practice have been overlooked or missed entirely due to a focus on regionally specific content. For instance, the application of mud to rice fields in the Menderes river valley in fourteenth-century western Anatolia resonates with the practice of using mud from canals in rice fields in China during "medieval times"; the former is a virtually unstudied agricultural practice noted in early fifteenth-century Ottoman tax surveys, while the latter has been treated extensively but solely within a Chinese historical context.[31] Historians have pointed to the importance of the diffusion of plants such as rice, sugarcane, and cotton from India to western Asia, North Africa, and across the Mediterranean in the first centuries following the emergence of Islam and during the Umayyad and Abbasid Caliphates.[32] Yet we tend to forget that a crucial aspect of knowing and moving plants was the soil:

30 For example, Sven Beckert, *Empire of Cotton: A Global History* (London: Penguin, 2014); Marcy Norton, *Sacred Gifts, Profane Pleasures: A History of Tobacco and Chocolate in the Atlantic World* (Ithaca, NY: Cornell University Press, 2008); Francesca Bray, Peter A. Coclanis, Edda L. Fields-Black, and Dagmar Schäfer, eds., *Rice: Global Networks and New Histories* (Cambridge: Cambridge University Press, 2017); among many others. For a critical roundup of popular nonfiction commodities histories, a particularly widespread genre in the 1990s and 2000s, see Bruce Robbins, "Commodities Histories," *Publication of the Modern Language Association of America* 120, no. 2 (2005): 454–463.

31 Cahit Telci, *Halil Beğ Defteri: Fetihten Sonra Aydın Sancağı'nın İlk Mufassal Tahrir Defteri 1425–1430* (İzmir: Kâtip Çelebi Üniversitesi, 2015); Francesca Bray, *The Rice Economies: Technologies and Development in Asian Societies* (Berkeley: University of California Press, 1994), 49.

32 Watson, "Arab Agricultural Revolution," 8–35.

that element foundational to the well-being of the living plant. In the early Arabic literary production of the newly expanding world that connected the Umayyad lands on the shores of the Atlantic Ocean with Tang China, it was the scent of the soil of one's homeland (*watan*) that provided healing qualities, a crucial element for defining and crafting home.[33] No wonder, then, that in this time when plants were becoming increasingly mobile, an Abbasid scholar from Basra, al-Jahiz (776–c. 869), grandson of a Black African camel driver, compared strangers to a "plant which has been removed from its soil and has lost the water that nourished it."[34] In other cases, these practices were directly related to the history of transatlantic slavery, as Judith Carney has shown in her discussions of the African origins of rice cultivation in the Americas. The introduction of rice from West Africa to the Carolinas depended heavily on knowledge about working with soil among the enslaved Africans working on plantations—ridging for soil aeration, building embankments, and other practices that were applicable in both the lowland mangrove swamps of the Guinean coast and the low-country salt marshes of South Carolina.[35] In each case, the focus on practice has revealed much about both the movement of people and knowledge and the similarity of solutions to persistent problems in analogous environments across the globe.

The movement of texts often reflected the movement of practices. Agricultural manuals prove to be a widespread phenomenon of virtually all literate Afro-Eurasian cultures. In these texts, authors conveyed agricultural techniques and traditional ecological knowledge, in some cases recorded textually for the first time, to describe the physical and chemical properties of soils. They recounted, among other things, how soils behaved in different climatic conditions, how they could be improved by agronomists, and how to test for properties like acidity, salinity, and fertility. In Europe, the Middle East, and China alike, the long tradition of recording agrarian practices morphed into new kinds of didactic texts designed to instruct farmers how to improve soils and provide new theoretical frameworks to explain how this was possible. In seventeenth-century Europe, novel alchemical theories concerning material

33 Zayde Antrim, *Routes and Realms: The Power of Place in the Early Islamic World* (Oxford: Oxford University Press, 2015), 17.

34 Franz Rosenthal, "The Stranger in Medieval Islam," *Arabica* 44, no. 1 (1997): 35–75, on 48.

35 On the importance of knowing soil in the process of rice farming, see Judith Carney, *Black Rice: The African Origins of Rice Cultivation in the Americas* (Cambridge, MA: Harvard University Press, 2002), 18. On the circulation of both West African plants and knowledge in this context, see Susan Scott Parrish, "Diasporic African Sources of Enlightenment Knowledge," in *Science and Empire in the Atlantic World*, ed. James Delbourgo and Nicholas Dew (New York: Routledge, 2008), 281–310.

transformation prompted agrarian reformers to develop new interpretive frameworks to explain how soils lost fertility and how chemical additives could transmute exhausted soils into new, productive varieties. In his *Discours admirable*, the sixteenth-century French craftsman and potter Bernard Palissy (c. 1510–c. 1589) described soil fertility in terms of a "vital spirit," which he identified with the alchemical quintessence.[36] In late Ming China, two prominent agronomic manuals—the *Nongshuo* 農説 (Explanations on agriculture) by Ma Yilong 馬一龍 (1499–1570) and the *Baodi quannong shu* 寶坻勸農 (Manual for encouraging agriculture in Baodi) by Yuan Huang 袁黄 (1533–1606)—employed the universal cosmological concepts of qi, to explain soil's vital properties, and yin and yang, to describe balance in nutrients, density, temperature, and moisture.[37]

Though produced within specific geographical settings with their own local ecologies during the late medieval and early modern era, agricultural manuals also harked back to many older traditions. Some of these were the result of very different cultural and linguistic milieus produced under distinct environmental conditions. Discrepancies based on these geographical or temporal distances could create tensions between traditional authorities and contemporary practitioners when the soil knowledge described by the former simply did not match the experience of the latter. The ancient world had bequeathed to early moderns across the globe numerous agricultural texts in the classical languages of Greek, Latin, and Sanskrit. Rarely were these passively received. Medieval intermediaries—such as the multiauthored tenth-century Byzantine *Geoponika* or the fourteenth-century Italian *Ruralia commoda* of Pietro de' Crescenzi—translated and adapted much of this knowledge, adding insights based on their experience and their own distinct climates. Abū Bakr Aḥmad Ibn Wahshiyya (fl. early 10th c.), author of the *Nabatean Agriculture* (*Al-filāḥah al-nabaṭīyah*) written in 903–904, most likely in Iraq, evoked his personal experience and knowledge he gained on his travels as well as earlier Greco-Roman and Syriac agricultural writings. Likewise, Ibn al-ʿAwwām's (?–1158) twelfth-century *Kitāb al-filāḥa* (*Book of Agriculture*), written in Andalusia, evokes the author's practical experience, especially with growing olive trees and saffron.[38]

36 See Justin Niermeier-Dohoney's chapter in this volume.

37 See Jörg Henning Hüsemann's chapter in this volume.

38 For the process in Europe, see, e.g., Mauro Ambrosoli, *The Wild and the Sown: Botany and Agriculture in Western Europe, 1350–1850*, trans. Mary McCann Salvatorelli (Cambridge: Cambridge University Press, 1997), esp. 12–95; for the Islamic world, see Karl W. Butzer, "The Islamic Traditions of Agroecology: Crosscultural Ideas, Experiences, and Innovations," *Ecumene* 1, no. 1 (1994): 7–50; Jaakko Hämeen-Anttila, *The Last Pagans of Iraq: Ibn*

Migrations, military conquests, the forging of new trade networks, and exploratory travel all contributed to a kind of agricultural pluralism, and the linguistic intermingling and competition associated with these movements created both discord and new fusions of soil knowledge. For example, both the Latin European West and the Muslim Middle East inherited the intellectual traditions of the ancient Greco-Roman Mediterranean. Late medieval and early modern authors of agricultural texts in many European vernacular languages often referenced the expertise of classical Greek and Latin writers, even when these writers wrote of soil types and agricultural practices significantly different from the ones they were attempting to promote. The rising scholarly interest in agriculture across Eurasia in the fifteenth century necessitated the development of literature to spread that knowledge. While most ancient Greek works on agriculture were lost, classical Roman agricultural writings survived in abundance and included the Latin works of Palladius, Columella, Cato, and Varro, among others.[39] Like other ancient texts, Greek and Roman works on agriculture became highly sought after among Renaissance Italian scholars. In 1472, the Italian humanists Georgius Merula (1430–1494) and Franciscus Colucia (fl. late 15th c.) compiled and edited the first edition of their collected agricultural texts in one volume, the *Scriptores rei rusticae*, in Venice. The competitive commercial environment of the Mediterranean and southwestern Asia, as well as Venetian maritime connections, assured that copies of this book spread widely, and there is evidence of both an uptick in the publication of agrarian manuals after this work appeared and of references to and citations of these ancient texts.[40] As deeper knowledge of agriculture gained social and economic currency across Afro-Eurasia, this web of textual knowledge became something that new agrarian writers could not afford to ignore.[41]

Waḥshiyya and his Nabatean Agriculture (Leiden: Brill, 2006), 88; Lucie Bolens, *Agronomes Andalous du Moyen-Âge* (Geneva: Droz, 1981), 30.

39 Joan Thirsk, "Making a Fresh Start: Sixteenth-Century Agriculture and the Classical Inspiration," in *Culture and Cultivation in Early Modern England: Writing and the Land*, ed. Michael Leslie and Timothy Raylor (Leicester: Leicester University Press, 1992), 15–34, on 18; G. E. Fussell, *The Classical Tradition in West European Farming* (Teaneck, NJ: Fairleigh Dickinson University Press, 1972).

40 On the cultural and commercial connections between Italy (especially Venice) and the Ottoman Empire in the early modern Mediterranean, see, e.g., Julian Raby, "A Sultan of Paradox: Mehmed the Conqueror as a Patron of the Arts," *Oxford Art Journal* 5, no. 1 (1982): 3–8; Cemal Kafadar, "A Death in Venice (1575): Anatolian Muslim Merchants Trading in the Serenissima," *Journal of Turkish Studies* 10 (1986): 191–219.

41 On the simultaneous global interest in agricultural writings around 1500, see Aleksandar Shopov, "Between the Pen and the Fields: Books on Farming, Changing Land Regimes, and Urban Agriculture in the Ottoman Eastern Mediterranean ca. 1500–1700" (PhD diss.,

Yet no soil practitioners took these ancient writings at face value, and all responded to their works and adapted their techniques in ways that best suited their specific situations.[42] Omissions, discrepancies, and gaps in soil knowledge were not always easily reconcilable. For example, Greco-Roman sources had very little to say about an exceptionally fertile soil variety called marl, found abundantly in northwestern Europe, because it existed only sparsely in the Mediterranean Basin. Later medieval and early modern European authors like Walter of Henley (fl. late 13th c.) and Gervase Markham (c. 1568–1637) constructed their own substantial body of knowledge about marl largely from scratch.[43] Similarly, though most of their textual soil knowledge derived from Roman sources through the medieval Italian intermediary Pietro de' Crescenzi (c. 1230–c. 1320), authors of early modern Polish agrarian manuals like Anzelm Gostomski (1508–1588) and Jan Herman (?–c. 1670) dispensed entirely with Roman advice on how to cultivate soil in high-lying and sloping land because they wrote largely for landowners who possessed estates in low-lying plains.[44]

Variations in precipitation levels, volcanic activity, flooding frequency, elevation, average temperature, and many other factors produced an extraordinary variety of soil types across the globe. Thus, even as agricultural manuals circulated widely, they often continued to refer to highly specific, locally produced knowledge. Textual advice for working the rich, volcanic soils of the Italian peninsula was largely useless for those working the rockier soils of the Scottish Highlands or the sandier soils of the Arabian Peninsula. Yet this is precisely the situation in which agrarian writers from Scotland to Arabia found themselves when they relied on soil classifications handed down through Pliny's *Historia naturalis*, a text widely referenced and employed by landowners in territories that had inherited classical Greco-Roman knowledge. Eighteenth-century Scottish agronomist Robert Maxwell complained of the "pompous and superficial readings of … Pliny" too common among professors, while the agrarian improver Adam Dickson, conversely, touted "the maxims of the ancient Roman farmers" as "the same with those of the best modern farmers in Britain."[45] It is tempting to view these debates as a sort of agronomic *Querelle des Anciens*

Harvard University, 2016), 66–67; Justin Niermeier-Dohoney, "A Vital Matter: Alchemy, Cornucopianism, and Agricultural Improvement in Seventeenth-Century England" (PhD diss., University of Chicago, 2018), 176–178.

42 Fussell, *Classical Tradition in West European Farming*.

43 Verena Winiwarter and Winfried E. H. Blum, "From Marl to Rock Powder: On the History of Soil Fertility Management by Rock Materials," *Journal of Plant Nutrition and Soil Science* 171 (2008): 316–324. See also Justin Niermeier-Dohoney's chapter for this volume.

44 See Monika Kozłowska-Szyc's chapter for this volume.

45 Robert Maxwell, *Selected Transactions of the Society of Improvers in Knowledge of Agriculture in Scotland*, vol. 5 (Edinburgh, 1743), xiv; Adam Dickson, *The Husbandry of the*

et des Modernes. A more nuanced and charitable reading is that farmers and agricultural writers across different cultures deftly synthesized soil knowledge from revered authorities with that of more recent experience, or used the lack of authoritative knowledge as a springboard for new explorations.

Sometimes, however, classical knowledge was rejected altogether. Medieval Islamic scholars seem to have completely abandoned the Hippocratic tradition according to which the underground soil (*batin al-ard*) was cooler in the summer than in winter, which can be seen as early as the writings of the Abbasid era scholar Abū Ḥanīfa Dīnawarī (828–895). When the almanac authors in Yemen in the thirteenth century began incorporating agricultural matters, they frequently noted the beginning of the warming of the soil on January 12 and its cooling after July 15–16.[46] The thirteenth century seems to have witnessed a reinvigoration of writings on agriculture in western Asia, contrary to the notion of an overwhelming agricultural decline that dominates the historiography.[47] Earlier agricultural writings had little to say on methods of treating the soil in mulberry tree orchards, but not the Ilkhanid agricultural writer and vizier Rashīd al-Dīn Hamadānī (1247–1318), as Himmet Taşkömür notes in his contribution to this volume, who went to great lengths to describe the process based on his observations made in the city of Yazd in Iran for his readers.

An increase in the trade along the Silk Road following the Mongol conquests required scholarly investigation of methods for improving soil to facilitate the growth of plants crucial for manufacturing goods such as silk.[48] In the twelfth century, sugarcane plantations were expanded on a large scale in Upper Egypt, and by the thirteenth century, a new type of large-sized plow (*muqalqila*) was invented. Ancient writers had little to say about soil and sugarcane growing, so Mamluk Egyptian scholars such as Shihāb al-Dīn Aḥmad bin ʿAbd al-Wahhāb al-Nuwayrī (1279–1333) wrote extensively on sugarcane's tendency to exhaust soil.[49] The effect of plants on soil and, in general, agricultural knowledge about

---

Ancients (Edinburgh, 1788), xv. On these, see T. C. Smout, "A New Look at the Scottish Improvers," *Scottish Historical Review* 91, no. 231, part 1 (2012): 125–149.

46 Daniel Martin Varisco, *Medieval Agriculture and Islamic Science: The Almanac of a Yemeni Sultan* (Seattle: University of Washington Press, 1994), 128.

47 Andrew M. Watson, "Agricultural Science," in *Science and Technology in Islam: Technology and Applied Sciences*, part II, vol. 2, ed. A. Y. Al-Hassan (Paris: UNESCO Pub, 2001): 38–39.

48 On Yazd as a center of manufacturing of silk in the thirteenth century, also reported to be such by Marco Polo, see A. K. Lambton, "Reflections on the Role of Agriculture in Medieval Persia," in *The Islamic Middle East, 700–1900: Studies in Economic and Social History*, ed. Abraham L. Udovitch (Princeton, NJ: Darwin Press, 1981), 293–294.

49 Sato Tsugitaka, *State and Rural Society in Medieval Islam: Sultans, Muqta's and Fallahun* (Leiden: Brill, 1997), 207–216.

soil in this period moved from Mamluk Syria and Egypt to Yemen, and most likely further to the opposite shores of the Indian Ocean such as East Africa.[50] The agricultural connection between the Arabian Peninsula and East Africa is exemplified by the transfer of coffee trees from Ethiopia to Yemen, which seems to have occurred in the late antique or medieval period, when coffee cultivation was on the rise and the coffee bean trade expanded.

These patterns were repeated across other regions. Climatological distinctions between drier, more temperate northern China and wetter, subtropical southern China were large enough that farming manuals that specialized in comparing their differences, like the *Nongshu* by Wang Zhen 王禎 (fl. 1290–1333), emerged during the Yuan dynasty. As the Mughals began to dominate much of the Indian subcontinent, Persian horticultural writings, which were already heavily influenced by ancient and Byzantine Greek models, began to vie with both the older *vṛkṣāyurveda* (plant medicine) of the Sanskrit tradition and contemporary agricultural writings in vernacular South Asian languages like Kashmiri and Mewari.[51] Farmers and agrarian reformers faced innumerable choices in this polyglot, multicultural, and methodologically diverse landscape. But these problems were not insurmountable. The early sixteenth-century Spanish writer Gabriel Alonso de Herrera (1470–1539), in his *Obra de agricultura*, explicitly rejected the argument that knowledge from other regions could not be applied to Iberian soils. His justification for this assertion relied on a Catholic vision of agriculture as ordained by God to be universally practicable, and soils of any sort to be workable through proper labor.[52] This attitude served Spanish and Portuguese agrarian experimenters well throughout the sixteenth and seventeenth centuries as Iberia became a kind of experimental farming laboratory for growing American and Asian plants on European soils for the first time. However, it also led to the belief that burgeoning European empires could bend colonial American soils to their will in economically extractive ways. These "agricultural regionalisms," as Jörg Henning Hüsemann argues in this volume, contributed to the heterogeneous nature of farming

50 On the transfer of Egyptian knowledge and classification of soils to Yemeni books on agriculture, see Hassanein Rabie, "Some Technical Aspects of Agriculture in Medieval Egypt," in Udovitch, *Islamic Middle East*, 73.

51 Richard Eaton, *India in the Persianate Age: 1000–1765* (Oakland: University of California Press, 2019); see also Nicolas Roth's and Zubair Khalid's chapters for this volume.

52 Gabriel Alonso de Herrera, *Obra de agricultura* (Alcalá de Henares: Arnao Guillen de Brocar, 1513), prologue; see also Alberto González Remuiñán and Dulce Freire's contribution to this volume.

manuals along with their sensitivity to soil variation.[53] As these manuals circulated, they also strengthened agronomists' comprehension of geographical, ecological, and climatic diversity, along with an understanding that knowledge about soil had undergone major changes over the centuries.

4 Language, the Body, the State, and Experimental Cultures

The chapters in the volume are organized under four broad themes related to the practices that generated soil knowledge and the material culture within which these practices operated: the translation and transmission of soil knowledge; soils, medicine, and bodies, both human and animal; the relationship between taxonomy and expertise in the soil practices valued by early modern states; and the increased specialization of soil practices within nascent experimental cultures. In Part 1, "Translation and Transmission of Soil Knowledge," we examine how soil imagery in poetry and rhetoric and the translations of soil terminologies across languages and cultures affected the dispersal and reception of specific types of agrarian knowledge. In Part 2, "Soil, Medicine, and the Body," we demonstrate how comparative analyses of soils and human and animal bodies created multidisciplinary knowledge across medical, pharmacological, veterinary, culinary, and agricultural domains. In Part 3, "Governing the Soil: Taxonomy, Expertise, and the State," we illustrate how the relationships between states, their political economies and agriculture, generated new soil taxonomies based on the economic concerns of elites and produced new political theories of social justice for farmers. Finally, in Part 4, "Soil, Specialization, and Experimental Culture," we show how the observational, empirical, and experimental engagement with soils in the contexts of transmutational alchemy, hydraulic engineering, flower breeding (*terbiye-i ezhār*), and *vṛkṣāyurveda* engendered highly versatile knowledge about soils that easily passed through porous disciplinary boundaries and contributed to new forms of expertise in burgeoning early modern sciences such as chemistry, botany, medicine, engineering, and agronomy.

4.1 *Translation and Transmission of Soil Knowledge*

Like geography and time, language could be both a barrier and a bridge. The translation of words was also the translation of concepts, and this could be

53 See Jörg Henning Hüsemann's chapter for this volume, which adapts the concept of "medical regionalism" found in Marta E. Hanson, "Northern Purgatives, Southern Restoratives: Ming Medical Regionalism," *Asian Medicine* 2, no. 2 (2006): 115–170.

an extraordinarily challenging endeavor, especially when two cultures with no historical relationship encountered one another. Spanish settler colonialism in the Americas provides us with numerous examples. For one, grammatical constructions in Nahuatl—one of the primary Indigenous languages of Central Mexico—emphasize qualitative over substantive notions, making mistranslations of concepts like *tlalcoztli* (literally, an "earth-yellow thing," meaning a yellow dye) into the Spanish *tierra amarilla* (literally, "yellow soil") quite common.[54] As Sarah Newman argues in her chapter on the pluralistic nature of soil taxonomies in sixteenth-century colonial Mexico, the use of different glyphs in the *Codex Vergara* and *Codex de Santa María Asunción* to refer to the same soils in the Indigenous logographic writing system may represent not scribal errors, as many anthropologists have presumed, but rather deliberate choices made to demonstrate intentional amendments to those soils. The authors' use of separate glyphs in these two cadastral codices linguistically reflect how natural and cultural processes intertwined in the care of soils by the "good farmers" of Tepetlaoztoc in Central Mexico. One-to-one translations of soil typologies from Nahuatl to Spanish from these cadastral codices remain problematic at least in part because many modern-day anthropologists have long sought stable referential categories rather than acknowledging soil as a dynamic material created through a combination of natural, cultural, and historical processes. The semantic content of words in many agglutinative Mesoamerican languages is incomplete without reference to accompanying social context, and thus crucial cultural information about soil knowledge is lost when literal translations into Spanish, English, or other non-agglutinative European languages are taken at face value.

Setting this potential linguistic and cultural irreconcilability aside, there was also remarkable comparability in the ways agrarian writers across Afro-Eurasia described soil qualities. Terminologies for fertility, acidity, density, and structure, among many other things, could be strikingly similar across these linguistic and cultural divides. The words "fat," "fatty," "plump," or "greasy" appear in many languages (e.g., Latin *pinguis*, Spanish *grasa/o*, Polish *tłusta* or *pulchna*, Chinese *fei* 肥, Turkish *et toprak*) to describe especially fertile soils or the fertilization process. Likewise, we find similar qualitative descriptions, often related to taste and smell, in multiple languages. Productive, nutrient-rich soils are described as "sweet" (e.g., Sanskrit *miṣṭa*), while exhausted, acidic, or nutrient-poor soils are referred to as "sour" (e.g., Sanskrit *amlaḥ*) or "salty"

54 Barbara J. Williams and H. R. Harvey, *The Códice de Santa María Asunción: Facsimile and Commentary; Households and Lands in Sixteenth-Century Tepetlaoztoc* (Salt Lake City: University of Utah Press, 1997), 31.

(e.g., Latin *salsam*, Polish *słona*, Sanskrit *lavaṇaḥ* or *aśubha*, Chinese *xian* 鹹). Some of these similarities are attributable to the ancient Roman *Scriptores rei rusticae*, agricultural writings that influenced many later cultures. However, many of these genuinely appear to be simultaneous, independent qualifications that emerged out of similar tactile engagement with soils across cultures.[55]

Several agricultural writers also used gendered language when referring to specific agricultural substances and practices. Soil or the earth in general were regularly regarded as feminine. English agrarian chymists often used the term "womb" to describe cultivated soil and "menstruum" to describe anything in which seeds could germinate.[56] The Mamluk book on farming *Selected Agriculture* (*Al-filāḥa al-muntakhaba*), written around 1400, claimed that fertile soil "becomes pregnant" by the sun's rays, which "gave birth with the help of water."[57] Chinese agronomists often called the heavens the "father" and the earth the "mother," which received the father's seed.[58] Agricultural laborers, in turn, were coded as masculine, and sexual metaphors often appeared to denote planting seeds and plowing fields.

Using rhetorical and analogical language like this to describe soils and using imagery and metaphors derived from agrarian knowledge in poetic traditions became increasingly common in many early modern cultures. As Zubair Khalid reveals in his chapter on the soil imagery employed in the mystical poetry of the fifteenth-century Kashmiri Sufi saint Nuruddin Rishi (c. 1377–c. 1440), agricultural allegories found natural homes in the works of several South Asian scholars and writers. In the *Dvitīyā Rājataraṅginī* by the Sanskrit court chronicler Jonaraja, we find Persian military incursions and Muslim migration into northern India equated with locusts devouring crops.[59] Conversely, in his own poetry, Nuruddin deploys soil imagery and presents agricultural practices as a natural process of crop growth aided by farmers in a way that mirrored

55 For comparisons between Greek and Latin terms with other languages, see Verena Winiwarter, "Prolegomena to a History of Soil Knowledge in Europe," in Winiwarter and McNeill, *Soils and Societies*, 177–215, on 202–206.

56 For example, Cressy Dymock, "A Discoverie for Division or Setting Out of Land," pamphlet (1653), 13.

57 Ṭaybaghā al-Jariglamishī al-Tamān Tamrī, *Al-filāḥa al-muntakhaba*, Bibliothèque national de France, Paris, MS no. 2808, fol. 3b.

58 See, e.g., Ma Yilong 馬一龍, *Yuantu dayan* 元圖大衍, in *Shuofu xu* 說郛續, ed. Tao Ting 陶珽 (China: s.n.), Shunzhi 3 [1646], *juan* 1, 1B, Harvard-Yenching Library, Chinese Rare Books Digitization Project-Collectanea: https://curiosity.lib.harvard.edu/chinese-rare-books/catalog/49-990067838030203941, accessed May 26, 2023.

59 See Zubair Khalid's chapter for this volume.

Nuruddin's own role as a knowledge broker and go-between facilitating the spread of Islam in northern and northwestern India.

4.2 *Soil, Medicine, and the Body*

The connection between soil and the human body is one of the oldest relationships acknowledged. As the Haudenosaunee creation narrative of Skywoman shows, many cultures observed a direct association between bodies and the soils out of which they were made. In Muslim, Christian, and Jewish traditions, there are frequent references to soil, the material from which Adam, the first man, was made.[60] Even as the cosmogonic contexts of the material culture of soil faded, the notion that the soil directly influenced the body did not. Hippocrates and the Hippocratic writers of fifth-century BCE Greece noted in *Airs, Waters, Places* that soil quality, moisture content, and the minerals contained within soil could directly impact human health by virtue of food consumed from it or the miasmas they exuded.[61] The second-century CE Roman physician Galen argued that disease was caused by an imbalance of the four humors (blood, phlegm, black bile, and yellow bile), and later Galenists suggested that an imbalance in bodily humors may reflect similar imbalances in the makeup of soil.[62] In the eleventh century, the Persian physician Ibn Sīnā (Avicenna) (c. 980–1037), whose work the *Canon of Medicine* was the foundational medical textbook in both the Islamic world and Christian Western Europe until the eighteenth century, argued that water and soil quality affected the prevalence of certain diseases.[63]

In some cases, the correlation between soil and body relied on a comparative framework that linked unlike bodies of knowledge through like practices. As Heba Mahmoud Saad Abdelnaby writes in her chapter, fourteenth- and fifteenth-century Mamluk writers of both agricultural manuals and medical treatises developed an analogical framework in which they imagined soil as the "skin of the earth" and skin as the "soil of the body," leading to major cross-disciplinary knowledge exchange concerning both prophylactic treatments

60 Genesis 2:7; Qur'an 15:28–29.

61 Hippocrates, "Airs, Waters, Places," in *Hippocratic Writings*, ed. G. E. R. Lloyd, trans. J. Chadwick and W. N. Mann (London: Penguin Classics, 1983), 148–153.

62 More specifically, they associated the humor black bile with earth, one of the four Empedoclean elements, both of which were presumed to be both cold and dry in the Aristotelian system of qualities. See G. E. R. Lloyd, *Greek Science after Aristotle* (New York: W. W. Norton, 1973), 136–153.

63 Avicenna, *A Treatise on the Canon of Medicine of Avicenna: Incorporating a Translation of the First Book*, trans. Cameron O. Gruner (Birmingham, AL: Classics of Medicine Library, 1984).

for crops and the use of certain types of fertilizers as remedies.[64] Notably, in Mamluk Egypt, Syria, and Palestine, bird droppings became extraordinarily important both as a soil improver and for treating diseases in human and animal medicine. How bird manure came to be seen as a "universal remedy" for soil infertility and the diseases of plants, animals, and the human body was a complicated, dialogical process that involved far more interaction between the authors of agrarian manuals and medical treatises than has previously been considered.[65]

Similarly, as Jörg Henning Hüsemann's chapter shows, agricultural manuals from the late Ming era of Chinese history demonstrate an increased interest in the relationship between soil types and their potential effects on the human body. Rather than the established practice of codifying technical knowledge and reproducing traditional accounts as was common among agrarian authors of the Ming dynasty, Hüsemann centers his discussion on the more innovative manuals of Ma Yilong and Yuan Huang, which focused on explaining the theoretical principles behind successful tillage. In doing so, they relied on traditional concepts of nature that were fundamental in numerous fields of systematic inquiry—especially the concepts of qi, yin, and yang—which, as this chapter argues, enabled them to link agriculture and medicine in novel ways. These scholars began to interpret agricultural practices in terms of ideas about balance in the human body and connected the agricultural treatment of soils to the medical treatment of diseases.

4.3 *Governing the Soil: Taxonomy, Expertise, and the State*

Following the Mongol conquests in the thirteenth century, agricultural texts, plants, and practices began to circulate more widely across Afro-Eurasia, and with these texts came new ways of ordering the economies of states that relied on agriculture for wealth. As early modern agrarian states developed more complex bureaucracies and agriculture developed into a source of wealth rather than the means of subsistence, soil was no longer a matter only for peasant farmers but began to attract the attention of the elites. A long tradition of recording agrarian practices transformed into new kinds of didactic texts designed to instruct farmers on how to improve soils and provide new

64 This phrase comes from the Mamluk Syrian author Raḍī al-Dīn al-Ghazzī (d.1529), who lived in Damascus and whose family hailed from Gaza (Ghazza) in Palestine. Al-Ghazzī, MS 8407, fol. 8, Al-Assad Library, Damascus, Syria. See Heba Mahmoud Saad Abdelnaby's chapter in this volume.

65 On manure, see Daniel Martin Varisco, "Zibl and Zirāʿa: Coming to Terms with Manure in Arab Agriculture," in *Manure Matters: Historical, Archaeological, and Ethnographic Perspectives*, ed. Richard Jones (New York: Routledge, 2012), 129–144.

theoretical frameworks to explain how this was possible. However, understanding the importance of these texts is more than a question of theory versus practice. Since agrarian manuals across Afro-Eurasia in this period were often written by elites, they often reflected their economic and political concerns. The *regional* emphasis on agricultural knowledge of these new treatises has been pointed out in the case of western Europe in the sixteenth and seventeenth centuries,[66] but this seems to have been a global phenomenon prior to the eighteenth century.

An early example of this is perhaps *Āṣār va aḥyā'* (Monumental traces and revival), a book on agriculture written by the Ilkhanid vizier Rashīd al-Dīn Hamadānī. In this work, which is discussed in Himmet Taşkömür's contribution to this volume, soil is central to the author's discussions of a variety of agricultural practices he observed across the sprawling Ilkhanid realm. Agricultural knowledge in late medieval western Asia has frequently been discussed within the framework of economic decline. Fewer books on agriculture were supposedly produced following the Golden Age that ended in the twelfth century. The case of Rashīd al-Dīn Hamadānī suggests otherwise. The author himself says that his writings improve upon earlier *filāḥa* books. His agricultural investments in Iraq, Iran, and Anatolia explain his work's emphasis on regional soil practices and support his claims to have observed these regional variations firsthand.

The political economy of empires and their metropoles became an important concern centering many agricultural manuals. As Alberto González and Dulce Freire argue in their chapter, writers of agrarian manuals in sixteenth- through eighteenth-century Iberia made distinct contributions to the arbitrismo economic reform movement, a widespread program recommending financial improvements to the Spanish and Portuguese crowns, in their case by upending received agricultural wisdom from both classical and Biblical sources.[67] Previous prevailing understandings of soil fertility in Iberia relied on specific Catholic interpretations in which land neither became exhausted nor lost its productive capacities because God had endowed soils with the necessary fertility to sustain individual nations. In the wake of Spanish and Portuguese imperial expansion into the Americas, new arguments emerged suggesting

66 Joan Thirsk, "Plough and Pen: Agricultural Writers in the Seventeenth Century," in *Social Relations and Ideas: Essays in Honour of R. H. Hilton*, ed. T. H. Ashton et al. (Cambridge: Cambridge University Press, 1983), 295–317.

67 For example, José Ignacio Fortea, "Economía, arbitrismo y política en la Monarquía hispánica a fines del siglo XVI," *Manuscrits: Revista d'història moderna* 16, no. 1 (1998): 155–176.

that true soil fertility depended on the labor and expertise of the farmers who worked it. New economic reinterpretations of soils touted by agrarian writers formed the basis of a new "universalist" agriculture that encompassed both the Spanish and Portuguese imperial visions *and* their position as strong, quintessentially Catholic (i.e., universal) monarchies.

Texts about soil and traditional agricultural practices come with certain epistemic problems. One major difficulty faced by historians is determining the degree to which actual farmers consulted these texts. Put differently, it is an open question whether agricultural manuals described what farmers *did* or prescribed what they *should* do.[68] This question is taken up by Monika Kozłowska-Szyc, who demonstrates in her chapter that the rise in demand for agricultural manuals in the late sixteenth-century Polish-Lithuanian Commonwealth reflected the need for the new landed gentry to have standardized, utilitarian reference points in order to extract the most wealth out of their property. Since these works were written with landowners and bureaucrats in mind, they dictated changes to agricultural production in service of these elites, often by classifying soil types and soil qualities in ways that provided economically useful information for commercial agriculture and taxation regimes.

Historians have long argued that the creation of taxonomies has played a vital role in political economy, reminding us that classifying nature often has far more to do with the practical uses of natural objects than with discovering some underlying, objective order. Indigenous taxonomic systems for bird species among the Foré people of New Guinea and of animal behaviors among the James Bay Cree of northern Quebec, for example, have reflected the social and economic needs of these societies.[69] Even Carolus Linnaeus's well-known hierarchical, binomial taxonomy—which served as the origin of modern-day

68 On the "descriptive-prescriptive" debate, see, e.g., Simon Schaffer, "The Earth's Fertility as a Social Fact in Early Modern Britain," in *Nature and Society in Historical Context*, ed. Mikuláš Teich, Roy Porter, and Bo Gustafsson (Cambridge: Cambridge University Press, 1997), 124–147; Winiwarter, "Prolegomena to a History of Soil Knowledge in Europe," 185. Winiwarter notes, for example, that George Duby fully believed that agrarian manuals influenced agrarian practice, while Adriaan E. Verhulst vehemently denied it. See George Duby, *The Fontana Economic History of Europe*, vol. 1, sec. 5, *Medieval Agriculture, 900–1500*, trans. Roger Greaves (London: Collins, 1969); Adriaan E. Verhulst, "Agrarian Revolutions: Myth or Reality?" *Sartoriana* 2, no. 1 (1989): 71–75.

69 Jared H. Diamond, "Zoological Classifications of a 'Primitive People,'" *Science* 151, no. 3714 (1966): 1102–1104; Colin Scott, "Science for the West, Myth for the Rest? The Case of James Bay Cree Knowledge Construction," in *The Postcolonial Science and Technology Studies Reader*, ed. Sandra Harding (Durham, NC: Duke University Press, 2011), 175–197.

biological systematics—began as a project to keep Sweden economically competitive following the failures of its empire.[70]

Soil classifications are, arguably, an even better example of this phenomenon. The late Zhou dynasty Chinese text "Yugong" 禹貢 (Tribute of Yu) of c. 500 BCE may have been the first attempt to systematically classify soils, while nearly every Greco-Roman agronomic text developed a classification system for soils based on tactile or sensory qualities like color, texture, density, mineral composition, and so on.[71] These taxonomies were often based on utilitarian concerns. The seventeenth-century English scholar John Evelyn famously declared that there were "one hundred seventy nine millions one thousand and sixty sorts of earths," eight or nine of which were useful for agriculture.[72] In Egypt, in the early years of the Ayyubid dynasty (1171–1260), comprehensive lists of soil types began to be recorded in which soil taxonomy was categorized according to the kinds of plants planted on the soil the previous year, as well as whether it was irrigated, left fallow, or exposed to flooding. One type, *sibah*, was described in terms of its mineral content as a soil high in salts that was also used as a fertilizer for flax. The idea of classifying soil by its previous uses may have reflected tax-collecting practices that encouraged a reassessment of soil quality on a yearly basis.[73] Indeed, soil taxonomies emerged within an economic and political context in which agricultural practices related to soil were negotiated or enforced in the agricultural fields. The reference to flax and not to any other plant in this taxonomy suggests the importance of flax exports to the Egyptian economy and the coercive role of debt in providing a stable supply of flax for merchants in Fatimid Egypt (969–1171).[74]

Just as transforming the soil in sixteenth-century America justified settler-colonial activities, soil taxonomies from many different groups around the world eventually played a similar role in the modern period. Hundreds of classification systems of soil developed among almost every cultural group in

---

70 Lisbet Koerner, *Linnaeus: Nature and Nation* (Cambridge, MA: Harvard University Press, 1999).

71 Zitong Gong, Xuelei Zhang, Jie Chen, and Ganlin Zhang, "Origin and Development of Soil Science in Ancient China," *Geoderma* 115, no. 1–2 (2003): 3–13; Winiwarter, "Prolegomena to a History of Soil Knowledge in Europe."

72 John Evelyn, *A Philosophical Discourse of Earth Relating to the Culture and Improvement of It for Vegetation, and the Propagation of Plants* (London: John Martyn, 1676), 11.

73 R. S. Cooper, "Ibn Mammati's Rules for the Ministers: Translation and Commentary of Qawanin al-Dawanin" (PhD diss., University of California, 1973), 32–34.

74 Lorenzo M. Bondioli, "Islam, Merchants, and Capitalism: Fifty-Five Years in the Socio-economic History of the Medieval Islamic World," *Capitalism: A Journal of History and Economics* 4, no. 2 (2023): 258–307, on 291.

West Africa.[75] Such classifications of soil had not existed in isolation prior to the nineteenth- and twentieth-century spread of what is called today "soil science," which emerged in northwestern Europe.[76] But this kind of knowledge about soil did play an important role in the earliest formation of European "soil science," as Indigenous knowledge was studied and utilized by colonial officials who sent reports, observations, and samples to scientists in the colonial centers.[77] In many of these cases, soil knowledge meant agricultural expertise. Agricultural expertise meant a well-fed population. And a well-fed population meant a strong state or empire.

4.4 *Soil, Specialization, and Experimental Culture*

If studying practices rather than merely content provides the key to a comparative understanding of soil across multiple bodies of knowledge, then experiment—and the specialization that it enables—provides the practical methodology by which the most innovative soil knowledge emerged. Soil experiments across the medieval and early modern globe unsettle the idea that the earliest systematic attempts to understand the nature of soil and its effect on plants are bound to a western European context. Though once seen as one of the quintessential features of the European "scientific revolution" of the seventeenth century, historians of science have noted that experimentation—meaning, broadly speaking, systematic, observational, and empirical trials to confirm or refute hypotheses about nature—is neither a European nor early modern innovation.[78] The Arab polymath Hasan ibn al-Haytham (c. 965–c. 1040) conducted intricate experiments (*iʿtibār muḥarrar*) with lenses and mirrors for his *Kitāb al-Manāẓir* (Book of optics), while the Song bureaucrat Shen Kuo 沈括 (1030–1095) recorded his own experiments with magnetic needles along with the experiments of others with a camera obscura and cast iron forging for his *Mengxi bitan* 夢溪筆談 (Dream brook brush talks).[79] In

75 Mensa Bonsu, "Indigenous Knowledge: The Basis for Survival of the Peasant Farmer in Africa," *Journal of Philosophy and Culture* 1, no. 2 (2006): 49–61.

76 Brevik and Hartemink, "Early Soil Knowledge," 23–33.

77 For example, "Soil and Cultivation in Yoruba-Land," *Bulletin of Miscellaneous Information (Royal Botanic Gardens, Kew)* 46 (1890): 238–244.

78 James Delbourgo, "The Knowing World: A New Global History of Science," *History of Science* 57, no. 3 (2019): 373–399.

79 The term *iʿtibār muḥarra* A. I. Sabra defines as "experiment with precision or exactness (or certainty)—but still without the implication of numerical quantification." On ibn al-Haytham's use of the term, see Ibn al-Haytham, *The Optics of Ibn al-Haytham: Books I–II: On Direct Vision*, trans. and ed. A. I. Sabra (London: Warburg Institute, University of London, 1989), 14–19.

premodern Europe, major figures in the history of natural philosophy, including Aristotle, Galen, and Roger Bacon all conducted recognizable experiments in zoology, anatomy, and alchemy respectively.[80]

Experiments show us how the practice of working with soil and the emergence of systematic attempts to understand its nature informed the development of early modern sciences. In his chapter, Justin Niermeier-Dohoney examines alchemical interpretations of a dense, clay-like, mineral rich soil called marl in early modern England. While medieval European sources had long touted the physical benefits of mixing marl with lighter soils to make working the ground easier, it was not until the sixteenth and seventeenth centuries that alchemical experiments explaining its fertility began to emerge. During this time, as English farmers grappled with the prospect of decreasing yields due to unfavorable climatic conditions, the propagation of experimental agrarian improvement projects and the rise of vernacular alchemical interpretations of natural change intersected for the first time. Alchemical theories about the transmutation of natural substances provided a conceptual bridge across both botanical and mineralogical domains to explain to early modern English agrarian experimenters how marl-based fertilizers could transform infertile soils into fertile ones.

Sometimes experiments developed less through applied theory and more through the necessity of ad hoc solutions to everyday problems encountered by laborers and technicians. In her chapter, Deniz Karakaş explores the construction of hydraulic works—canals, pipes, sewage systems, drainage ditches, aqueducts—and shows how the "water technicians" and "earth workers" responsible for these projects in early modern Istanbul engaged with and gained extraordinarily specialized knowledge by evaluating the soil removed for these projects.[81] Through their practical expertise, these laborers became the principal observational and experimental experts on the types of soil, mud, and hydrogeological practices that had previously largely been the purview

80 David C. Lindberg, *Theories of Vision from Al-Kindi to Kepler* (Chicago: University of Chicago Press, 1976), 60–67; Nathan Sivin, *Science in Ancient China: Researches and Reflections* (Brookfield, VT: Ashgate Publishing, 1995), 21, 27, 34; David C. Lindberg, *The Beginnings of Western Science: The European Scientific Tradition in Philosophical, Religious, and Institutional Contexts, Prehistory to A.D. 1450*, 2nd ed. (Chicago: University of Chicago Press, 2007), 60–64, 124–131, 357–369.

81 On the social and epistemic status of "earth workers" and "water technicians," see Lydia Barnett, "Showing and Hiding: The Flickering Visibility of Earth Workers in the Archives of Earth Science," *History of Science* 58, no. 2 (2019): 245–274; Pietro Daniel Omodeo, "Hydrogeological Knowledge from Below: Water Expertise as a Republican Common in Early-Modern Venice," *Berichte zur Wissenschaftsgeschichte* 45, no. 4 (2022): 538–560.

of elites. However, the authority and needs of the Ottoman state, and more specifically the water channel superintendent's office (*ṣu nāẓırı*), often took precedence over the knowledge made by these water technicians, who typically came from more marginalized social backgrounds. Nevertheless, the practical authority of these technicians helped boost the epistemic value of experience and experiment in Ottoman culture.

While water technicians had become experts in certain types of soil in early modern Istanbul, an entirely separate social group of practitioners specialized in gaining knowledge about different types of soil through experimentation. In their chapter, Aleksandar Shopov and Himmet Taşkömür examine the earliest Ottoman books on flower breeding and the concurrent rise of a science of soil in mid-seventeenth-century Istanbul. They argue that these twin developments reflected the emergence of a community of practitioners in the city and its surroundings, and an experimental culture among flower breeders who saw soil as one of the crucial elements in the creation of new signature varieties of flowers. As flower breeding became recognized by the Ottoman state as a discrete body of knowledge (*ʿilm*), methods of improving soil were described, for the first time, as belonging to a specific "school of thought" (*mezheb*), a term originating in Islamic law. In this living discourse, both new floral varieties themselves and the techniques used to generate them became associated with individual breeders.

Experimental practices could also forge common ground between distinct traditions that otherwise had little in common. For example, northern India during the sixteenth and seventeenth centuries could boast of two distinct traditions of technical agricultural literature that coexisted and at times influenced each other. One, composed in Sanskrit, was developed entirely in the subcontinent, while another, in Persian, included texts written in Iran and Afghanistan—often drawing on Byzantine Greek sources—as well as adaptations, recompilations, and original works produced in India. One of the more striking features found in both traditions was an emphasis on taste as an essential mode of knowing soils and judging their qualities. In the final chapter of this volume, Nicolas Roth argues that, as empirical knowledge gained greater prominence vis-à-vis received tradition, these methodologies of experimental and experiential knowing of the soil by way of ingesting it and "savoring" its flavors became essential in these texts. Analyzing this shift thus not only sheds light on a sensory dimension of soil that is often neglected today, but also speaks to evolving ways of knowing in early modern South Asia.

•••

In the Sufi mystical poem "I climbed the plum tree branch to eat grapes," an absurdist listing of various material paradoxes ("the stork gave birth to a donkey foal") meant to "hide the face of meaning from the faces of hypocrites," the Anatolian mystic poet Yunus Emre (d. 1320?) claims:

> I put mudbrick [*kerpiç*, i.e., mud mixed with hay] in a cauldron and boiled it with the northeast wind. When someone asked me what it was, I scooped it and gave it to him.[82]

These verses frame the mystical experience of the Sufi and his relationship with God and self within the material world of Anatolia. Mud cannot be cooked with wind; yet these lines convey an understanding of mud as subject to material transformations. Sun-dried mud mixed with hay was used for building houses. Indeed, the material coming from ruined buildings was regarded by Yunus Emre's contemporary Ilkhanid and Mamluk agricultural writers as the best fertilizer.[83] According to an ownership deed from 1290, piles of dirt and garbage in central Anatolia had economic value and were considered private property.[84] Mud was also moved and taxed across fields to serve as fertilizer in newly established rice fields in western Anatolia.[85] Yet, by handing the person who asked about the dried mud the mixture itself, Yunus Emre was suggesting that he, the inquirer, and all of us, come from the soil and rely on it for everything. By focusing on practices related to soil between 1300 and 1750, this volume shows the fluctuating and changing understanding of soils across Afro-Eurasia and the Americas. In the process of changing soil, those who worked with it also changed history.

---

82 Yūnus Emre, *Yûnus Emre dîvânı*, ed. Mustafa Tatcı (Istanbul: Millî Eğitim Bakanlığı, 1997), 428–430, authors' translation.

83 See Himmet Taşkömür's contribution to this volume; see also Anthony T. Quickel and Gregory Williams, "In Search of Sibākh: Digging Up Egypt from Antiquity to the Present Day," *Journal of Islamic Archaeology* 3 no. 1 (2016): 89–108.

84 Halil Sahilioğlu, "Ikinci Keykâvüs'ün bir Mülknamesi," *Vakıflar Dergisi* 8 (1969): 62.

85 See Telci, *Halil Beğ Defteri*.

PART 1

Translation and Transmission of Soil Knowledge

CHAPTER 1

Soil Practices in the Kashmiri-Language Mystical Poetry of Shaykh Nuruddin (c. 1377–1440) in the Persian Textual Tradition

Between Agriculture and Sufi Mysticism

*Zubair Khalid**

1 Introduction

> Why would you sow grains in a sandy soil?
>
> NURUDDIN

Nuruddin, originally named Nund, was born in c. 1377 at Khee in the Kulgam area of south Kashmir. This son of a migrant family, with no formal education, gained recognition in his lifetime for his Kashmiri-language poetry and for founding a local mystic order, the Rishi silsila.[1] Nuruddin used highly technical agricultural metaphors to explain a range of Islamic beliefs and rituals to a primarily rural audience of agriculturists with little background in Islam.[2] A testimony to his credentials as a scholar-saint and an erudite agriculturist lies in the fact that the specialized terminology of agricultural, specifically soil-related, and mystical knowledge that he gathered and used in his vernacular poetry later migrated into the Persian textual tradition. While he did not

* ORCID: 0009-0003-8710-5197

1 For more on Nuruddin's religious life and his poetry, see G. N. Gauhar, *Sheikh Noor-ud-Din Wali* (*Nund Rishi*) (Delhi: Sahitya Akademi, 1988), 27–42. Some modern scholars see Nuruddin as a reviver, rather than the founder, of the Rishi order. For a discussion, see Basher Bashir, "Arz-i Nashir" [Publisher's introduction], in *Noor-Nama: Baba Naseebudin Ghazi key Farsi Noor-Nama ka Urdu Tarajuma* [Noor-Nama: An Urdu translation of Baba Naseebudin Ghazi's Persian Noor-Nama], trans. and ed. Marghoob Banihali (Srinagar: Markaz-i Noor Centre for Sheikh-ul Alam Studies, Kashmir University, 2013), i–vii.

2 As Mohammed Ishaq Khan argues, his poetry played a decisive role in Islamizing the rural non-Persian-speaking population of premodern Kashmir, a group the Persian-speaking immigrant Sufis could not otherwise address due to the language barrier. See Mohammed Ishaq Khan, *Kashmir's Transition to Islam: The Role of Muslim Rishis, Fifteenth to Eighteenth Century* (New Delhi: Manohar, 1994).

DOI:10.1163/9789004748484_003

commit his poetry to writing, others recorded it during his lifetime. Given his status as the preeminent local Sufi mystic of his time, local hagiographers in medieval and early modern Kashmir compiled, translated, and interpreted his Kashmiri-language poetry in Persian-language texts, which enabled a movement of knowledge across linguistic and literary traditions. This chapter demonstrates how Nuruddin's poetry played a vital role in the transfer of soil-related metaphors from the Kashmiri language to Persian-language hagiography. Nuruddin's use of soil metaphors in his poetry points to larger patterns of engagement with soil knowledge in premodern South Asia, a theme also discussed in Nicolas Roth's chapter in this volume.

I approach Nuruddin as a broker, a go-between navigating knowledge networks, and this chapter underlines how his poetry helped the movement of knowledge across various traditions and languages: from oral to textual, from Kashmiri to (among others) Persian.[3] Using this approach makes it possible to reinterpret the regional Sufi *tazkira* genre and to move beyond the dichotomous frame of fact versus fiction through which these hagiographies have often been analyzed.[4] This Kashmiri-language form of biographical anthology dedicated to poetry proved to be a potent medium of exchange for scientific and material knowledge in medieval and early modern Kashmir and, by extension, the larger Persianate world. Due to favorable soil and climate conditions, agriculture prospered in the region and was widely practiced in medieval Kashmir. Given the pivotal role of agriculture in the regional economy and the significance of agricultural products in different socioreligious occasions, it comes as no surprise that soil-related metaphors featured prominently in local literature. Nuruddin, however, used them to an unprecedented degree in his vernacular Sufi poetry. The following chapter explores the birth of a new genre of Persian-language Sufi hagiography, the *Rishi-Nama* or *Noor-Nama*, which, by recording Nuruddin's vernacular poetry and the history of his Sufi order, helped incorporate these highly technical Kashmiri-language metaphors of soil into the Persian textual tradition.

3 On the role of "brokers" and "go-betweens" in the functioning of early modern knowledge networks, see Harold J. Cook and Sven Dupré, introduction to *Translating Knowledge in the Early Modern Low Countries*, ed. Harold J. Cook and Sven Dupré (Zurich: Lit Verlag, 2012), 3–17; Simon Schaffer, Lissa Roberts, Kapil Raj, and James Delbourgo, eds., *The Brokered World: Go-Betweens and Global Intelligence, 1770–1820* (Sagamore Beach, MA: Science History, 2009).

4 For the dichotomous historiographical approach of myth versus history, see Mufti Mudasir, "Holy Lives as Texts: Saints and the Fashioning of Kashmir's Muslim Identity," *Philological Encounters* 1, nos. 1–4 (2016): 288–312.

The Himalayan region of Kashmir has a long agricultural history, and sources confirm that soil cultivation was practiced by its earliest inhabitants.[5] A network of rivers and streams, regularly fed by rain and snow, runs through the valley, ensuring fresh water for the local population and plentiful irrigation for crops.[6] Just as crucial to agriculture is the abundance of fertile soil beds created by the Jhelum River with its numerous contributory streams as it wound its course through the plains of the valley.[7] This advantageous combination of fertile soil and ample water resources not only facilitated the cultivation of crops in general but also of water-intensive crops such as rice. Some of the oldest extant regional texts attest to the enduring cultivation of rice in Kashmir.

References to agricultural activities in the region can be found in texts written as early as the seventh century CE, and others from late medieval Kashmir trace the emergence of agriculture in the valley to the time of its first settlers. One of the earliest extant texts from Kashmir, a seventh-century Sanskrit text, the *Nilamata Purana* (The instructions of Nila), describes Kashmir as a land "filled with rows of rice-fields," "endowed with the qualities (of producing) all grains" and "with good fruits."[8] The same text also contains references to the region's agricultural calendar and describes several agricultural ceremonies associated with different agricultural practices in the region.[9] The anonymously composed late sixteenth-century Persian chronicle, *Baharistan-i-Shahi* (Abode of the spring of kings), one of the earliest Persian chronicles from Kashmir, specifically credits a person named Bekadarat with initiating agriculture in Kashmir soon after it had emerged as a piece of land from a primordial lake, and describes him as having "sowed many kinds of seed [*anwāʿ takẖm*] in

5 The history of agriculture in the Himalayas has been traced to as early as the Neolithic period. See K. P. Nautiyal, Pradeep M. Saklani, and Vinod Nautiyal, "Agriculture in the Garhwal Himalayas: An Ethnographic Perspective," in *History of Agriculture in India (up to c. 1200 AD)*, ed. Lalanji Gopal and V. C. Srivastava (New Delhi: Concept Publishing Company, 2008), 159–170.

6 Aparna Shukla and Iram Ali, "Major River Systems of Jammu and Kashmir," in *The Indian Rivers: Scientific and Socio-Economic Aspects*, ed. Dhruv Sen Singh (Singapore: Springer, 2018), 383–411.

7 For an account of the river systems, soil types, and climate of Kashmir, see M. A. Stein, *Memoir on Maps Illustrating the Ancient Geography of Kashmir* (Calcutta: Baptist Mission Press, 1899), 82–120.

8 Ved Kumari [Ghai], *The Nilamata Purana*, vol. 2, *A Critical Edition & English Translation* (Srinagar: J&K Academy of Art, Culture and Languages, 1994), 5–6.

9 For a summary, see Ved Kumari, *The Nilamata Purana*, vol. 1, *A Cultural and Literary Study of a Kashmiri Purana* (Srinagar: J&K Academy of Art, Culture and Languages, 1968), 125–128.

the muddy soil [*be gil-ha kandeh*] and raised crops."[10] Another Persian chronicle authored by the early seventeenth-century Mughal administrator of Kashmir, Haidar Malik Chadurah's *Tarikh*, completed in 1030 AH (1620–1621 CE), traces the emergence of agriculture in the region to the first group of its inhabitants, who initiated agriculture (*ḥiraṣ wa g̱hiras*) after the Islamic prophet Solomon ordered the draining of what was originally an uninhabited lake.[11]

Agriculture was widely practiced in medieval Kashmir, and agricultural products formed a significant part of a range of socioeconomic practices in the region. This is especially true of rice (*orizum sativum*), often labeled the staple produce of the region, and described as *dhânyã*[12] or *śāli*[13] in the Sanskrit chronicles of Kashmir.[14] Sanskrit and Persian chronicles from early modern Kashmir confirm this, with references to feasts of rice organized by the contemporary Shahmiri sultans of Kashmir, to granaries of rice located in villages, as well as mentions of storehouses of paddy and the cooking and serving of rice at local Sufi hospices in fifteenth-century Kashmir.[15] One of the only

10 For the Persian terms, see Akbar Haidari, ed., *Baharistan-i-Shahi* [Abode of the spring of kings] (Kashmir: s.n., 1982), 248. For the English translation of the terms, see K. N. Pandit, trans., *Baharistan-i-Shahi: A Chronicle of Mediaeval Kashmir* (Calcutta: Firma KLM Private Limited, 1991), 1.

11 Haidar Malik Chadurah, *History of Kashmir*, ed. and trans. Raja Bano (Srinagar: JayKay Books, 2016), 7, 23. The medieval Persian texts of Kashmir Islamized the Sanskritic legend of Satisara, which referred to the origin story of Kashmir emerging from a lake. For the original legend, see Stein, *Ancient Geography of Kashmir*, 65–67. For a recent interpretation, see Khalid Bashir Ahmad, *Kashmir: Exposing the Myth behind the Narrative* (New Delhi: Sage Publications, 2017), 5–6.

12 For *dhânyã* as grain, see Arthur Anthony Macdonell, *A Practical Sanskrit Dictionary with Transliteration, Accentuation, and Etymological Analysis throughout* (London: Oxford University Press, 1929), 132.

13 For *śāli* as rice, see Macdonell, *A Practical Sanskrit Dictionary*, 312.

14 For rice as "the largest and most important produce of the valley," see M. Stein's note in his translation of Kalhana's *Rajatarangini*. M. A. Stein, trans., *Kalhana's Rajatarangini: A Chronicle of the Kings of Kaśmīr* (Westminster: Archibald Constable, 1900), 2:427n78. According to him, Sanskrit chronicles of Kashmir referred to rice as "the grain." For an account of the prevalence of extensive rice cultivation in nineteenth-century Kashmir by the then Settlement Commissioner of Kashmir, see Walter Lawrence, *The Valley of Kashmir* (London: Oxford University Press Warehouse, 1895), 319.

15 The fifteenth-century Sanskrit chronicler of Kashmir, Pandit Shrivara, mentions rice feasts and granaries during the reign of Sultan Zain (1420–1470 CE). See Pandit Shrivara, *Kings of Kashmira Third Series (Shrivara)* [*Rajatarangini*] in *The Rajatarangini of Jonaraja*, trans. Jogesh Chunder Dutt (1898; repr., Delhi: Gian Publishing House, 1986), 98–336, on 139–140. The early seventeenth-century Persian chronicle *Baharistan-i-Shahi* refers to the daily cooking of rice and storage of paddy in the storehouses at Shamsuddin Araki's Zadibal hospice in Srinagar, Kashmir. See Pandit, *Baharistan-i-Shahi*, 47. Another sixteenth-century Persian hagiography, *Tohfat-al Ahbab*, describes similar practices of

extant contemporary Sanskrit chronicles from this period, the *Rajatarangini* (*River of Kings*) of Pandit Jonaraja (d. 1459), which he wrote as a continuation of a twelfth-century Sanskrit chronicle of the same name by Kalhana (d. ca. 1150), gives us an idea about the regional specialization in agricultural production in medieval Kashmir.[16] According to Jonaraja, Kashmir was "adorned ... by the vine, and the saffron, the grain, and the *sháli* rice," kings would organize feasts of rice, and it was used as a currency as well.[17] *Ain-i Akbari* (Mirror of governance of Akbar), one of the most important sixteenth-century Mughal administrative documents, authored by court historian Abul Fazl ibn Mubarak (d. 1602), details the revenue system and crop cultivation patterns in sixteenth-century Kashmir, describing rice as the staple crop.[18] Another early seventeenth-century source, the *Tuzuk-i Jahangiri* (Memoirs of Jahangir) of Mughal ruler Nur al-Din Jahangir (r. 1605–1627), similarly attests to the wide-scale prevalence of rice cultivation in Kashmir, noting that rice was cultivated on 75 percent of total cultivable land in the valley.[19] Descriptions of economic transactions in pre-twelfth-century Kashmir by the famous twelfth-century Sanskrit chronicler of Kashmir, Kalhana, indicate that grain, especially rice, formed one of the primary media of currency in the region, a trend still in vogue as late as the nineteenth century, according to the British Settlement Commissioner of Kashmir at the time, Walter Lawrence.[20] Extraction of land

daily cooking and serving of rice to the Sufis of the Khanqah, besides storage of paddy at one of the most famous Sufi hospices in Kashmir, the Hamadanniyeh hospice. See Muhammad Ali Kashmiri, *Tohfat-al Ahbab* [Gift for friends], trans. K. N. Pandit (New Delhi: Voice of India, 2009), 145, 183, 207, 246.

16 *Rajatarangini* is a genre of Sanskrit chronicles from Kashmir written between the twelfth and the sixteenth centuries that derive their name from the *Rajatarangini* of Kalhana, written in the twelfth century during the reign of King Jayasimha (r. 1128–1155). He was followed by Jonaraja (d. c. 1459), who, while in the court of Shahmiri ruler Sultan Zain-ul Abidin (1420–1470), complied a similar chronicle. Three more chroniclers, including Shrivara, followed Jonaraja and compiled their versions of the *Rajatarangini*. See Satoshi Ogura, "Rajatarangini," *Perso-Indica: An Analytical Survey of Persian Works on Indian Learned Traditions*, ed. F. Speziale and Carl W. Ernst, June 12, 2020, http://www.perso-indica.net/work/rajatarangini_%28zayn_al-abidin_translation%29.

17 For Jonaraja's references to agricultural practices and rice cultivation in Kashmir, see *Rajatarangini of Jonaraja*, trans. Dutt, 430.

18 Abul Fazl Allámi, *The Ain-i Akbari* [The mirror of governance of Akbar], trans. H. S. Jarret (Calcutta: Asiatic Society of Bengal, 1891), 2:366–371.

19 Nur al-Din Jahangir, *The Tuzuk-i Jahangiri; or, Memoirs of Jahangir*, trans. Alexander Rogers, ed. Henry Beveridge (Delhi: Munshiram Manoharlal, 1968), 2 vols., 2:146.

20 Stein, *Kalhana's Rajatarangini*, 2:313. Lawrence mentions that in the 1880s he was offered oil seeds as his salary and that others were also paid salaries in grain. Lawrence, *Valley of Kashmir*, 243.

revenue (taxes) from cultivators by the state in the form of crops instead of cash was a well-established practice in Kashmir, as evidenced by textual references from, and even before, medieval times.[21] Similarly, assigning revenue-villages to persons and institutions was a well-established mode of state patronage in medieval Kashmir, with references in locally produced Sanskrit and Persian sources emerging from the medieval period. For instance, *Tarikh-i-Sayyed Ali* (Chronicle of Sayyed Ali), a local Persian history written around the 1560s, refers to the allotment of revenue-land to Sufi personalities and institutions by the Shahmiri ruler Sultan Sikander (r. 1389–1413) to provide an income to cover their personal and institutional expenses.[22]

Given its centrality to the regional economy, agriculture has historically witnessed significant state intervention through the extension of irrigation facilities and the building of new canals. In fact, one of the main reasons for the prominence of the ninth-century Kashmiri king Avantivarman (r. 855–883) was his massive project to construct an extensive network of canals in the region with the help of his engineer-minster Suyya, who, in turn, is described as having accumulated religious merit on account of his tremendous efforts on these irrigation schemes.[23] Similarly, the Sanskrit and Persian chronicles from medieval and early modern Kashmir make a note of a significant number of agricultural projects initiated by the regional rulers, the sultans, including the extension of a network of irrigation canals and attempts to bring new areas of land under cultivation.[24] Agriculture also directly impacted the regional political dynamics, especially between the eleventh and the thirteenth centuries, when an agriculturist warrior group, the Damaras, played a role in deciding the political fortunes of contenders to the throne of Kashmir.[25]

---

21 For pre-Sultanate-era references, see Stein, *Kalhana's Rajatarangini*, 2:6, 96. For references regarding the Sultanate period, see Shrivara, *Rajatarangini*, 156.

22 According to the text, the pargana Nagama was assigned to a Central Asian Sufi saint, Sayyid Taj'ud Din, who had recently migrated to Kashmir. Similarly, villages were assigned for the maintenance of the famous Sufi hospice, the Khanqah-I Mou'lla, located in Srinagar. See Sayyed Ali, *Tarikh-i-Sayyed Ali: History of Kashmir; 1374–1570*, trans. Zubida Jan (Srinagar: Jay Kay Book Shop, 2009), 30, 47.

23 Stein, *Kalhana's Rajatarangini*, 1:200–201.

24 For references to Sultan Zain's canal building projects and expansion of cultivable area by his contemporary Sanskrit chroniclers, see Jonaraja, *Rajatarangini*, 87. See also Ali, *Tarikh-i-Sayyed Ali*, 58.

25 The translator of Kalhana's *Rajatarangini*, M. A. Stein, describes them as "feudal landholders," most of whom belonged to the tribal division of the Lavanyas in pre-Sultanate Kashmir. Kalhana also describes them as having won victories against the twelfth-century Kashmiri king Harsha, even as the same king is subsequently described as persecuting them. Stein, *Kalhana's Rajatarangini*, 1:19.

Various agricultural products, including rice, had an essential function in different socioreligious festivals and rituals in medieval Kashmir. Kalhana's twelfth-century *Rajatarangini* refers to several such festivals and rituals involving the use of rice or other grains. For instance, there are references to grains being thrown at sacrificial oblations, rice and ghee offered for *shraddha* fortnights, as well as sesamum grains at the *Tiladvadasi* festival.[26] A mid-sixteenth-century Persian chronicle of the Mughal governor of Kashmir, Mirza Haidar Dughlat (r. 1541–1551), *Tarikh-i Rashidi* (*Chronicle of Rashid*), similarly attests to the association of rice with several ritual practices in the valley.[27]

Given the importance of agriculture in the region, metaphors of soil and other agricultural terms came to be used as literary and poetic devices in a range of works from Kashmir, even before Nuruddin's time. For instance, Kalhana used the metaphor of sowing "corn in barren soil" in his *Rajatarangini* to refer to the appointment of undeserving persons to positions of authority, whereas "planting trees in the right soil" refers to a proper upbringing.[28] Metaphors employing technical terms for soil, such as *shāṭh* (stony and sandy soil)[29] and *trāg* (a pond overgrown with vegetation),[30] are used similarly by Nuruddin's senior contemporary, the famous fourteenth-century mystic poet Lalla Ded (d. 1392), in her Kashmiri-language poetic verses.[31] Nuruddin's contemporary, Jonaraja, similarly used several agricultural allegories in his fifteenth-century Sanskrit chronicle to describe the changing socioreligious configurations in Kashmir in the wake of increased immigration into the region from Persianate regions and the subsequent spread of Islam in the following manner: "As the

26 See Stein, *Kalhana's Rajatarangini*, 1:155, 228, 2:326.

27 Mirza Muhammad Haidar Dughlat, *A History of the Moghuls of Central Asia: The Tarikh-i Rashidi of Mirza Muhammad Haidar, Dughlat*, trans. E. Denison Ross (Delhi: Gyan, 2022), 428. The text was written by Dughlat as a Mughal governor of Kashmir toward the end of the first half of the sixteenth century.

28 Kalhana, *Rajatarangini*, 1:95, 145.

29 See Sir George Grierson, *A Dictionary of the Kashmiri Language, Compiled Partly from Materials Left by the Late Pandit Isvara Kaula* (Calcutta: Asiatic Society of Bengal, 1916), 898.

30 See Grierson, *Dictionary of the Kashmiri Language*, 1020.

31 For the two terms, see Ghulam Qadir Bedar, *Lal Ded aur Lal Vaakh: Mua Tarjama Urdu* [Lal Ded and Lal Vaakh: With Urdu translation] (Srinagar: Sheikh Usman and Sons Tajiran-i Kutb, 2016), 103, 107. For other agricultural metaphors in Lalla's poetry, see Ranjit Hoskote, *I, Lalla: The Poems of Lal Ded* (New Delhi: Penguin Books, 2011), 36, 239, 254. Lalla Ded is known as one of the most famous spiritual and literary figures of Kashmir. Her Kashmiri-language poetry, even though compiled at a later stage, is among the earliest known manifestations of a Kashmiri literature.

wind destroys the trees, and the locusts the shāli crop, so did Yavanas destroy the usages of Kashmīra."[32]

More specifically, Jonaraja used a metaphor of stony versus fertile soil to highlight the alleged innate political incompetence of a twelfth-century ruler of Kashmir, Jassaka (r. 1180–1198): "The parrot that imitates the human is caught, but not the crow; stony soil is not ploughed and dug like fertile land; and stones are not powdered to dust like rock-salt."[33]

References to agricultural terms commonly appear in the local hagiographic tradition, because, as historian Chitralekha Zutshi has argued, these works sought to portray Sufi saints as agriculturists who productively transformed the regional landscape and thereby supported their protagonists' claims of authority over it.[34] However, the language, context, and manner in which Nuruddin employed agricultural metaphors, including those related to soil, has led others to label Nuruddin's case as essentially unique.[35] As the following section makes clear, Nuruddin employed metaphors of soil in the Kashmiri language in a didactic sense, often alongside Islamicate and sometimes explicitly Qur'anic terminology at a time when Islam was a recent phenomenon in Kashmir, earning him the title of earliest "translator of the Qur'an" in the region, and his poetry the title of *Koshur Qur'an* (literally, the "Kashmiri Qur'an").[36] Subsequently, local Persian chronicles and hagiographies revered Nuruddin and his order, the Rishi silsila, with several works, including that of the sixteenth-century Suharwardy Sufi Baba Dawud Khaki (d. 1585), who praised Nuruddin's poetry specifically for its religious discourses.[37] One of the

32 Jonaraja, *Kings of Kashmira Second Series* (*Jonaraja*) [*Rajatarangini*], in *Rajatarangini of Jonaraja*, trans. Jogesh Chunder Dutt, 1–97, on 57. "Yavanas" is a Sanskrit term used by Jonaraja and other Sanskrit chroniclers of Kashmir to denote non-Sanskritic western immigrants to Kashmir. "The usages of Kashmīra" refer to the customs of Kashmir.

33 Jonaraja, *Kings of Kashmira Second Series* (*Jonaraja*), 7. For a discussion on Nuruddin's poetry as representing aspects of contemporary poetic tradition, see Shafi Shouq, *Sheikh-ul Alam te tasund zamane* [Sheikh-ul Alam and his time] (Kapran: Bazm-i Adab, 1978).

34 Chitralekha Zutshi, *Kashmir's Contested Pasts: Narratives, Sacred Geographies, and Historical Imagination* (New Delhi: Oxford University Press, 2014), 70.

35 See, for instance, Khan, *Kashmir's Transition to Islam*; Asadullah Afaqi, *Aayin-e haq ya'ni Kulliyat-i Sheikh-ul Alam* [The mirror of truth, that is, complete works of Sheikh-ul Alam] (Srinagar: Life Foundation, 2008), 35–40.

36 For a history of Islam in the region, see Khan, *Kashmir's Transition to Islam*; Muhammad Ashraf Wani, *Islam in Kashmir: Fourteenth to Sixteenth Century* (Srinagar: Oriental Publishing House, 2004). His poetry is termed as *Koshur Qur'an* since the bulk of it is a rendition of the Qur'an and Hadith. See Gauhar, *Sheikh Noor-ud-Din Wali*, 54.

37 Baba Dawud Khaki, *Dastur-us Salikin Sharh-e Wirdul Murideen* [A manual for the divine seekers, commentary on Wirdul Murideen] (Srinagar: Ghulam Muhammad Nur Muhammad Tajiran-i Kutb, n.d.), 213–214.

earliest instances of Nuruddin's life sketch and poetry entering the Persian textual tradition is in the seventeenth-century hagiography, *Noor-Nama* (Book of light), authored by a disciple of Khaki, Baba Naseebudin Ghazi (d. 1637).[38] This set the stage for a new trend in the local Persian textual tradition, which continued well into the nineteenth century, whereby either original Kashmiri versions or Persian translations of Nuruddin's poetry were incorporated into Persian-language hagiographies, along with a short biography.

2 Between Agriculture and Sufi Mysticism: Soil Imagery in Nuruddin's Mystic Poetry

Shaykh Nuruddin Rishi (or, more commonly, Nund Rishi) was born in 1377 in the Khee village of south Kashmir.[39] Soon after his marriage, Nuruddin is said to have left his family and spent most of his life traveling and in solitary meditation across Kashmir.[40] During these travels and sojourns, Nuruddin composed and narrated his poetry, with certain verses and poems directly narrating his encounters with various people, including farmers, administrators, and Sufi saints, among others.[41] Nuruddin died around 1440 and is buried at a place called Tsrar, now known as *Tsrar-i shareef* (the holy Tsrar) and revered as one of the most prominent Sufi shrines in Kashmir.

Nuruddin's birthplace of Khee was in Kulgam, an area that, owing to its location around the basin of the Jhelum River and the consequent abundance of alluvial soil, was—and still is—considered very suitable for agricultural activities.[42] This may be why the sixteenth-century *Tarikh-i Rashidi* described Devsar, the then larger administrative division that incorporated areas to the west of upper Jhelum, including Kulgam, as one of the important districts of

38 *Noor-Nama* is the title of Ghazi's Persian-language work. At the same time, *Noor-Nama*, alternatively called *Rishi-Nama*, also refers to a hagiographic genre of writings about Nuruddin. See Bashir, "Arz-i Nashir."

39 Nuruddin was born as Nund. Persian hagiographers mentioned him as Shaykh Nuruddin, while in Kashmiri he came to be called Nund Rishi, the latter term signifying his association with the local mystic Rishi order. His sobriquets include Alamdar-i Kashmir (The standard-bearer of Kashmir), Noorani (The illuminated), Shaykh-ul Alam (The shaykh of the world), Shazanand (One who has attained the ultimate truth), and Wali (The friend of God). See Gauhar, *Sheikh Noor-ud-Din Wali*, 20.

40 For an account of Nuruddin's life sketch, including his travels, see Gauhar, 16–43.

41 For an explanation of Nuruddin's verses in English along with the mention of "stories" associated with them, see Nund Rishi, *Unity in Diversity*, trans. B. N. Parimoo (Srinagar: J&K Academy of Art, Culture and Languages, 1984).

42 Rattan Lal Hangloo, "Agricultural Technology in Kashmir (AD 1600 to 1900)," *Medieval History Journal* 11, no. 1 (2008): 63–99.

Kashmir.[43] This area was significant in terms of agricultural output. According to Abul Fazl's *Ain-i Akbari*, Kulgam accounted for a substantial share of the overall revenue from the administrative division of the Marraj tract of which it was a part.[44] Not surprisingly then, Nuruddin described himself as being quite familiar with agriculture, a justified claim, one would say, given the depth of his agricultural knowledge, which the following passages from Nuruddin's poetry make clear:

> Nund Rishi is conversant with agriculture;
> one who celebrates the Gongol will surely celebrate the Kraav.[45]

Nuruddin used several soil-related agricultural metaphors in the context of discourses on a wide range of mundane everyday realities and more complex mystic doctrines, thereby bringing together the agricultural and the mystical. Soil-related metaphors are often used alongside terms referring to a wide range of Islamic and specifically Sufi doctrines, rituals, and practices—at times even within a single verse. As such, references to almost the entire agricultural cycle, from selecting the most suitable soil to preparing the soil for cultivating different crops, find repeated mention in his verses.

As the legend goes, Nuruddin, sent by his mother to learn the craft, was perplexed by the alternate upward and downward movement of the feet of the weaver he saw working on a loom. Hence, he came up with a deeper mystic explanation of the cyclic movements of the weaver's feet: he understood one foot to represent the spurning of soil, while the other insisted on the importance of soil to human life, thus emphasizing the inappropriateness of spurning it.[46] This, according to his seventeenth-century Kashmiri hagiographer Ghazi, was the context in which Nuruddin expressed the significance of soil by highlighting its centrality to the cycle of human life, such as its creation, sustenance, and the eventual merger of the human body with the very material of its creation after its death:

---

43 Dughlat, *Tarikh-i Rashidi*, 428.

44 According to the *Ain-i Akbari*, with an annual revenue of 85,644 *kharwars* in kind, Devsar was one of the significant contributors among the eleven *Mahals* in the southeastern part of the Marraj tract. See Abul Fazl Allámi, *Ain-i Akbari*, 2:369.

45 Moti Lal Saqi, *Kulliyat-i Sheikh-ul Alam* [The complete works of Sheikh-ul Alam] (Srinagar: J&K Academy of Art, Culture and Languages, 1985), 163. Gongol and Kraav are two local festivals. On Nuruddin's names, see n. 39 above.

46 For the legend, see Ghazi, *Noor-Nama*, trans. Banihali, 37–38 (see n. 1); Nund Rishi, *Unity in Diversity*, 33–41.

> Adam was made out of soil,
> and his frame was cast in soil;
> out of soil grow all the bounties;
> in pots of soil, we cook our food;
> into the soil is the body consigned after its death,
> and the soil gets mixed with soil at the end.[47]

As such, when Nuruddin alluded to the Qur'anic story of the creation of Adam from the soil to highlight its centrality to cosmology and human life, he was also referring to more mundane and material uses of soil. Nuruddin also used different soil types as metaphors to explain matters of Islamic faith and spirituality. For instance, the line of verse that opens this chapter uses the term *sekh-shāṭh* (literally, "sandy soil")[48] in this context:

> Why would you sow grains in a sandy soil,
> bow to an illiterate, or scratch an elephant?[49]

A different couplet employs terms used for two different soil types—*vīr-kôn*[50] and *bangī-kôn*[51]—in order to emphasize the futility of looking for things in the wrong place:

> Don't expect heaps of rice from a land choked by the hemp,
> grains won't grow in a land choked by the willow tree.[52]

Nuruddin uses the metaphor of *vīr-kôn*, which refers to a willow-choked land. Because of their dense root system, which helps prevent soil erosion, willow trees are still a common sight around the paddy fields in Kashmir. However, their roots invade the wet fields, making any land within their reach unfit for cultivation.

---

47 Nund Rishi, *Unity in Diversity*, 36, translation modified; see also Saqi, *Kulliyat*, 35; Ghazi, *Noor-Nama*, 37–38.

48 For the meaning of the term, see Grierson, *Dictionary of the Kashmiri Language*, s.v. "sekh-shāṭh."

49 Afaqi, *Kulliyat*, 260.

50 See Grierson, *Dictionary of the Kashmiri Language*, s.vv. "vīr," "kôn."

51 *Bangī-kôn* means a piece of land choked up by hemp (*bangī*), which grows naturally in wastelands otherwise left uncultivated. For *bangī* and *kôn*, see Grierson, *Dictionary of the Kashmiri Language*, 113, 452.

52 Saqi, *Kulliyat*, 41.

Similarly, the cultivation of hemp (*bangī*) on a piece of land prepared for the cultivation of saffron (*kong-khah*)[53] is used as a metaphor for tragedy in the following line:

> Alas! Who cultivated hemp in my saffron field![54]

In Kashmir, saffron is cultivated in the elevated tablelands, known locally as *karewas*. The crop requires scant irrigation, and saturation can cause significant damage.[55] Thus, Nuruddin used the unsuitability of watering the soil intended for saffron cultivation as a metaphor for a heedless life wasted in worldly pleasures in the following line:

> water won't suit a saffron field.[56]

Referring to the practice of soaking seeds before they were sown in the field to make them sprout quickly, locally known as *kah*,[57] Nuruddin used the metaphor of an unsoaked seed to describe a follower (*murid*) without a spiritual master (*pir*), along with a host of other metaphors:

> A disciple without a preceptor is a collar without a yoke,
> . . . .
> A sheep without wool, and a seed that does not sprout.[58]

Similarly, the practice of erecting embankments, or dykes, around fields to regulate the flow of water in them, *dongd gandun*,[59] is used metaphorically to refer to the regulation of biological mechanisms such as appetite and sleep and is described as a potent way of gaining closeness to God:

> He who gains knowledge of the soul within,
> Is redeemed from hell by God:

53 See Grierson, *Dictionary of the Kashmiri Language*, s.vv. "kong" and "khah."
54 Saqi, *Kulliyat*, 110.
55 Latief Ahmad, Sabah Parvaze, Saqib Parvaze, and R. H. Kanth, "Crop Water Requirement of Saffron (*Crocus sativus*) in Kashmir Valley," *Journal of Agrometeorology* 19, no. 4 (2017): 380–381.
56 Afaqi, *Kulliyat*, 119.
57 For the meaning, see Grierson, *Dictionary of the Kashmiri Language*, s.v. "kah."
58 Nund Rishi, *Unity in Diversity*, 252, translation modified.
59 For the meaning of the term, see Parimoo's note in Nund Rishi, *Unity in Diversity*, 266n7.

> He who contains hunger, sleep, and attachments
> Gets admittance to God, spontaneously.[60]

Some of Nuruddin's metaphors incorporate procedures or equipment involved in preparing the soil for planting seedlings into an exposition of several Islamic rituals, most of them specifically related to Islamic prayer. For example, soil-related agricultural terms such as *alafal* (the plowshare) and *vaan* (the tilling of the soil) are equated with *ghusl* (the ritual bath and the recitation of the Qur'an) in the following verses:

> Thy plowshare is thy bath and ablutions: perform them well,
> . . . .
> Tilling the soil is reading the Qur'an: read it correctly.[61]

The extensive use of specialized agricultural and soil-related terminology in Nuruddin's poetry indicates that he was well versed in the knowledge and practice of agriculture in Kashmir. Marshalling this knowledge, his poetry vividly used soil-related terminology as potent metaphors to explain and elaborate otherwise complex Sufi mystic doctrines to an audience who may have lacked formal education but were well versed in agricultural practices. His ability to appeal to a wider audience by evoking well-known agricultural knowledge may explain why many modern historians consider him the preeminent saint-poet of the Kashmiri language.

3 The Journey of Nuruddin's Poetry from Kashmiri to the Persian Textual Tradition

By the time Nuruddin's poetry was written down, a vibrant culture of multilingual translations was well established in Kashmir, which played an important role in knowledge exchange across different linguistic traditions and cross-cultural interactions. One example, from the second half of the fifteenth century, is Kashmiri scholar Shrivara's Sanskrit translation of the allegorical Sufi romance *Yusuf-wa-Zulaykha* by his contemporary, the Persian poet Jami.[62] Abul Fazl's *Ain-i Akbari* recounts another, when, in the late sixteenth

60 Nund Rishi, 266–267.

61 Nund Rishi, 193, translation modified.

62 Shrivara completed his translation around the third quarter of the fifteenth century. For the translation, see Luther Obrock, "Srivara's *Kathakautuka*: Cosmology, Translation, and

century, the Mughal emperor Akbar (r. 1542–1605) ordered his courtier, Shah Muhammad Shahabadi, to translate the Sanskrit text *Rajatarangini* into Persian.[63] Instances of Kashmiri-language soil terminology in nonmystical Persian works such as historical chronicles are not in themselves surprising. For example, Abul Fazl's *Ain-i Akbari* mentions procedures involved in the preparation of soil for saffron cultivation, including references to the tilling and plowing of soil and a local agricultural unit of measurement, the *trakh*. The latter also appears in Nuruddin's poetry.[64] While we cannot definitively trace the source of Abul Fazl's use of this particular term, there is little doubt that Nuruddin's use of agricultural imagery in his mystic poetry would have contributed to linguistic interaction between Persian and Kashmiri. This is supported by references in the eighteenth-century Persian chronicle *Gauhar-i Alam*, which indicate that Nuruddin's poetry was translated into Persian during his lifetime by Mulla Ahmad Kashmiri in the form of a work titled *Mirat-ul Awliya* (Mirror of the saints).[65] As a matter of fact, metaphors of soil, along with other agricultural metaphors, continued to be used within the Persian Sufi textual tradition in early modern Kashmir well after Nuruddin's time, even if their relation to the use of similar metaphors by Nuruddin has not yet been confirmed. For instance, metaphors of soil are similarly used in discussions on several Sufi doctrines in *Dastur-us Salikin* (Manual of the divine seekers), a late sixteenth-century Perso-Arabic hagiography of the famous sixteenth-century Sufi saint Hamza Makhdum by his disciple Baba Daud Khaki (d. 1585). Quoting a Sufi text, the *Zayn al Mu'taqad* of Kubrawi saint Ala al-Dawla Simnani (d. 1336), Khaki referred to one of the eight preconditions before a spiritual master could accept a new follower, as proposed by the famous Sufi Junayd al-Baghdadi (d. 910): "*Dhikr* is a seed [*tukhm*], and the heart of a seeker [*dil-e saalik*] is the soil [*zameen*], so an assessment of the condition of the soil [*haal-e zameen*] should be done so that the fresh seed does not go to waste."[66]

Although Khaki did not use Nuruddin's Kashmiri-language verses in his Persian Sufi manual, Khaki's contribution to the historiography of regional Sufi mysticism is evident from his compilation of Persian poetical works, which

the Life of a Text in Sultanate Kashmir," in *Jami in Regional Contexts: The Reception of 'Abd al-Rahman Jami's Works in the Islamicate World, ca. 9th/15th–14th/20th Century*, ed. Thibaut d'Hubert and Alexandre Papas (Leiden: Brill, 2018), 752–776.

63 Abul Fazl Allámi, *Ain-i Akbari*, 1:106.

64 For *trakh* as a unit, see Grierson, *Dictionary of the Kashmiri Language*, s.v. "trakh."

65 *Gauhar-i Alam* is an eighteenth-century work written by Badi-ud Din Abdul Qasim. See Gauhar, *Sheikh Noor-ud-Din Wali*, 11.

66 Khaki, *Dastur-us Salikin*, 99.

recounts the lives and thoughts of some of the most prominent Rishi saints in Kashmir.[67] Khaki's literary endeavors may well have inspired his disciple Ghazi, whose work *Noor-Nama* (Book of light) represented one of the foremost examples of the migration of Nuruddin's Kashmiri-language mystical poetry into the Persian textual tradition. Composed toward the middle of the seventeenth century, Ghazi's work details aspects of beliefs and practices of the Rishi order, including those of Nuruddin and some of his most prominent disciples. More importantly, his work summarizes portions of Nuruddin's poetry in the Persian language, while other parts, including those containing metaphors of agriculture and soil, are noted in the Kashmiri language itself.[68] Ghazi's work marked an important milestone in the regional Persian literature, particularly in Rishi hagiography, as similar works patterned on his *Noor-Nama* style soon followed. These included Baba Muhammad Kamal's late eighteenth-century Rishi hagiography, *Rishi-Nama Ambar Shimama*, which recorded the bulk of Nuruddin's original verses along with a commentary in Persian prose, followed by another Persian work in *mathnavi* form, his own *Noor-Nama*, in which he translated certain portions of Nuruddin's poetry into Persian.[69] A slightly later Persian Rishi hagiography written by Baba Muhammad Khalil around the mid-nineteenth century, *Rauzat-ur Riyaz* (Garden of the meadows), similarly produced a versified Persian translation of portions of Nuruddin's poetry, while retaining others in their original Kashmiri.[70] In a fascinating display of multilingualism, these works record Nuruddin's poetic verses in the Kashmiri language along with explanatory passages in Persian and Arabic.[71] For instance, in *Noor-Nama Ghazi* cites certain verses of Nuruddin's poetry in the original Kashmiri, commenting on them in Arabic and Persian with frequent references to Qur'anic quotations.[72] As such, Nuruddin's Kashmiri-language verses, including those containing imageries of soil-related agricultural practices, came to be part of locally produced Persian-language Sufi hagiographic texts.

---

67 Besides his *Dastur*, these included *Rishi-Nama Lamiyya* (the *lam* panegyric), a Persian panegyric recounting the life of sixteenth-century Rishi saint of Kashmir Baba Hyder Rishi (b. 1504).

68 Baba Naseebudin Ghazi, *Nur-Nama*, Markaz-i Noor Centre for Sheikh-ul Alam Studies, University of Kashmir, Acc. No. 105. For a discussion on the history of composition of Nuruddin's poetry, see Gauhar, *Sheikh Noor-ud-Din Wali*, 12–14; Saqi, *Kulliyat*, 17–20.

69 Gauhar, *Sheikh Noor-ud-Din Wali*, 13.

70 Gauhar, 14.

71 See, for instance, the manuscript of Baba Khalil's *Rauzat-ur Riyaz* in the Markaz-i Noor Centre for Sheikh-ul Alam Studies, University of Kashmir, fol. 6, Acc. No. 135.

72 See, for instance, Ghazi, *Noor-Nama*, 42–43, 117, 267–268.

The transfer of Nuruddin's Kashmiri-language verses into the Persian tradition can be understood as a multidimensional phenomenon, which involved factors such as Nuruddin's stature as a prominent local saint-poet, the didactic nature of his poetry, the proliferation of the Persian language and its literature in Kashmir, and intra-Sufi competition. Clearly, Nuruddin had become a prominent mystic and poet before his death. Based on references in later hagiographies and collections of his poetry, such as Ghazi's seventeenth-century *Noor-Nama* and Mir Abdullah Baihaqi's eighteenth-century *Kulliyat-i Sheikh-ul Alam*, there is a consensus among modern historians that Nuruddin's poetry was written down during his lifetime in the Sharda script of the Kashmiri language by his disciple and Sanskrit scholar Kati Pandit.[73] Near-contemporary writings also made a prominent mention of his life and his mystic Rishi order. For instance, some of the earliest extant textual sources from early modern Kashmir, such as the sixteenth-century Persian chronicle *Tarikh-i-Sayyed Ali*, detail the life events of Nuruddin and other prominent personalities associated with his mystic order based on earlier, now nonextant texts. Notable among these are the Persian hagiography *Maqamat-i Auliya Kashmiri* (An anthology of Kashmiri saints) of Baba Haji Adhami and a Persian chronicle, *Tarikh-i Kashmir* (A history of Kashmir), by a local historian, Qazi Ibrahim, both from the fifteenth century.[74] We get the same impression from the earliest extant works from and on Kashmir, as these similarly attest to the stature of Nuruddin as a prominent mystic and describe the Rishis as forming a significant mystic order. This equally applies to works styled as historical chronicles and hagiographies of some of the most well-known Sufi saints associated with the spread of Islam in medieval Kashmir. Included in the latter category of texts are Persian chronicles such as the sixteenth-century *Tarikh-i-Sayyed Ali*, the early seventeenth-century *Baharistan-i Shahi* and *Tarikh-i Haidar Malik* (The chronicle of Haidar Malik), and the eighteenth-century *Waqiat-i Kashmir* (The history of Kashmir), *Gauhar-i Alam* (Pearl of the world), and *Bagh-i Sulaiman* (The garden of Solomon).[75]

There were two complex historical processes, both of which had a direct and yet complex relationship with the spread of Islam in medieval

73 Baba Khalil states that Kati Pandit had written down the verses of Nuruddin on birch paper. For details about Kati Pandit, see Pir Ghulam Hassan Khuihami, *Tarikh-i Hassan* [The chronicle of Hassan], vol. 3, *Tazkira-i Auliya-i Kashmir Asrar-ul Akhyar* [History of the saints of Kashmir secrets of the righteous], 142; Khan, *Kashmir's Transition to Islam*, 188; Gauhar, *Sheikh Noor-ud-Din Wali*, 11.

74 For an account of Nuruddin in *Tarikh-i-Sayyed Ali*, see Ali, *Tarikh-i-Sayyed Ali*, 81–85.

75 For a reference to the mention of Nuruddin and his Rishi order in these works, see Khan, *Kashmir's Transition to Islam*, 11–12.

Kashmir: intra-Sufi competition and the proliferation of Persian language and literature.[76] The latter was a long and slow process linked to Kashmir's connections with certain areas in Central Asia and Iran.[77] By the fifteenth century, well before it came under the Mughal rule, Persian was the official language of administration under the Shahmiri sultans of Kashmir.[78] Thus, Persian, and not Kashmiri, was the language of *Rishi-Namas* and *Noor-Namas*, a local genre of hagiographic literature that focused on recording the life of Nuruddin and his Rishi order. Given that no earlier works about his life and order survived, these are the primary texts that preserve portions of Nuruddin's Kashmiri-language poetry. However, while *Noor-Namas* and *Rishi-Namas* may emerge as primary sites of trilingualism in the regional literary context due to their combination of Persian, Arabic, and Kashmiri languages, they are not exceptional. Rather, they represent the advanced stage of a more extensive and older process, which likely started with the arrival of Arabic- and Persian-speaking immigrants into Kashmir after they settled and set up Arabic and Persian learning institutions in the region.[79] This meant that Kashmiris soon took up Persian as one of the main languages of their literary expression; so much so that, as Zutshi puts it, "some of the most potent expressions of regional belonging from the late-eighteenth and early-nineteenth centuries are, in fact, in Persian."[80]

These processes provide the general context in which Nuruddin's Kashmiri-language verses came to be part of Persian hagiographic texts. A key milestone in this linguistic interaction was the making of Persian-language hagiographic works in the *Noor-Nama* and *Rishi-Nama* genre that featured verses of Nuruddin's Kashmiri-language Sufi poetry, including those containing soil-related agricultural metaphors. Nuruddin's act of bringing together the mystical and the agricultural in his poetry already enabled the transfer of knowledge of soil across literary and linguistic traditions. This transfer was, however, facilitated by hagiographers of the saint and members of his Rishi order, the earliest of whom, such as Ghazi and Baihaqi, were prominent saints

76 For a discussion, see Zutshi, *Kashmir's Contested Pasts*, 21–71.

77 For an overview of the spread of the Persian language in Kashmir, see Abdul Qadir Sarwari, *Kashmir mein Farsi Adab ki Tarikh* [A history of Persian literature in Kashmir] (Srinagar: Sheikh Mohammed Usman and Sons Tajiran-i Kutb, 2012).

78 Persian as a state language is usually attributed to the reign of Sultan Zain-ul Abidin (r. 1420–1470), even as earlier contributions to Persian literature, including those by the previous sultans, were equally remarkable. See Sarwari, *Kashmir mein Farsi Adab*, 42–55.

79 For a discussion of the earliest phase of this development, see Sarwari, *Kashmir mein Farsi Adab*, 33–41.

80 Chitralekha Zutshi, *Languages of Belonging: Islam, Regional Identity, and the Making of Kashmir* (London: Hurst, 2004), 27.

and members of litterateurs of their time. Thus it was the medium of hagiography—a genre modern historians of the region have dismissed as fictitious and of little historiographical value—that acted as a conduit for the transmission of local agricultural knowledge of soil into the Persian textual tradition.[81] The making of these bilingual Persian-Kashmiri hagiographic texts may also have set a general precedent in the Persian literary tradition in Kashmir, as indicated by the frequent use of Kashmiri-language words and phrases in one of the most comprehensive Persian-language works on the history of Kashmir ever produced, the multivolume *Tarikh-i Hassan* by the nineteenth-century Kashmiri chronicler Ghulam Hassan Shah Khuihami (d. 1898). This work recorded Kashmiri-language terms for a range of "products" from the region,[82] illustrating not an isolated instance but rather an advanced stage of an extensive and older process of multilingual interactions between the Sanskritic and Persianate literary cultures, the earliest of which could be traced back to the beginning of the second millennium CE.

4 Conclusion

Examples of agricultural metaphors, including some specifically related to the soil, abound across various multilingual textual and oral compositions in Kashmir before, during, and after Nuruddin's time. These included descriptions of changing sociopolitical configurations in Sanskrit chronicles and the expositions of mystic discourses in Persian Sufi hagiographies and Kashmiri-language poetry. While there are earlier examples of the use of Kashmiri-language words in Persian as well as Sanskrit texts from Kashmir, it is important to note that the works belonging to the genre of *Noor-Nama* and *Rishi-Nama* helped bring Kashmiri-language vocabulary in general and agricultural and soil-related metaphors in particular into the Persian tradition at an unprecedented scale. This was a direct result of the fusion of the agricultural and the mystical in Nuruddin's Kashmiri-language poetry.

Well-known agricultural metaphors in the Kashmiri language, including those related to soil, are extensively used in Nuruddin's mystic poetry, serving as highly effective communicative devices to expound complex Islamic—and specifically Sufi—beliefs and practices in a region where Islam was a recent

81 For an appraisal, see Mudasir, "Holy Lives as Texts."

82 Khuihami, *Tarikh-i Hassan*, vol. 1, *dar bayan-i Jugrafiyya Kashmir* [The geography of Kashmir] (Srinagar: Research and Publications Department, Govt. of Jammu and Kashmir, 1954), 163–190.

phenomenon. With the proliferation of Persian language and literature in early modern Kashmir, Nuruddin's Kashmiri-language poetry, laden with extensive references to local agricultural knowledge of soil, was compiled by local hagiographers in the form of Persian-language texts. This helped directly transfer soil-related terminology and knowledge from the Kashmiri language into the Persian textual tradition.

The use of Kashmiri-language terms in Persian works from Kashmir aroused interest among Persian litterateurs outside Kashmir. For instance, the seventeenth-century Persian poet Saib Tabrizi (d. 1677), on a visit to Kashmir, sought out his contemporary, the Persian poet of Kashmir Mulla Tahir Ghani Kashmiri (d. 1669), to ask for an explanation of the Kashmiri-language term *krāl-pan* (potter's string) that Ghani had used in his Persian-language poetry.[83] Further research would be required to address the reception of Kashmiri-language soil-related knowledge in the wider Persian literary tradition. The composition and compilation of Nuruddin's poetry in early modern Kashmir occurred at a time when the region's ties to the Persianate world were firmly established. Thus, analyses of the reception of Nuruddin's poetry and the search for signs of transfer of soil-related metaphors from Sanskrit-dominated Kashmiri- to Persian-language works beyond Kashmir could potentially provide new perspectives on the broader patterns in the exchange of soil knowledge between the Sanskritic and Persianate spheres in the medieval and early modern world.

83 R. K. Parmu, *A History of Muslim Rule in Kashmir: 1320–1819* (Delhi: People's Publishing House, 1969), 322–323. For the meaning of the term, see Grierson, *Dictionary of the Kashmiri Language*, s.v. "krāl-pan."

CHAPTER 2

Soil Transformations in the *Codex Vergara* and the *Codex de Santa María Asunción*

The "Good Farmers" of Sixteenth-Century Mexico

Sarah Newman

In the mid-sixteenth century, a few decades after Spaniards had reclaimed the Aztec capital of Tenochtitlán as the city of Mexico, the Franciscan friar Bernardino de Sahagún began working with Indigenous students and collaborators at the newly founded Colegio de la Santa Cruz de Tlatelolco to document Aztec life and culture. Together, they compiled a three-volume, twelve-book bilingual (Nahuatl-Spanish) illustrated encyclopedia known as the *Florentine Codex*. There, an entry in the tenth book, dedicated to descriptions of people, includes a list of what makes a "good farmer." The good farmer

> is bound to the soil; he works—works the soil, stirs the soil anew, prepares the soil ... levels the soil, makes separate furrows, breaks up the soil. ... He sets the landmarks, the separate landmarks; he sets the boundaries, the separate boundaries; he stirs the soil anew during the summer; he works [the soil] during the summer; he takes up the stones; he digs furrows; he makes holes.[1]

Around the same time that Sahagún was directing the work of the *Florentine Codex*, scribes in the nearby region of Tepetlaoztoc (Map 2.1) were recording two pictorial census-cadastral documents, now known as the *Codex Vergara* and the *Codex de Santa María Asunción* (hereafter, the CV and the CSMA). The CV and the CSMA convey a wealth of information about landowners, landholdings, and soil types or qualities, primarily through the combined alphabetic and logographic hieroglyphic writing system used by Indigenous Nahuatl-speaking communities of Central Mexico. The glyphs in the codices are occasionally accompanied by glosses or annotations in alphabetic Nahuatl and Spanish (see Figure 2.1). Although characters written with the alphabet represent only

1 Bernardino de Sahagún, *Florentine Codex*, bk. 10, *The People*, trans. Charles E. Dibble and Arthur J. O. Anderson (Santa Fe: The School of American Research and the University of Utah, 1961), 41.

 | DOI:10.1163/9789004748484_004

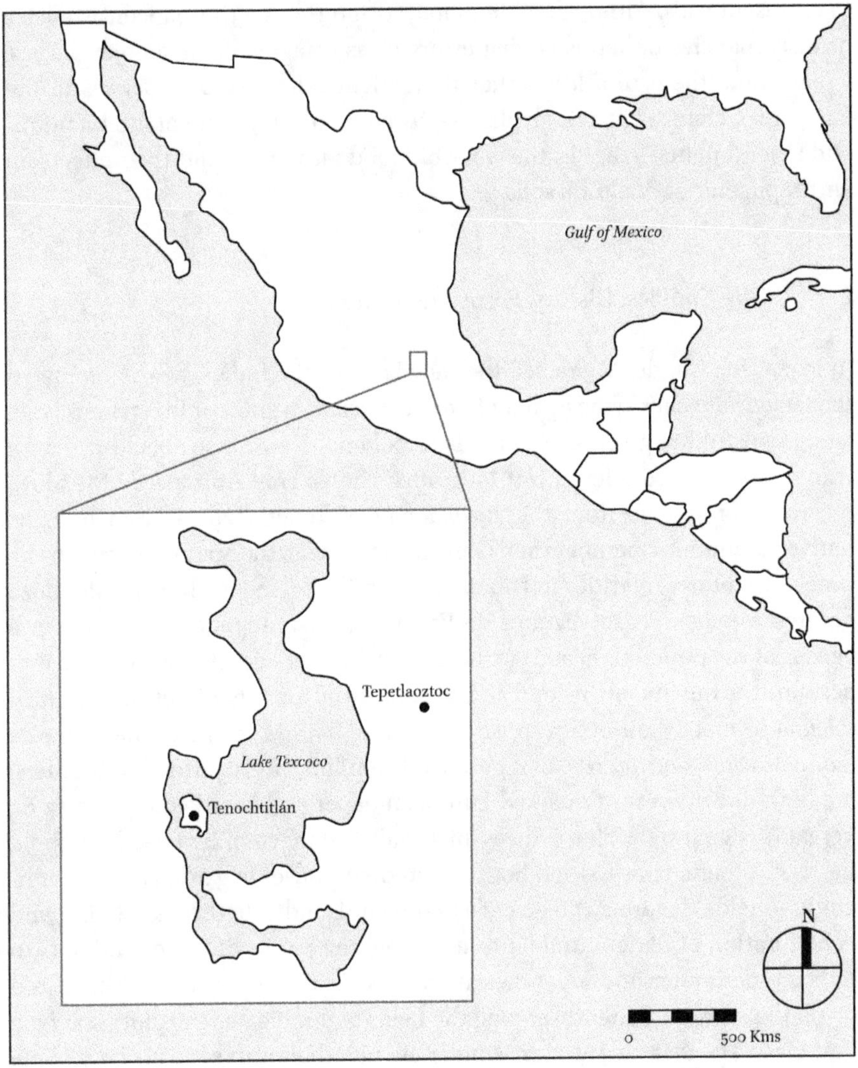

MAP 2.1 Map showing Tepetlaoztoc's location in the Valley of Mexico and its proximity to the former Lake Texcoco and Aztec island capital of Tenochtitlán

sounds, Nahuatl glyphs can stand for sounds and/or concepts: thus, a drawing of a tooth can stand for the word *tlantli* (tooth) or, in rebus writing, for the common place-ending *tlan*. Since the nineteenth century, scholars have proposed translations and taxonomies for individual glyphs in the *CV* and the *CSMA*, including those that appear to indicate soil types. Building upon that long history of work, I focus on certain inconsistencies or discrepancies that have been noted by scholars who have examined the two codices and the glyphs used to represent soils. Although earlier analysts interpreted those inconsistencies or discrepancies either as scribal errors or as disagreements among scribes, I reexamine them to ask whether they might reflect actual changes in the land. If so, changes in soil glyphs over time record intentional interventions and amendments—that is, the work of "good farmers"—and their effects on anthropogenic agricultural soils.

1 The Codices: History, Form, and Content

In 1542, King Charles v enacted the New Laws of the Indies, which stipulated that an *encomendero* (the owner of an *encomienda*, a grant of Indigenous land, labor, and tribute given to Spaniards in colonial New Spain) could not raise tribute above a fixed level. Not long after, the viceroy Antonio de Mendoza received complaints from the *encomienda* of Tepetlaoztoc—from both its native community members and Gonzalo de Salazar, the Spanish owner at the time—about irregularities in tribute and service payments. In 1543, Mendoza ordered a judge, Pedro Vázquez de Vergara, to visit Tepetlaoztoc, conduct a census of the population, and establish new tribute levies that would take into account the quality of the land and the ability of the inhabitants of the *encomienda* to meet their obligations.[2] The *Codex Vergara* (*CV*) and the *Codex de Santa María Asunción* (*CSMA*), produced to fulfill Viceroy Mendoza's orders, are both drawn on watermarked European paper and bear Pedro Vázquez de Vergara's signature.[3] The codices pictorially connect individual household heads in Tepetlaoztoc to their houses, and their houses to gardens and/or agricultural fields (Figure 2.1). The cadastral records indicate the shape, size, and configuration of agricultural lands and label the plots with lines and dots to indicate perimeter and areal measurements. At the center of each field is a glyph that varies frequently among the fields and indicates the plot's soil type or quality. The *CV* and *CSMA* are similar enough that most scholars have come

2 Perla Valle Pérez, *Memorial de los indios de Tepetlaoztoc o Códice Kingsborough, a cuatrocientos cuarenta años* (Mexico City: Instituto Nacional de Antropología e Historia, 1993); see also Barbara J. Williams and H. R. Harvey, *The Códice de Santa María Asunción: Facsimile and Commentary; Households and Lands in Sixteenth-Century Tepetlaoztoc* (Salt Lake City: University of Utah Press, 1997), 3.

3 Jerome A. Offner, "Household Organization in the Texcocan Heartland: The Evidence in the *Codex Vergara*," in *Explorations in Ethnohistory: Indians of Central Mexico in the Sixteenth Century*, ed. Herbert R. Harvey and Hanns J. Prem (Albuquerque: University of New Mexico Press, 1984), 127–146, on 127–128.

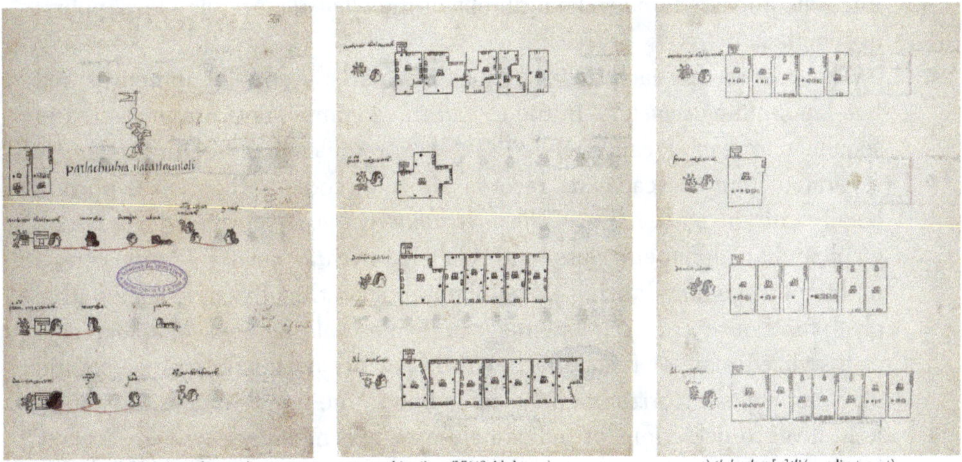

a) *tlacatlacuilol[l]i* (pictorial census) b) *milcocol[l]i* (field shapes) c) *tlahuelma[n]tli* (an adjustment)

FIGURE 2.1, A–C Households from the *altepemaitl* (subdistrict or hamlet) of Patlachiuhca shown across the three registers of the *Codex Vergara*: a) in the household census (*tlacatlacuilol[l]i*), b) in the cadastral register showing field shapes and perimeter measurements (*milcocol[l]i*), and c) in the "adjustment" cadastral record showing areal measurements (*tlahuelma[n]tli*)
SOURCE: *CODEX VERGARA*, 1539, BIBLIOTHÈQUE NATIONALE DE FRANCE, DÉPARTEMENT DES MANUSCRITS, MEXICAIN 37–39, FOLS. 3R, 34V, 36V. SCAN FROM WORLD DIGITAL LIBRARY, LIBRARY OF CONGRESS

to believe they were once a single manuscript.[4] At the very least, they appear to have been produced by the same school of manuscript painting.[5]

Although there are difficulties in capturing political-spatial categories with English (or Spanish) terms, the CV and the CSMA are organized according to local political hierarchies.[6] Individual households are presented in groups according to the specific *altepemaitl* (subdistrict or hamlet) where they are located. Those *altepemaitl* together comprise a wider *tlaxilacalli* (ward or neighborhood): Chimalpan in the case of the CV and Cuauhtepoztlan in the case of the CSMA. Chimalpan, Cuauhtepotzlan, and other *tlaxilacalli* were, in

4 Charles Gibson, *The Aztecs under Spanish Rule: A History of the Indians of the Valley of Mexico, 1519–1810* (Stanford: Stanford University Press, 1964), 269; Barbara J. Williams and Janice K. Pierce, "Evidence of Acolhua Science in Pictorial Land Records," in *Texcoco: Prehispanic and Colonial Perspectives*, ed. Jongsoo Lee and Galen Brokaw (Boulder: University Press of Colorado, 2014), 147–164, on 148; but cf. Williams and Harvey, *Códice de Santa María Asunción*, 2.

5 Donald Robertson, *Mexican Manuscript Painting of the Early Colonial Period: The Metropolitan Schools* (New Haven, CT: Yale University Press, 1959), 153–154.

6 Benjamin D. Johnson, *Pueblos within Pueblos: Tlaxilacalli Communities in Acolhuacan, Mexico, ca. 1272–1692* (Boulder: University Press of Colorado, 2017), 54.

turn, encompassed under the umbrella of the *altepetl* (sovereign local polity) of Tepetlaoztoc.

The codices are each divided into three parts: a population census and two landholding registers. In the CV, all three components are presented in sequence for each locality, while in the CSMA, the census records are largely separated from the cadastral records.[7] In both codices, the census portion begins with the place name of the *altepemaitl* being surveyed (given as both a glyph and an alphabetic gloss) and is labeled *tlacatlacuilolli* (pictorial census or, literally, "people-drawing"). Each household is then listed, labeled with a glyph for "house" (*calli*) and an image of the head of that household, depicted as a stylized male or female face with his or her personal name glyph and alphabetic gloss. Horizontal lines extending to the right connect the household head to depictions of individual household members, who are usually labeled with their names (Figure 2.1a). These census records offer precise demographic information (including age, gender, marital status, and even ethnicity) about the sixteenth-century communities they depict.[8]

The CV and CSMA appear to have been first drawn between 1543 and 1544 and then updated around 1546.[9] Not long after Vázquez de Vergara conducted his initial census, the Audiencia (the highest tribunal of the Spanish crown in New Spain) ordered a reassessment to take into account what was called in Nahuatl a *cocolitztli* (pestilence), an epidemic that swept through Mexico in 1545. According to the friar Domingo de Betanzos, founder of a local Dominican convent in 1529, the *cocolitztli* killed more than 14,000 people in Tepetlaoztoc alone.[10] Additional annotations, including narrative descriptions of territorial boundaries and land transfers overseen by Indigenous officials, were added to both documents throughout the mid-1570s.[11]

In both codices, the first cadastral register begins, like the census, with the place glyph and gloss of the *altepemaitl*, usually accompanied by the label *milcocol(l)i* (field shapes).[12] The *milcocoli* cadastre follows the same sequence

7 Barbara J. Williams and Herbert R. Harvey, "Content, Provenience, and Significance of the *Codex Vergara* and the *Códice de Santa María Asunción*," *American Antiquity* 53, no. 2 (1988): 337–351, on 338.

8 See, e.g., Herbert R. Harvey, "Household and Family Structure in Early Colonial Tepetlaoztoc," *Estudios de cultura náhuatl* 18 (1986): 272–294; Offner, "Household Organization."

9 Williams and Pierce, "Evidence of Acolhua Science," 148.

10 Joaquin García Icazbalceta, *Colección de documentos para la historia de México, tomo segundo* (Mexico City: Antigua Librería, 1866), 200.

11 Williams and Harvey, "Content, Provenience, and Significance," 339.

12 This term is most often spelled *milcocoli* in both the CV and the CSMA and I follow that spelling in referring to the sections of the codices labeled as such. Joseph Marius Alexis

and style as the census, but only the household head appears in each entry. The lines from each individual lead to his or her house plot or garden (indicated by a house glyph) and agricultural fields (Figure 2.1b). The *milcocoli* cadastre shows the shape and configuration of the fields, which are labeled with lines and dots along the sides of each field to indicate perimeter measurements,[13] using a local standard of linear measurement known as the *tlalcuahuitl*, or "a measure for land, a rod" (literally, "land stick"),[14] suggested to be equivalent to roughly 2.5 m.[15] "Cutouts" from field depictions suggest that residential structures, trees, rocks, and paths were removed from the field areas, leaving only the arable lands quantified and qualified in the cadastral surveys. In the center of each parcel is a glyph that is presumed to connote soil type or quality.

The third section is labeled *tlahuelma(n)tli* (an adjustment),[16] understood as a reassessment of the corresponding *milcocoli* cadastre after the 1545

Aubin and René Simón defined "milcocolli" (in French and Spanish, respectively) as "contour, shape of the lands, of properties." See *Dictionnaire de la langue Nahuatl Classique*, ed. Alexis Wimmer, s.v. "milcocolli," accessed May 26, 2023, http://sites.estvideo.net/malinal/nahuatl.page.html (hereafter cited as *DNC*). Williams and Harvey thus gloss the *milcocoli* used in the codices as "field shapes." Williams and Harvey, *Códice de Santa María de Asunción*, 25. *Milli* (Spanish *milpa*) has the meaning of "cultivated land, field," but *cocolli* could come either from the verb *coloa*, meaning "to twist, to change direction; to bend, fold something; to detour around something" or from the verb *cocoa*, meaning "to hurt," from which is derived *cocolli*, glossed in Alonso de Molina's sixteenth-century dictionary as *riña* (quarrel) and as *carga, obligación* (charge, obligation). See Frances Karttunen, *An Analytical Dictionary of Nahuatl* (Austin: University of Texas Press, 1983), s.vv. "milli," "coloa" (hereafter cited as *ADN*); as well as the *DNC*. The first possibility suggests *milcocolli* refers to the shapes of fields, the second implies either some kind of dispute over land (*milpas*) or some kind of obligation involving land.

13 Eduard Seler, "Alexander von Humboldt's Picture Manuscripts in the Royal Library at Berlin," in *Mexican and Central American Antiquities, Calendar Systems, and History*, trans. Charles P. Bowditch (Washington, DC: Smithsonian Institution Bureau of American Ethnology, 1904), 123–229, on 203.

14 *Online Nahuatl Dictionary*, ed. Stephanie Wood, s.v. "tlalcuahuitl," accessed May 26, 2023, https://nahuatl.uoregon.edu/content/tlalcuahuitl (hereafter cited as *OND*).

15 Williams and Harvey, *Códice de Santa María Asunción*, 27; Williams and Pierce, "Evidence of Acolhua Science," 149–153.

16 This term is spelled *tlahuelmatli* in both the *CV* and the *CSMA* and, as with *milcocoli*, I follow the spelling used in the codices throughout this chapter. It seems that scribes did not always write the final "n" (e.g., in the *CV*, the letter "n" is also not marked at the end of place names, such as "Teocaltitla" in place of "Teocaltitlan"). See Frances Karttunen and James Lockhart, *Nahuatl in the Middle Years: Language Contact Phenomena in Texts of the Colonial Period* (Berkeley and Los Angeles: University of California Press, 1976), 8–14. *Tlahuelmantli* is glossed as "an adjustment, something evened out or leveled off" (cosa igualada o allanada) by Alonso de Molina, but it is otherwise unattested beyond the Tepetlaoztoc codices. See *OND*, s.v. "tlahuelmantli."

epidemic. It, too, follows the household sequence from the census and connects owners to their fields, once again with soil glyphs at their centers, but the plots in the *tlahuelmatli* cadastre are drawn as consistent, stylized rectangles. The lines and dots in the *tlahuelmatli* are placed differently than in the *milcocoli*: they appear in a "tab" in the upper right-hand corner, along the bottom margin, or in the center of a field near the soil glyph (Figure 2.1c). As opposed to the linear perimeter measurements of the *milcocoli*, the lines and dots in the *tlahuelmatli* are recorded by positional notation and can be used to calculate the surface areas of the fields depicted.[17]

2 The Semantics of Sand-Spitting Stones

Scholars from a variety of disciplinary backgrounds have plumbed the Tepetlaoztoc codices for information on precolonial demographics and social organization, local systems of measurement and mathematical notation, and evidence of ancient soil classification systems. More than a century ago, anthropologist Eduard Seler compared pages from the CV to another cadastral document brought to the Royal Library of Berlin by Alexander von Humboldt (known as Humboldt Fragment VIII) and suggested that the glyphs painted at the center of each agricultural plot in Humboldt Fragment VIII and the CV were intended to indicate soil type.[18] To decipher these glyphs, Seler looked to the *Florentine Codex*, specifically to its eleventh volume, which is dedicated to "Earthly Things," including different types of soils, their material qualities, factors involved in their formation, and their agricultural potential. For example, the Nahuatl text, in Charles E. Dibble and Arthur J. O. Anderson's translation, glosses and describes a soil called *xalatoctli* as "sand borne by water. It is very loose. Borne by water, it is porous, very porous"; *quauhtlalli* as "rotten wood or oak leaves; humus; or silt with [rotten] wood. It is black or yellow. It is

17 Williams and Harvey, *Códice de Santa María Asunción*, 26–29. For a detailed explanation of land surveying, mathematical units, and perimeter and area calculations in the codices, see María del Carmen Jorge et al., "Mathematical Accuracy of Aztec Land Surveys Assessed from Records in the *Codex Vergara*," *Proceedings of the National Academy of Science* 108, no. 37 (2011): 15053–15057; Williams and Pierce, "Evidence of Acolhua Science"; María del Carmen Jorge, Clara Eugenia Garza-Hume, and Ramiro Chávez, "A Mathematical Description of the Agricultural Fields of the Acolhua Codices Vergara and Santa María Asunción," *Latin American Antiquity* 32, no. 24 (2021): 835–849.

18 Seler, "Humboldt's Picture Manuscripts," 204–206.

a) Macuilcoatl
macuilli (five) + ***coatl*** (snake)

b) Macuilcoatl
macuilli (five) + **co**mitl (jar) + ***atl*** (water)

FIGURE 2.2 Pablo Macuilcoatl's personal name glyphs: a) the name spelled using two logograms as a rebus in the *tlacatlacuilolli* register, and b) spelled using three logograms as a rebus in the *milcocoli* and *tlahuelmatli* registers. Note that in b) the reading order (top, bottom, center) is not linear.

a fertile place"; and another, called *callalli*, as "the land upon which a house has rested, and also the surrounding houses. It is fertile, it germinates."[19]

The codices from Tepetlaoztoc are notable for their highly phoneticized logo-syllabic writing, which is somewhat distinct from other well-known Central Mexican documents, such as the *Codex Mendoza* or the *Codex Boturini*.[20] The *CV* and the *CSMA* very often use glyphs as parts of rebuses, in which the phonetic value of a logogram is used to evoke the same or a similar sound with a different meaning.[21] The alphabetic glosses that accompany personal names make this clear. For example, the head of a household in Cuauhtepuztitla, Pablo Macuilcoatl, appears in the *CSMA* identified by two different name glyphs (Figure 2.2). The first, in the census register, combines the glyph for *macuilli* (five), a group of five lines joined by a horizontal line above (probably representing a highly schematized hand), and the logogram for *coatl* (snake). In the *milcocoli* and *tlahuelmatli*, Pablo Macuilcoatl's name is written using the same five-line symbol for *macuilli*, but instead of using the snake logogram for *coatl*, the second half of Pablo's surname is produced by the combination

19 Bernardino de Sahagún, *Florentine Codex*, bk. 11, *Earthly Things*, trans. Charles E. Dibble and Arthur J. O. Anderson (Santa Fe: University of Utah Press, 1963), 251–255.

20 Alfonso Lacadena, "Regional Scribal Traditions: Methodological Implications for the Decipherment of Nahuatl Writing," *PARI Journal* 8, no. 4 (2008): 1–22, on 2.

21 Lacadena, 2; cf. Gordon Whittaker, *Deciphering Aztec Hieroglyphs: A Guide to Nahuatl Writing* (Oakland: University of California Press, 2021), 130.

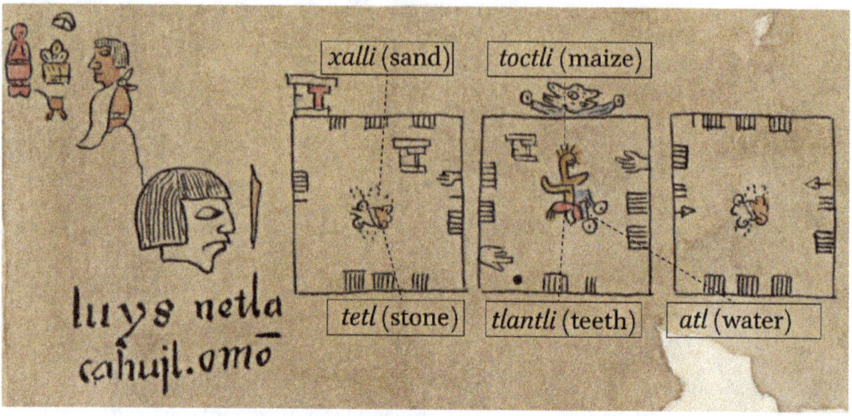

FIGURE 2.3 Seler's proposed readings of soil glyphs from Humboldt Fragment VIII as *tetlalli xallalli* and *atoctlan* or *atoctlalli*. Facsimile of the Humboldt Fragment VIII drawn by Agostino Aglio, 1825–1831
SOURCE: NATIONAL ANTHROPOLOGICAL ARCHIVES, SMITHSONIAN INSTITUTION, NAA MS 2205, INV 08882500

of the phonetic complement *co-*, cued by an image of a *comitl* (jar), and the logogram for *atl* (water).

Assuming the soil glyphs also functioned as rebuses (in the same way as personal name glyphs), Seler interpreted the soil glyphs depicted in the Humboldt Fragment VIII and the *CV* as sequences of multiple graphemes, each providing a phonetic value (as *comitl* served to produce the syllable *co-* in the example of Pablo Macuilcoatl's name). Seler then searched the text of the *Florentine Codex* for a comparable soil classified by Sahagún—a term with a name that matched his phonetic reading and a description with qualities that matched those suggested by the graphic images (Figure 2.3). In a basic example, an image of a stone with a series of fine dots preceding it was read by Seler as *tetl* (stone) combined with *xalli* (sand). "Combining these descriptors with the word for 'earth,' *tlalli*," Seler wrote, "this hieroglyph, then, would denote stony, sandy soil, which the Mexicans called *tetlalli xallalli*." In another example, an image of a stream of water (*atl*) is accompanied by that of a maize (*toctli*) plant with both tassels and ear. Together they express the passive verb *atoco* (meaning "to be carried by water"). Below the water and maize, a row of teeth (*tlantli*) adds *-tlan* (place of). All three images together produce *atoctlan*, which Seler translated as "land rich in *atoctli* (fertile vegetable mold)," though Seler also suggested the row of teeth might be intended to express only *-tla* for *tlalli* (rather than *-tlan*), to produce *atoctlalli* (literally, "land carried by water"). Seler also identified *quauhtlalli* ("earth which has been manured with rotten wood ... soft, rich, and golden") and *contlalli* ("clay in this earth for making tiles and pots") using this method of rebus transliteration and comparison with Sahagún's encyclopedia.[22]

One of the most common glyphs in the *CV* and the *CSMA* is a stone pierced by a pointed object or spine (see Figure 2.4, T6: *tezoquitl* [clay]). The stone glyph is well known—it appears in a wide number of Central Mexican pictorial documents—and its translation as *tetl* (along with its use to provide the phonetic complement *te-* for other words) is secure thanks to the range of sources where it is employed, often with alphabetic glosses.[23] The *pierced* stone glyph, however, appears *only* in the Tepetlaoztoc codices, which precludes a reading informed by comparison with other contemporary pictorial sources. Geographer Barbara Williams, who is single-handedly responsible for the bulk of existing scholarship on the *CV* and the *CSMA*, argued that the pierced stone glyph should be read as *tezoquitl*, translated as "hard clay."[24] In her reading of the glyph, a rebus-based interpretation like Seler's, *te-* comes from the stone, *tetl*, while the *zo* sound is communicated by the spine piercing it, as *zo* is a verbal root meaning "to pierce oneself, to draw blood by piercing someone."[25] According to the *Florentine Codex*, the name *tezoquitl* explicitly "comes from *tetl* (rock) and *zoquitl* (mud), because it is firm, gummy, hard; dark, blackish, bitumen-like."[26]

Zoquitl, however, is a specific kind of mud, as archaeologist Aurelio López Corral has noted.[27] The description of *zoquitl* is not included among Sahagún's list of soils in the Nahuatl text of the *Florentine Codex*, but rather in a separate section, one dedicated to the kinds of earth used to make pottery, where it is further glossed as "clay" and described as "moist, wet, thick, hard, firm. It is water and earth mixed together."[28] The term *tezoquitl* is also found there, but not as an independent entry. It is used in the explanation for *contlalli*, the

---

22 Seler, "Humboldt's Picture Manuscripts," 204–206.

23 *Gran Diccionario Náhuatl*, ed. Instituto de Investigaciones Bibliográficas / Instituto de Investigaciones Históricas, Universidad Nacional Autónoma de México, s.v. "tetl," accessed June 10, 2022, https://gdn.iib.unam.mx/diccionario/tetl/11571 (hereafter cited as *GDN*). See also *Visual Lexicon of Aztec Hieroglyphs*, ed. Stephanie Wood, accessed June 10, 2022, https://aztecglyphs.uoregon.edu/fulltext-quick-search?search_api_views_fulltext=tetl (hereafter cited as *VLH*).

24 Barbara J. Williams, "Pictorial Representation of Soils in the Valley of Mexico: Evidence from the *Codex Vergara*," *Geoscience and Man* 21 (1980): 51–62, on 54; see also Williams and Harvey, *Códice de Santa María Asunción*, 51–62.

25 *ADN*, 347.

26 Sahagún, *Florentine Codex*, bk. 11, 252.

27 Aurelio López Corral, "Los glifos de suelo en códices acolhua de la Colonia temprana: Un reanálisis de su significado," *Desacatos* 37 (2011): 145–162, on 155.

28 Sahagún, *Florentine Codex*, bk. 11, 257.

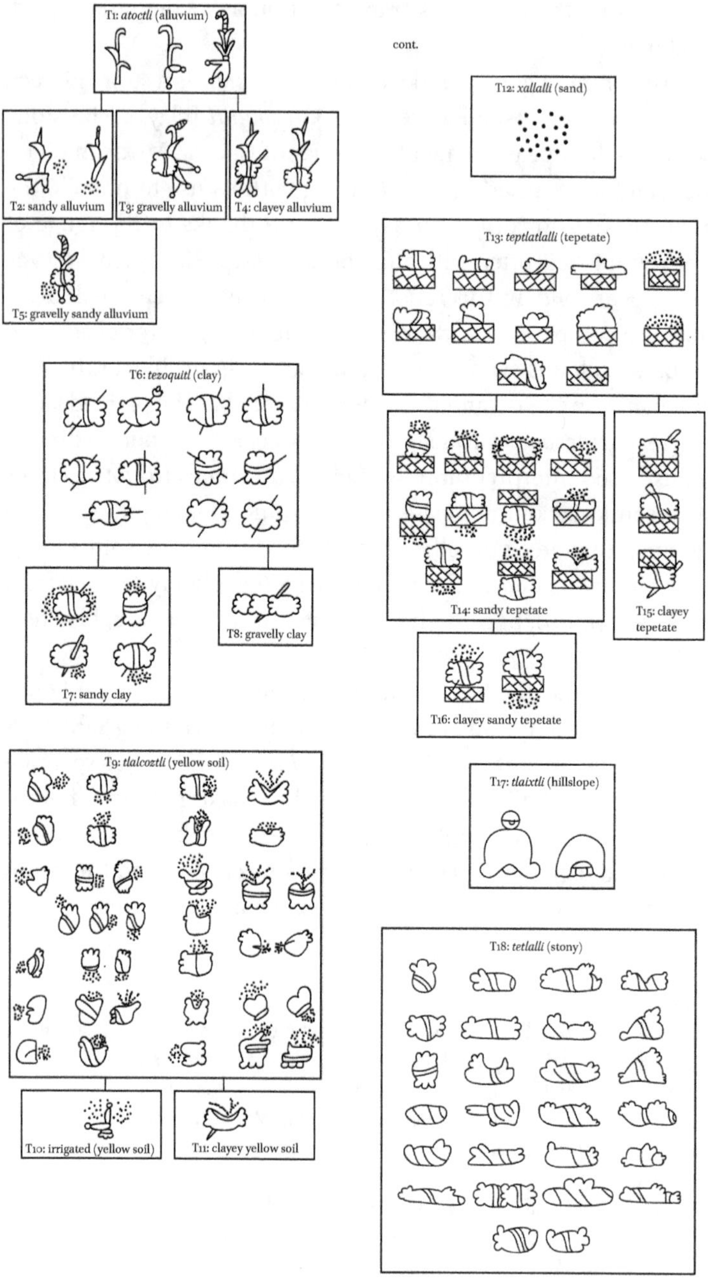

FIGURE 2.4 Williams's proposed taxonomic structure of soil classes recorded in the *Codex de Santa María Asunción* (listed in order of probable agricultural productivity)
SOURCE: ADAPTED FROM BARBARA J. WILLIAMS AND H. R. HARVEY, *CÓDICE DE SANTA MARÍA ASUNCIÓN: FACSIMILE AND COMMENTARY; HOUSEHOLDS AND LANDS IN SIXTEENTH-CENTURY TEPETLAOZTOC* (SALT LAKE CITY: UNIVERSITY OF UTAH PRESS, 1997), 31–32, FIGURE 16. REPRINTED COURTESY OF THE UNIVERSITY OF UTAH PRESS

earth from which water jars, bowls, and basins are made, where it is described as "sticky, gummy ... it is that which is kneaded, which is tempered with reed stem fibers."[29] Don Hernando Alvarado Tezozomoc's *Crónica Mexicayotl*, a late sixteenth-century Nahuatl-language history of the Aztec Empire, translates *tezoquitl* as *lodo de piedra* (stone mud) and describes how it was used to construct the strong pipes of aqueducts carrying water to Tenochtitlán.[30] As López Corral has pointed out, these terms are specific in referring to dense, hard clays used for ceramics, not for agriculture. He argues that the spine should instead be read as *tezoni*, an instrument or lance for drawing blood, and that the stone, *tetl*, serves as a complement to confirm and clarify the intended *te-* syllable.[31] He thus reads the glyph as *tezontli*, or *tezontle* in Spanish, which is the name for a light and porous volcanic rock found throughout the Central Mexican highlands (a possibility originally proposed [as *teçontlalli*] by Williams).[32] Today, tezontle soil is a preferred growing medium in Mexico, especially for large-scale production of greenhouse tomatoes, as the presence of the porous tezontle rock in soil reduces the amount of both water and fertilizer needed by crops.[33] The differences in Williams's and López Corral's readings are important: one effectively reduces a field to an agricultural dead zone, the other communicates desirable productive potential.

Williams not only proposed translations for the soil glyphs of the *CV* and *CSMA*, but, together with ethnohistorian Herbert Harvey, organized the glyphs of the *CSMA* and her translations of them into a folk soil classification, which consists of eighteen taxa at three taxonomic levels, described in order of productivity: seven generic, nine specific, and two varietal (Figure 2.4).[34] In the later facsimile of the *CV*, Williams and ethnohistorian and anthropologist Frederic Hicks made use of the same generic categories, but no longer presented

29 Sahagún, 257.

30 *GDN*, s.v. "tezoquitl."

31 Note that *tetl* and *tezoni* have different vowel lengths: the vowel is short in *tetl*, long in *tezoni*.

32 Barbara J. Williams, "Nahuatl Soil Glyphs from the *Códice de Santa María Asunción*," in *Actes du XLIIe congrès international des américanistes, volume VII* (1976) (Paris: Société des américanistes, 1979), 27–37, on 32; Williams, "Pictorial Representation of Soils," 54. Williams also suggested that "another interpretation of the pierced stone glyph results if the spine indicates 'worn out,' as in *zozoltic*, yielding worn out stone land (*tetlalli*, *tlaçolli*), that is, degraded soil (or perhaps land in fallow)." Williams, "Nahuatl Soil Glyphs," 32. See also Tlachia en CEN, Compendio Enciclopédico Náhuatl, s.v. "tezontli," accessed July 31, 2024, https://cen.sup-infor.com/home/lachia/.

33 See, e.g., J. Z. Castellanos, J. L. Ojodeagua, P. Varga-Tapia, and J. J. Muñoz-Ramos, "Water and Nutrient Use of Greenhouse Tomatoes Grown in 'Tezontle' Media and in Soil in Central Mexico," *Acta Horticulturae* 659 (2004): 565–568.

34 Williams and Harvey, *Códice de Santa María Asunción*, 30–34.

them as a hierarchical taxonomy.[35] In terms of classifications, Williams was a "lumper." Her soil categories combine glyphs that she considered similar looking, despite both subtle and striking morphological variations (see, e.g., the variety of glyphs lumped together under Williams's classifications for T9: *tlalcoztli* [yellow soil] or T18: *tetlalli* [stony] in Figure 2.4).

3 Error or Intention?

Given the structure of the CV and CSMA, each household and its corresponding agricultural fields are represented twice: once in the *tlahuelmatli* section and once in the *milcocoli* section (see Figure 2.1). Although the soil glyphs are consistent across the two registers most of the time, differences between the registers are not infrequent. Of the 607 fields belonging to 110 households depicted in the *milcocoli* and *tlahuelmatli* sections of the CV, there are 96 individual instances in which the soil glyph changes from one register to the other (~16%). In the CSMA, which depicts a total of 823 individual fields belonging to 258 households, there are 253 differences across the two registers (~31%).[36]

As noted above, Williams tended toward minimizing potential meaning in glyphic variation. She attributed the differences between the two cadastral sections of the codices to mistakes made by the sixteenth-century scribes who painted them. In explaining the differences in soil glyphs across the registers of the CV, she wrote:

> At least on the subject matter of soils, Náhuatl glyphics were standardized, but still permitted some variation in glyph depiction. If one considers that a scribe recorded a single soil type dozens of times at various places in the document, then stylistic variation might be expected because of fatigue, haste, boredom, or caprice. … Further, the drafting task was analogous to record statistics, and "typos" might also be expected. … Small details in glyphs portraying other information may be significant, but in soil glyphs, they are not.[37]

35 Barbara J. Williams and Frederic Hicks, *El Códice Vergara: Edición facsimilar con comentario: pintura indígena de casas, campos, y organización social de Tepetlaoztoc a mediados del siglo XVI* (Mexico City: Universidad Nacional Autónoma del México, 2011), 32–37.

36 Williams and Pierce, "Evidence of Acolhua Science," 148; cf. Williams, "Pictorial Representation of Soils," 53–55.

37 Williams, 58.

Williams and Janice K. Pierce similarly stated of the *CSMA* that

> soil glyphs drawn on *milcocolli* fields do not always match those of the *tlahuelmantli* record. Explanation of the discrepancies may involve recording procedures. For example, the two field registers of a community may not have been drawn at the same time, causing a mismatch in *milcocolli-tlahuelmantli* paired fields. Or discrepancies may stem from varying classification judgements between different classifiers or even a single classifier.[38]

López Corral understands divergences between the *milcocoli* and *tlahuelmatli* not as mistakes but as evidence that the registers were made by different scribes in different years, during which time certain fields had changed hands.[39] For him, each record is an accurate assessment, but some fields assessed in the *milcocoli* were no longer owned by the same individual when the *tlahuelmatli* was recorded. He details an illustrative example of such changes between the *milcocoli* and *tlahuelmatli* in the *CSMA*: two of the four fields of Damián Hecacozca, in the *altepemaitl* of Tlaltecahuaca. The first two fields attributed to Hecacozca are identical in the two registers, but his third and fourth fields change. In the *milcocoli* (Figure 2.5a), Hecacozca's third field is shaded gray. The gray shading is understood by Williams to show that a field is abandoned (i.e., available for and awaiting redistribution);[40] it is also marked with a glyph that López Corral sees as a combination of an eye, smoke, and spiraling air particles, which he reads as *ixpouhqui* and translates as "stale, withered, altered, faded, discolored."[41] The fourth field contains the glyph López Corral reads as *tezontli*, but also a glyph that is widely attested and read as *zacatl* (grass), an indicator of terrain that is covered in weeds, that is, allowed to lie fallow.[42] In the *tlahuelmatli* register (Figure 2.5b), Damián Hecacozca's third field is marked with the pierced stone glyph, read as *tezontli* by López Corral, and the

38 Williams and Pierce, "Evidence of Acolhua Science," 159.

39 López Corral, "Los glifos de suelo," 161.

40 Williams and Harvey, *Códice de Santa María Asunción*, 35.

41 López Corral, "Los glifos de suelo," 153. López Corral's translation draws from Alonso de Molina's sixteenth-century dictionary, where *ixpouhqui* is glossed as "damaged, or faded in color" (dañado, o rebotado color). See *OND*, s.v. "ixpouhqui."

42 *GDN*, s.v. "zacatl." See also the *VLH*. All three scholars who have proposed readings of the glyphs in the *CV* and/or *CSMA* likewise read the glyph as *zacatl*. López Corral, 156; Seler, "Humboldt's Picture Manuscripts," 201; Barbara J. Williams, "Aztec Soil Knowledge: Classes, Management, and Ecology," in *Footprints in the Soil: People and Ideas in Soil History*, ed. Benno Peter Warkentin (Oxford: Elsevier, 2006), 17–42, on 26.

a)
b)

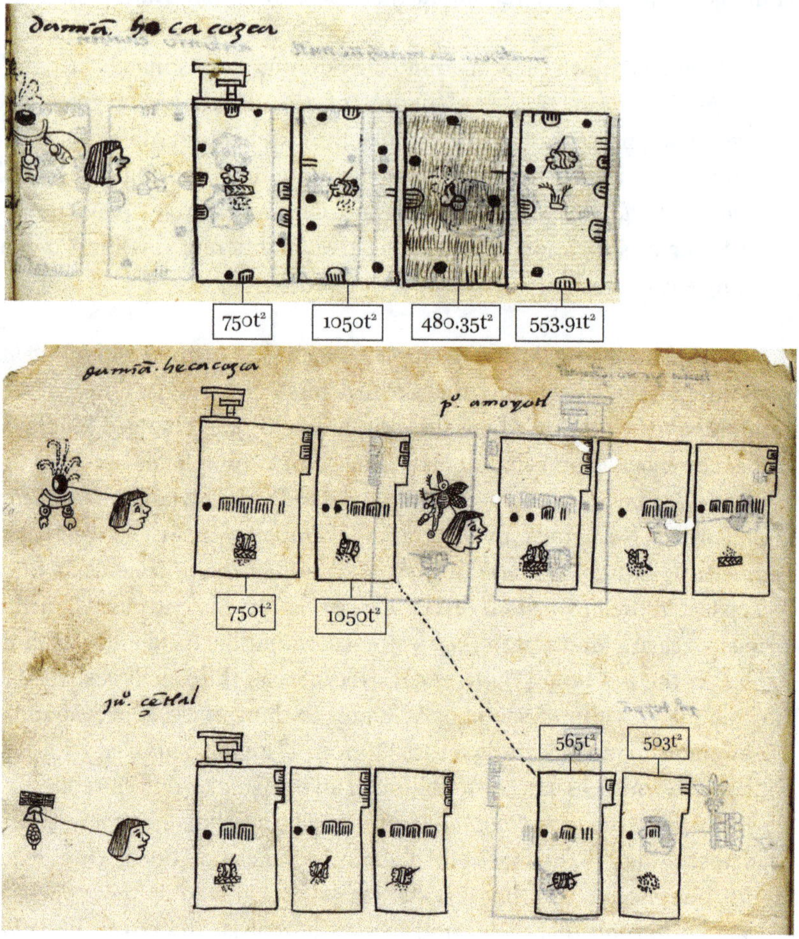

FIGURE 2.5, A–B Areal measurements (in square *tlalcuahuitl* [T]) of the fields of Damián Hecacozca from the CSMA: a) Hecacozca's fields in the *milcocoli*, and b) Hecacozca's fields in the *tlahuelmatli*
SOURCE: DETAIL FROM MS 1497BIS *CÓDICE DE SANTA MARÍA ASUNCIÓN* (NO. DE SISTEMA BNM 645351), FOLS. 66R, 67V, BIBLIOTECA NACIONAL DE MÉXICO, CIUDAD DE MÉXICO. AREAL CALCULATIONS FROM CLARA E. GARZA-HUME AND MARÍA DEL CARMEN JORGE, "AGRIMENSURA AZTECA-ACOLHUA," ACCESSED JUNE 1, 2022, HTTPS://AGRIMENSURAAZTECA.IIMAS.UNAM.MX

fourth with a glyph that López Corral, Williams, and Seler (and others) all read as *xalli* (sand).[43] Comparing the perimeter and areal measurements for the divergent fields, López Corral shows that the agricultural plots with glyphs that

43 GDN, s.v. "xalli." See also the VLH; Whittaker, *Deciphering Aztec Hieroglyphs*, 154, figure 5.17.

differ between the *milcocoli* and the *tlahuelmatli* also change in size—that is, unlike the fields with glyphs that remain constant, the fields with changes in their soil glyphs between the *milcocoli* and the *tlahuelmatli* also differ in their overall areas.

Dimensional comparisons, however, pose the same issues as glyphic ones: when and to what extent is variation important? The fields that remain the same and those that differ with respect to their central glyphs and sizes align neatly in the case of Damián Hecacozca's territories, but this does not hold for the rest of the *CSMA* or the *CV*. Antonio Tlapalzayol, for example, owned two fields in the *altepemaitl* of Conçotlan that are illustrated in the *CSMA* (Figure 2.6). Tlapalzayol's second field remains exactly the same size in both registers, but the soil glyph changes between the *milcocoli* and the *tlahuelmantli*. In the *milcocoli*, the second field displays a pierced stone glyph atop a woven mat, surrounded by sand (translated as a "'clayey sandy' form of *tepetlatlalli*," a cemented soil horizon exposed on the surface through erosion, by Williams, and as *xalteçonpetlatl*, a layer or stratum of tezontle gravel, by López Corral). In the *tlahuelmatli*, however, it is marked by a stone glyph atop a woven mat with sand below (translated as "sandy tepetate (volcanic tuff)" by Williams,[44] and as *xaltepetlatl*, a layer or stratum of stone gravel, by López Corral).[45] Similarly, the four fields of Mateo Calhua in the *altepemaitl* of Topotitla, depicted in the *CV*, include a field that again remains the same size in both registers, but is marked by the "sandy tepetate"/*xaltepetlatl* glyph in the *milcocoli* register and by a different glyph, translated by Williams as *tlalcoztli* (yellow soil) and by López Corral as *tepitzactli*, thin or elongated stones, in the *tlahuelmatli*.[46]

It is unlikely that fields in Tepetlaoztoc would have changed hands as frequently as soil glyphs and areal measurements differ between the registers of the *CV* and the *CMSA*. Some instances of land transfer are explicitly indicated, either by textual additions in Spanish or alphabetic Nahuatl and/or fields shown subdivided by solid or dotted lines.[47] Moreover, Aztec agriculture was characterized by highly intensive techniques—primarily terracing, but also

44 Williams and Harvey, *Códice de Santa María Asunción*, 32.

45 For *tepetlatlalli* and "sandy tepetate," see Williams and Harvey, *Códice de Santa María Asunción*, 32; for *xalteçonpetlatl* and *xaltepetlatl*, see López Corral, "Los glifos de suelo," 154. Note, however, that López Corral (who generally attributes meaning to subtle morphological variations in soil glyphs) does not provide a translation for the specific variant illustrated for Tlapalzayol's *tlahuelmatli* field, where the sand is depicted below the stone and mat, but only a variant where sand surrounds a stone atop a mat.

46 For *tlalcoztli*, see Williams and Harvey, *Códice de Santa María Asunción*, 31; for *tepitzactli*, see López Corral, 151.

47 Williams and Harvey, *Códice de Santa María Asunción*, 35.

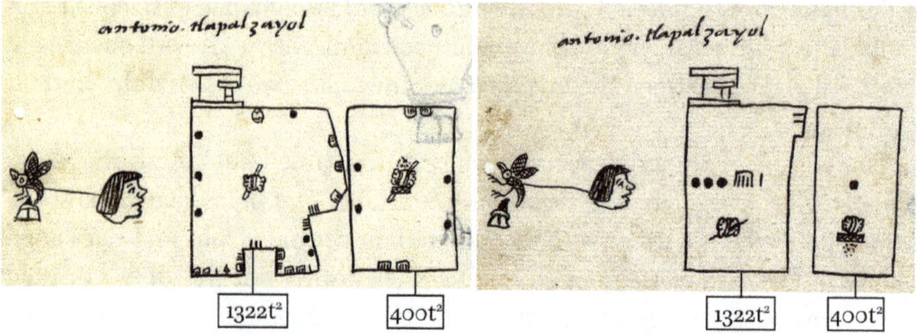

FIGURE 2.6 Areal measurements (in square *tlalcuahuitl* [T]) of the fields of Antonio Tlapalzayol in the *milcocoli* (L) and *tlahuelmatli* (R) registers of the CSMA. Note that the area of Tlapalzayol's second field in both the *milcocoli* and *tlahuelmatli* is exactly 400 square *tlalcuahuitl*, but the field's central glyph changes between registers.
SOURCE: DETAIL FROM MS 1497BIS *CÓDICE DE SANTA MARÍA ASUNCIÓN* (NO. DE SISTEMA BNM 645351), FOLS. 77V, 78V, BIBLIOTECA NACIONAL DE MÉXICO, CIUDAD DE MÉXICO. AREAL CALCULATIONS FROM GARZA-HUME AND JORGE, "AGRIMENSURA AZTECA-ACOLHUA," ACCESSED JUNE 1, 2022, HTTPS://AGRIMENSURAAZTECA.IIMAS.UNAM.MX

canals, dams, and *chinampas* (raised plots constructed above water level and separated by narrow canals). By the end of the fifteenth century, most of the arable land available in the Basin of Mexico was covered by terrace systems and actively being farmed.[48] Such landscape modifications were bottom-up strategies undertaken by individual households or communities, time- and labor-consuming efforts that bound producers in place "via complex and nested systems of investment, entitlement, and conveyance."[49] The continual work of maintenance and cultivation required farmers to situate their houses on their agricultural holdings, producing distinctive patterns of dense

48 Deborah L. Nichols, "Agricultural Practices and Environmental Impacts of Aztec and Pre-Aztec Central Mexico," in *Oxford Research Encyclopedias: Environmental Science* (Oxford: Oxford University Press, 2018), 9; William T. Sanders, Jeffrey R. Parsons, and Robert S. Santley, *The Basin of Mexico: Ecological Processes in the Evolution of a Civilization* (New York: Academic Press, 1979), 243.

49 Christopher Morehart, "Aztec Agricultural Strategies: Intensification, Landesque Capital, and the Sociopolitics of Production," in *The Oxford Handbook of the Aztecs*, ed. Deborah L. Nichols and Enrique Rodríguez-Alegría (Oxford: Oxford University Press, 2017), 263–280, on 269. See also Michael E. Smith and T. Jeffrey Price, "Aztec-Period Agricultural Terraces in Morelos, Mexico: Evidence for Household Level Agricultural Intensification," *Journal of Field Archaeology* 21, no. 2 (2013): 169–179.

yet dispersed settlements that have been documented by archaeologists in the Basin of Mexico, including in Tepetlaoztoc itself.[50]

Both Williams and López Corral attempt to explain the discrepancies between the two registers of the *CV* and the *CSMA* by assuming that the actual soil remains constant: it is either marked incorrectly, assessed differently by the same scribe or by different scribes, or belongs to a new household (and thus is no longer depicted, or at least not in the same place). The idea that researchers can infer what people *intended* to say or write (versus what they actually said or wrote) is not uncommon in efforts to relate Indigenous and Western knowledge.[51] Textual critics, however, often heed a principle attributed to the nineteenth-century classical philologist Ulrich von Wilamowitz-Moellendorff: *einmal ist keinmal, zweimal ist immer* (once is never, twice is always)—meaning that if more than one instance of a discrepancy is found, it should be carefully considered as a possible established usage, rather than an error.[52] Following Wilamowitz-Moellendorff, what if, rather than errors or exchanges of fields, the discrepancies among soil glyphs are taken to record actual changes made to the land between Vázquez de Vergara's initial census and the postepidemic reassessment? If that is the case, tracking the changes in soil glyphs indexes farmers' intentional amendments to the agricultural landscape of sixteenth-century Mexico and their effects.

4 Anthropogenic Transformations in Tepetlaoztoc

For the Nahuatl-speaking informants interviewed by Sahagún and his assistants, certain types of soils were not only worked but actively *made* by humans. In addition to the description of the many activities carried out by a "good farmer" at the beginning of this chapter, the *Florentine Codex*'s entry for

50 See, e.g., Morehart, "Aztec Agricultural Strategies," 266; Michael E. Smith et al., "Aztec Period Houses and Terraces at Calixtlahuaca: The Changing Morphology of a Mesoamerican Hilltop Urban Center," *Journal of Field Archaeology* 38, no. 3 (2013): 227–243. For Tepetlaoztoc, see William T. Sanders, *Tepetlaoztoc Project: Archaeological Investigations* (n.p.: Foundation for the Advancement of Mesoamerican Studies, Inc., 2002), http://www.famsi.org/reports/95047/index.html.

51 See, e.g., Franz Boas, "On Alternating Sounds," *American Anthropologist* 2, no. 1 (1889): 47–54.

52 See, e.g., Tom Mackenzie, "A Canonical Author: The Case of Hesiod," in *Liddell and Scott: The History, Methodology, and Languages of the World's Leading Lexicon of Ancient Greek*, ed. Christopher Stray, Michael Clarke, and Joshua T. Katz (Oxford: Oxford University Press, 2019), 105–123, on 117.

FIGURE 2.7 Fertilizing soil with excrement in the *Florentine Codex*
SOURCE: DETAIL FROM MEDICEO PALATINO 220, FOL. 378V. SCAN FROM WORLD DIGITAL LIBRARY, LIBRARY OF CONGRESS, REPRINTED WITH PERMISSION: FIRENZE, BIBLIOTECA MEDICEA LAURENZIANA, MS MED. PALAT. 220, FOL. 378V. SU CONCESSIONE DEL MIC. E' VIETATA OGNI ULTERIORE RIPRODUZIONE CON QUALSIASI MEZZO

tlalauiyac (i.e., *tlalahhuiyac*), a descriptor of soil fertilized with manure, reads: "This is the land which is good, which produces, which is mellow. I fertilize it. I add humus to it. I make it mellow. I make it good."[53] The description of *tlalauiyac* is accompanied by an illustration showing a man crouched and working the soil, marked with a lump of excrement, with his hands (Figure 2.7).

The CV and *CSMA* make note of intensive agricultural investments. For example, various plots in the codices are marked with glyphs formed by elongated or repeating *tetl* (stone) glyphs. Williams grouped most of these within a generic "stony" taxon, except for a series of three overlapping pierced stone glyphs, which she translated as "gravelly [i.e., extra stony] clay" (see Figure 2.4, T18: *tetlalli* [stony]). López Corral however, suggests that the elongation expresses the Nahuatl word *pantli*, which he translates as "line" or "strip," with the stone *tetl* adding the prefix *te-*, making the reading of the stretched stones *tepantli* or "line or wall of stone."[54] López Corral's interpretation of elongated stones as *pantli*, combined with his reading of the pierced stone glyph as *tezontli*, leads him to suggest that the overlapping pierced stone glyphs should not be read as "gravelly clay," as Williams proposed, but instead as a wall or boundary marker

53 Sahagún, *Florentine Codex*, bk. 11, 252.

54 López Corral, "Los glifos de suelo," 157. *Pantli* is glossed by Karttunen as "row, wall," with the additional note that *pantli* is "abundantly attested in *tepantli*, 'stone wall.'" *Pantli* is glossed by Wood as "a ridge, a furrow, an agricultural row" (i.e., it is a "line" that has width), with *tepantli* defined as "boundary; wall; fence; construction." See *ADN* and *OND*, s.vv. "pantli" and "tepantli."

constructed from tezontle stone, a building material commonly used in pre-Columbian and colonial-period architecture.[55]

Certain land parcels in the cadastres are marked with a glyph featuring an eye atop a hill, which both Williams and López Corral interpret as an indication of a field's placement on a hillslope. Williams read the glyph as *tlaixtli* (hillside lands), and López Corral as *tepetlixpan* or *tepeixpan* (on the face of the hill).[56] Williams and Harvey suggested that the hillside fields were located on the slopes of Tetepayo, a ridge which forms the eastern boundary of Tepetlaoztoc, and that they were unterraced, dry-farmed plots.[57] López Corral, however, notes that "where there are hills, there are terraces"—that is, that the "hillside" and "terrace" glyphs both occur only in the sloping *altepemaitl* of Cuauhtepuztitla and Huiznahuac.[58] Archaeological and historical surveys have revealed that the piedmont hillsides of the Basin of Mexico were once covered with terraces built from rock and earth and "semi-terraces" of stabilizing plants (such as maguey), which retain both soil and moisture in rain-fed agricultural zones.[59]

Smaller-scale interventions and other forms of management, such as manuring, burning, fallowing, and crop rotations, may also be recorded in the *CV* and *CSMA*. Returning to Damián Hecacozca's fields (Figure 2.5), comparison of the glyph at the center of his third *milcocoli* field with other contemporary pictorial documents (some of which include Spanish or Nahuatl glosses to confirm their meaning) suggests the image is a lump of excrement (read as *cuitlatl*), rather than the smoke curl suggested by López Corral.[60] Marked with manure,

55 See, e.g., Ethelia Ruiz Medrano, "The Lords of the Land: The Historical Context of the Mapa de Cuauhtinchan No. 2," in *Cave, City, Eagle's Nest: An Interpretive Journey through the Mapa de Cuauhtinchan No. 2*, vol. 2, ed. Davíd Carrasco (Albuquerque: University of New Mexico Press, 2007), 91–120, on 100.

56 For *tlaixtli*, see Williams and Harvey, *Códice de Santa María Asunción*, 34. Note that *tlaixtli* is glossed by Karttunen as "uphill stretch, rise." See *ADN*, s.v. "tlaixtli." For *tepetlixpan* or *tepeixpan*, see López Corral, "Los glifos de suelo," 156–157.

57 Williams and Harvey, 34.

58 López Corral, 159. See also María del Carmen Gutiérrez Castorena et al., "Los suelos del área de influencia del Códice Santa María Asunción y su representación pictórica," *Terra Latinoamericana* 35, no. 2 (2017): 101–111, on 104, fig. 3.

59 See, e.g., R. A. Donkin, *Agricultural Terracing in the Aboriginal New World* (Tucson: University of Arizona Press, 1979); Sanders, "Tepetlaoztoc Project"; Sanders, Parsons, and Santley, *Basin of Mexico*, 251; Michael E. Smith and T. Jeffrey Price, "Aztec Period Terraces in Morelos, Mexico: Evidence for Household-Level Agricultural Intensification," *Journal of Field Archaeology* 21, no. 2 (1994): 169–179; Thomas M. Whitmore and B. L. Turner, "Landscapes of Cultivation in Mesoamerica on the Eve of the Conquest," *Annals of the Association of American Geographers* 82, no. 3 (1992): 406–408.

60 *GDN*, s.v. "cuitlatl." See also the *VLH*; the *DNC*; *Tlachia en CEN*; Williams and Harvey, *Códice de Santa María Asunción*, 35, 40.

the field thus appears to be undergoing fertilization with night soil (*cuitlatl*), as described and depicted by Sahagún's collaborators in the *Florentine Codex* (see Figure 2.7). Similarly, as mentioned above, the *zacatl* grass depicted in Damián Hecacozca's fourth *milcocoli* agricultural plot is an indicator not only of a fallow field but also, specifically, of pioneer plants, the first to grow after fields have been cleared and burned.

Fields that lay fallow or were being manured when the *milcocoli* registers of the CV and the *CSMA* were recorded may have been transformed by the time the *tlahuelmatli* registers were made. Historical records indicate that sixteenth-century farmers in the Basin of Mexico practiced relatively short cultivation intervals, allowing fields to lie fallow for four to five years for each year of cultivation.[61] Indeed, the Spanish lawyer and royal official Alonso de Zorita, in his *Breve y sumaria relación de los señores de la Nueva España*, wrote that the elder of each *calpulli* (territory)

> is responsible for guarding and defending the *calpulli* lands. He has pictures on which are shown all the parcels, and the boundaries, and where and with whose fields the lots meet, and who cultivates what field, and what land each one has. The paintings also show which lands are vacant, and which have been given to Spaniards, and by and to whom and when they were given. *The Indians continually alter these pictures according to the changes worked by time* and they understand perfectly what these pictures show.[62]

Rather than mistakes, differences of opinion or assessment, or new fields altogether, the changing glyphs at the center of individual agricultural fields across the *milcocoli* and *tlahuelmatli* registers of the CV and *CSMA* may identify actual transformations resulting from interactions between natural and cultural processes (see also Roth, this volume, on Cakrapāni Miśra's Sanskrit gardening

61 Teresa Rojas Rabiela, "La technología agrícola mesoamericana en el siglo XVI," in *Historia de la agricultural época prehispánica—siglo XVI*, ed. Teresa Rojas Rabiela and William T. Sanders (Mexico City: Instituto Nacional de Antropología e Historia, 1985), 129–232, on 154; Teresa Rojas Rabiela, *Las siembras de ayer: La agricultura indígena del siglo XVI* (Mexico City: Secretaría de Educación Pública y Centro de Investigaciones y Estudios Superiores en Antropología Social, 1988), 62; see also Whitmore and Turner, "Landscapes of Cultivation," 407.

62 Alonso de Zorita, *Life and Labor in Ancient Mexico: The Brief and Summary Relation of the Lords of New Spain*, trans. Benjamin Keen (New Brunswick: Rutgers University Press, 1963), 110; italics added for emphasis.

treatise, which similarly records the impacts of specific and changing farming practices on soil composition).

5 Conclusions

Since at least the late nineteenth century, the *CV* and *CSMA* have been probed by modern scholars as unique sources of insight into precolonial and early colonial systems of knowledge and classifications of land in Central Mexico. The heavily rebus-based logosyllabic writing used in the codices has incited multiple attempts, by scholars from various backgrounds, to translate and taxonomize glyphs found at the centers of the agricultural plots in the two cadastral manuscripts. Disputes over some of those proposed translations remain unresolved, but different meanings for the same hieroglyphs are important for more than just semantics or soil science. Whether a field was dominated by sand or alluvium, full of clay or porous rock, would have made a meaningful difference in the daily lives, agricultural yields, and tribute expectations of the sixteenth-century farmers of Tepetlaoztoc.

Discrepancies between the two registers of the *CV* and the *CSMA* may similarly reflect real and meaningful differences on—or perhaps in—the ground. Although they have been explained as variations of a particular hieroglyph, or "typos" due to "fatigue, haste, boredom, or caprice,"[63] as errors in the recognition or assessment of soil types, or as a change in ownership of a particular field, a simpler and more straightforward explanation for changes in soil glyphs from one cadastral survey to another is also possible: that they correspond to actual changes in the soils themselves. Those changes, moreover, may have been—and also may have been understood as—the result of intentional anthropogenic activities. Based on the information in the codices themselves and the linguistic, historical, and cultural context in which they were made, I argue that the *tlahuelmatli* registers of the *CV* and the *CSMA* should be understood as literal "adjustments"—that they reflect actual changes in the people and the earth that *together* make up the community of Tepetlaoztoc. Rather than errors, changes in glyphs between registers of the *CV* and *CSMA* encode soils in flow. They are evidence of a history that is simultaneously natural, cultural, and always in motion.[64]

63 Williams, "Pictorial Representation of Soils," 58.

64 See James Maffie, *Aztec Philosophy: Understanding a World in Motion* (Boulder: University Press of Colorado, 2014), 12–14.

PART 2

Soil, Medicine, and the Body

CHAPTER 3

Improving Soil and Healing the Body in Mamluk Egypt

Bird Droppings as a Universal Remedy

Heba Mahmoud Saad Abdelnaby

1 Introduction

Agriculture, which was practiced along river valleys, around oases, and in areas receiving rainfall, has formed the economic foundation of many Islamic societies. It provided the nutritional needs of the population, raw materials used in trade, and shaped the construction of settlements and urbanization movements within societies. Yet practicing agriculture was not an easy task. Agricultural land was exposed to various natural and man-made hazards and threats such as variations in rainfall and river flow; excessive heat or cold; infestations of vermin, including rodents, locusts, and other insects; and outdated farming methods. These challenges called for specialized agricultural knowledge that would enable farmers to respond to them adequately.

Islamic societies inherited a wealth of literature in agricultural science from older cultures, especially the Roman and Nabataean. In the early ninth century, their major writings were translated into Syriac, Persian, and, most importantly, Arabic. Ancient texts were not only translated but also interpreted and critically studied by Muslim scholars, and from that platform of inherited knowledge, they made important progress in many sciences, surpassing the achievements of the ancient world. Islamic agricultural science was catalyzed by two developments: the first was the late eighth-century significant advances in botany, medicine, and mechanics, as sciences related to agriculture; the second, from the tenth century, was the emergence of the Kutub al-filāḥa (Books of Husbandry). This genre of agricultural manuals compiled by Arab agronomists disseminated both primary source material and know-how accumulated by the Arabs, including knowledge about soil, irrigation, cultivation, planting of different crops and trees, grafting, harvesting, the use of fertilizers, measures to fight insect pests, and many other topics essential for successful farming. Although this genre continued to attract the interest of authors who used it as a channel to contribute their knowledge and experience up until

DOI:10.1163/9789004748484_005

the eighteenth century, the period in which it most flourished was during the Mamluk Sultanate (1250–1517).[1]

Because of its importance for ameliorating the soil and intensifying the productivity of crops and trees, fertilizer production was an essential topic in most Arabic agricultural manuals. The manuals contained rich details such as the different types of fertilizer, their means of production, their effects on soil and plants, and their usage in treating certain diseases of crops and trees. Bird droppings are one of the most important fertilizers used from the eleventh to the sixteenth century, according to the manuals of that period. Nevertheless, fertilizers in general remain understudied,[2] and bird droppings in particular, though of high importance, have never been studied in this historical context. Therefore, this chapter examines how bird droppings were discussed in Mamluk agricultural manuals in order to highlight their importance, usage, application, and benefits. This study illustrates that Mamluk agricultural manuals provided very detailed information about the classification of bird droppings among the different categories of fertilizers, their preparation, and application at different stages of the agricultural process based on the type of soil or plant. Moreover, it highlights both the agricultural and medical importance of bird droppings by showing how different sciences informed each other, considering that pigeon droppings were used not only as fertilizer but also as remedies for plant diseases.

Understanding the properties and characteristic features of plants, animals, and minerals helped physicians to understand their effects on the human body when used in medication. This enabled them to develop a system of knowledge based on the usage of different plants and different parts of animals—and their extracts and excrements—to produce single and compound medicines. I argue in this chapter that medical knowledge also benefited from the advancement of agriculture, reflecting a relation or dialogue between the two sciences about which little has been written. This chapter shows how bird droppings were

---

1 For more information on these manuals and the genre in general, see Muhammad Zuhir al-Bābā, "ʿIlm al-filāḥa fi Bilād al-Shām," *Majallat al-turāth al-ʿArabī* 10 (1990): 31–63; A. H. Fitzwilliam-Hall, "An Introductory Survey of the Arabic Books of Filāḥa and Farming Almanacs," The Filāḥa Texts Project: The Arabic Books of Husbandry, accessed August 30, 2023, http://www.filaha.org/introduction.html; Bethany Walker et al., "Did the Mamluks Have an Environmental Sense? Natural Resource Management in Syrian Villages," *Mamluk Studies Review* 20 (2017): 179–181.

2 One important historical work about fertilizers, *Manure Matters*, includes a chapter by Daniel Varisco, which is the only study on that topic based on Arabic manuals of agriculture. See Daniel Varisco, "Zibl and Zirāʿa: Coming to Terms with Manure in Arab Agriculture," in *Manure Matters: Historical, Archaeological and Ethnographic Perspectives*, ed. Richard Jones (London: Routledge, 2012), 132–133.

seen as a "universal remedy" for soil infertility and for diseases of plants, animals, and the human body, attesting to the interaction between the authors of agrarian manuals and medical treatises. The stimulus for this approach was provided by an observation made by Raḍī al-Dīn al-Ghazzī (d. 935 AH/1529)[3] about fertilizers. He stated that "fertilizers open the pores of the soil, improve them, and give space for roots to permeate them."[4] In this metaphorical passage, he compares soil to skin, noting that the soil's "pores" are "opened" by fertilizers. This observation raised the question of whether bird droppings were used in human medicine, and this led to further investigation of other Mamluk sources. My investigation revealed that bird droppings were used not only in traditional medicine for the treatment of symptoms and diseases but also for the treatment of social and behavioral problems. It also revealed that the study of fertilizers in general, and bird droppings in particular, can demonstrate the relationship between sciences, as will be explained on the following pages.

2 Taxonomy of Soil and Fertilizers

Understanding soil, its nature and its properties, is of prime importance in agriculture because this knowledge guides decisions made by farmers. This includes knowing how to work the soil and how to water it, what suitable fertilizers to use, how to plow it, what to grow in it, and in which rotation. The discussion of soils and the compatible type of fertilizer for each soil has been indispensable for understanding which kind of crop could be paired with each soil. Since historical agrarian authors discussed soil from very different perspectives, Mamluk manuals contain considerable variation in the taxonomy of soil. For example, in the second chapter of the work of al-Waṭwāṭ (d. 718 AH/1318),[5]

3 Raḍī al-Dīn Abū al-Faḍl Muḥammad ibn Raḍī al-Dīn Abū al-Barakāt Muḥammad ibn Aḥmad ibn ʿAbd Allāh ibn Badr al-Ghazzī al-Dimashqī. He was a judge in Damascus whose life spanned the end of Mamluk and the beginning of Ottoman rule in Syria, as he died in 935 AH/1529. He was famous for his agricultural manual *Jāmiʿ farāʾiḍ al-milāḥa fī jawāmiʿ fawāʾid al-filāḥa*, which received much attention during the seventeenth and eighteenth centuries. He also visited Egypt and Palestine and recorded his observations about their plants and agriculture. ʿĪsā Aḥmad, *Tārīkh al-Nabāt ʿinda al-ʿArab* (Cairo: Matbaʿt al-iʿtmād, 1944), 110–111; al-Bābā, "ʿIlm al-filāḥa," 55–58.

4 "واعلم أن الزبل يفتح مسام الأرض ويجودها وينفشها لولوج العروق."
MS 8407, fol. 8, Al-Asad National Library, Damascus. Unless otherwise noted, all translations by the author.

5 Muḥammad ibn Ibrāhīm ibn Yaḥya al-Waṭwāṭ, known also as al-Kutubī, was born in Egypt in 632 AH/1235 and lived there all his life. He worked as a bookseller, and through that profession

Mabāhij al-fikar wa manāhij al-ʿibar (The pleasures of thoughts and the ways of the lesson),[6] the author surveys "Which soil and fertilizers are appropriate for which plants" (*Ma ywāfiq al-nbāt min al-arḍīn wal-sirqīn*). He discusses the types of soil revealed by the annual receding of the Nile and classifies them into thirteen types. He lists four types of soil that could not be repaired and ten types that could be cultivated following repair and improvement.[7] An anonymous Syrian author—who is believed to have lived during the first half of the eighth/fourteenth century AH/CE[8]—discussed soil taxonomy in the second chapter of his text *Miftāḥ al-rāḥah li-ahl al-filāḥa* (The key of comfort for the people of agriculture),[9] using the same chapter title as al-Waṭwāṭ. He presented a taxonomy similar to al-Waṭwāṭ's, with identical names of soil types and slight differences in details.[10] In fact, neither developed original ideas about the taxonomy of soil, but both were strongly influenced by Ibn Mamātī's book *Qawānīn al-dawāwīn*,[11] Ibn Waḥshiya's (fl. 318 AH/930–931) *Al-filāḥa al-Nabaṭīya* (Nabataean agriculture)—a foundational, highly influential work

he was well read in the arts and sciences and was widely regarded a clever author. His most important works are an encyclopedia (see n. 6) and *Ghurar al-khaṣāʾis al-wāḍiḥa*. He died in 718 AH/1318. Jurjis Manash, "Al-Manāhij fi Waṣf al-Mabāhig by Jamālalddin al-Watwāt," in *Publications of the Institute for the History of Arabic-Islamic Science*, ed. Fuat Sezgin (Frankfurt am Main: Institute for the History of Arabic-Islamic Science at the Goethe University, 1994), 1–22, originally published in the journal *Al-Mashriq* 16 (1907); ʿAbd al-Rāziq Aḥmad al-Ḥarbī, *Mabāhij al-Fikar wa Manāhij al-ʿIbar: Taʾlīf Muḥammad ibn Ibrāhīm al-Waṭwāṭ* (Lebanon: al-Dār al-Arabiya lil-Mawsuʿāt, 2000), 77–79; Ilham al-Kurkī, *Manhajīyat al-Waṭwāṭ al-Kutubī fi Mawsuʿtuh Mabāhij al-fikar wa Manāhij al-ʿIbar* (MA thesis, Mu'tah University, 2005), 35–43.

6 This is an encyclopedic book of natural science and geography divided into four arts, the fourth of which is devoted to plants and agriculture. MS Arabe 2809, Bibliothèque nationale de France, Paris (hereafter BnF); MS Arabe 6745, BnF; MS Wetzstein II 1241, Staatsbibliothek zu Berlin. It is worth mentioning that the latter two copies referred to this part as the second art, not the fourth.

7 *Mabāhij al-fikar*, MS Arabe 6745, fols. 21–30, BnF.

8 Muḥammad ʿĪsā Ṣāliḥīyah and Iḥsān Ṣidqī al-ʿAmad, *Miftāḥ al-rāḥah li-ahl al-filāḥa (li-muʾallif majhūl min al-qarn al-thāmin al-hijrī)* (Kuwait: Al-Majlis al-Waṭanī lil-Thaqāfa wa-l-Funūn wa-l-Adāb, 1984), 7–14; al-Bābā, "ʿIlm al-filāḥah," 52–55.

9 MS Ahlwardt 6208, Staatsbibliothek zu Berlin. The copy has been misattributed to Ibn Waḥshiya.

10 Ṣāliḥīyah and al-ʿAmad, *Miftāḥ al-rāḥah*, 101–111.

11 This is one of the most important sources about Egypt's geography and agricultural affairs during the sixth/twelfth century AH/CE. It was written by al-Asʿad ibn Mamātī, who was the head of several diwans and a notable statesman and writer. ʿAziz Surial Attia, *Kitāb Qawānīn al-dawāwīn* (Cairo: Madbulī Library, 1991), 5–10.

that is considered the earliest agricultural treatise written in Arabic[12]—and Ibn Baṣṣāl's *Dīwān al-filāḥa*.[13]

The Syrian agronomic manual *Al-Durr al-multaqiṭ fī ʿilm filāḥatay al-Rūm wa-al-Nabaṭ* (Pearls gleaned from the science of the two agricultures of the Romans/Byzantines and the Nabataeans),[14] attributed to al-Dimashqī (d. 725 or 728 AH/1325 or 1328),[15] followed a different methodology. The fifth chapter of this work shows that al-Dimashqī avoided the debate about soil taxonomy altogether in that he only mentions the "good" and "corrupt" types of soil for planting.[16] On the other hand, both Ṭaybughā al-Jariklamishī (d. 797 AH/1394–1395),[17] in the third chapter of his text *Al-filāḥa al-muntakhaba* (Selected agriculture),[18] and Raḍī al-Dīn al-Ghazzī, in the first chapter of his manual *Jāmiʿ farāʾiḍ al-milāḥa fī jawāmiʿ fawāʾid al-filāḥa* (Complete rules for elegance in all the uses of farming),[19] agreed on the classification of soil into suitable and

12 The book was based on earlier works by several Syriac authors and remained influential for centuries. Al-Bābā, "ʿIlm al-filāḥah," 37; Saʿīd Ismāʿīl ʿAlī, *Al-Nabāt wa al-filāḥa wa al-rayy ʿind al-ʿArab* (Cairo: ʿAlam al-Kutub, 2006), 45.

13 Muḥammad Ibn Baṣṣāl al-Andalūsī was a leading Andalusian agronomist born in the mid-eleventh century in Toledo. He was known for his unique agronomic treatise because it contains no reference to earlier agronomists but was based exclusively on the author's personal knowledge and experience. Ibn Baṣṣāl, *Kitāb al-filāḥa*, ed. and trans. Khosy Maryeh and Mohammad ʿAziman (Tétouan: Mawlāy al-Hasan Institute, 1955), 12–18, 29; ʿAlī, *Al-Nabāt*, 39.

14 MS Zirāʿah 21, Dār al-Kutub, Cairo.

15 Shams al-Dīn Abū ʿAbd Allāh Muḥammad ibn Abī Bakr ibn Abī Ṭālib al-Anṣārī al-Dimashqī, also called Ibn Shaykh Ḥiṭṭīn and Shaykh al-Rabwah. He was born in Damascus in 654 AH/1256, and during his life, he traveled to Egypt and Palestine. He was famous for his encyclopedic book *Nukhbat al-dahr fī ʿajāʾib al-barr wa-al-baḥr* in addition to the aforementioned filāḥa book, which was a selected abridgment of two ancient Byzantine and Nabataean works. He died in 725 AH/1325 or 728 AH/1328. Aḥmad, *Tārīkh al-Nabāt*, 100; al-Bābā, "ʿIlm al-filāḥa," 49.

16 *Al-Durr al-multaqiṭ*, MS Zirāʿah 21, fols. 40–47, Dār al-Kutub. Two copies of this manuscript are preserved in Dār al-Kutub in Cairo.

17 We do not know much about the author, and several versions of his name were mentioned in extant copies of his manual; yet we know that he was a Circassian Mamluk who lived in Egypt during the second half of the eighth/fourteenth century AH/CE and died in 797 AH/1395. He was also familiar with Syria. See Aḥmad, *Tārīkh al-Nabāt*, 109–110; Muḥammad ʿĪsā Ṣālihīyya, "Mulāḥaẓāt ʿala makhṭūṭāt al-filāḥa al-taṭbīqīya al-maḥfūẓa fī al-maktabāt al-ʿArabīya wal-ajnabīya," *Majallāt Mujmaʿ al-Lugha al-ʿArabīya* 59, no. 3 (1984): 572.

18 MS Arabe 2805, BnF; MS Arabe 2807, BnF. The second copy mentions a long title for the treatise and records the name of the author as Ṭaybaghā al-Jariklamishī al-Tumāni Tamurī.

19 MS 8407, Al-Asad National Library, Damascus; MS Zirāʿah Taymūr 42, Dār al-Kutub, Cairo.

unsuitable soils for planting and those that could be improved and repaired.[20] They also agreed on the five types of soil suitable for planting, which they marked by their color, while al-Ghazzī surveyed more details about the other two categories. These differences among Mamluk agricultural manuals suggest that classifying soils was by no means straightforward and soil taxonomy was a lively debate in Mamluk Egypt.

In Mamluk Egypt and Syria, many types of soil were considered good soil suitable for cultivation, especially since the annual flooding of the rivers in the two territories deposited a layer of silt on their respective valleys. This river silt was rich in various nutrients such as phosphoric acid and potash, but deficient in nitrogen.[21] Soil characteristics, however, were not immutable, and some authors recognized that the continuous cultivation of soil could lead to degradation or at least alteration of its features. On this issue, al-Waṭwāṭ quotes Ibn Baṣṣāl as follows: "when the soil is used, its moisture is depleted and its properties are weakened; consequently, it needs to be strengthened with fertilizers because of what they contain of heat and moisture."[22] Generally, Mamluk agrarian authors recognized the importance of fertilizers, and al-Ghazzī emphasized this importance by explaining that

> fertilizers open the pores of the soil, improve them, and give space for roots to permeate them. Fertilizers also increase the quality of soil that is good, repair and improve soil that is corrupt, and strengthen soil that is moderate, weak, and soft.[23]

He also added that "the soil that is not fertilized is cold, and that which is over-fertilized is burnt,"[24] which means that the use of fertilizers was an essential practice to balance, enhance, or repair the properties of soil, and consequently, understanding the effects of fertilizers on soil was crucial.

The sources agreed on their classification of fertilizers into two main groups: simple or single fertilizers and compound fertilizers. The first group comprised

---

20 *Al-filāḥa al-muntakhaba*, MS Arabe 2805, fols. 5–7, BnF; *Jāmiʿ farāʾiḍ al-milāḥa*, MS 8407, fols. 2–6, Al-Asad National Library.

21 Richard S. Cooper, "Agriculture in Egypt, 640–1800," in *Geschichte der islamischen Länder*, ed. Bertold Spuler (Leiden: Brill, 1977), 1:188–204, on 192; Anthony T. Quickel and Gregory Williams, "In Search of Sibākh: Digging up Egypt from Antiquity to the Present Day," *Journal of Islamic Archaeology* 3, no. 1 (2016): 91–108, on 94.

22 *Mabāhij al-fikar*, MS Arabe 6745, fol. 32, BnF.

23 See n. 4 above.

24 *Jāmiʿ farāʾiḍ al-milāḥa*, MS 8407, fol. 5a, Al-Asad National Library; MS Zirāʿah Taymūr 42, fol. 20, Dār al-Kutub.

three categories: *azbāl*, *atbān*, and *armidah*. The first category, *azbāl*, referred to single or simple fertilizers made of animal or human excrement, which included various types such as manure of cows, deer, pigs, sheep, buffaloes, horses, donkeys, and pigeons, as well as night soil.[25] The manuals imply that the word *zibl* (plural *azbāl*) refers to animal or human manure after it has been processed or has at least dried or rotted. The second category, *atbān*, referred to hay and leaves of plants such as broad bean, barley, wheat, gourd, blackberry, mallow, marshmallow, and roses, in addition to carrot leaves, fig stalks and leaves, and palm fronds. The third category, *armidah*, referred to the ashes of all the previous types of hay and leaves, and it was recommended that the ashes of each tree or crop be used on the same type of tree or crop.[26] Authors of agricultural manuals were generally keen to discuss the properties of each type of simple fertilizer and how each could be used successfully. They also used specific vocabulary to distinguish between various types of dung and excrement of animals and humans: *akhthā'* for cows; *zibl* for deer, pigs, horses, and donkeys; *ba'r* for sheep; *dharq*, *zibl*, or *khar'* for birds; and *zibl* or *khar'* for humans.[27]

The second group, compound fertilizers, also included three categories: *zibl muḍāf*, *zibl muwallad*, and *tarkibāt al-sirqīn*. The first category, *zibl muḍāf*, referred to animal manure plus the ashes from hot baths. Bath ashes are marked by dryness, salinity, and lack of moisture; consequently, they were not usually employed as a sole fertilizer without any other additions. However, when mixed with dung, they absorbed moisture and could be used. It seems that this category of fertilizer was applied when there was an insufficient or limited amount of animal dung available. It was considered one of the most compatible types of fertilizer for soil and water. And, being a harsh fertilizer, only a small amount of it should be used.

When no animal dung was available, the second category of compound fertilizers, *zibl muwallad*, was prepared out of several varieties of herbage, grass, hay, and ashes. Al-Waṭwāṭ and the anonymous Syrian author, both quoting from Ibn Waḥshiya, describe the preparation process of three types of fertilizer within this category.[28]

25 *Mabāhij al-fikar*, MS Arabe 6745, fol. 32, BnF; *Jāmi' farā'iḍ al-milāḥa*, MS 8407, fols. 5a, 5b, Al-Asad National Library; MS Zirā'ah Taymūr 42, fols. 20–21, Dār al-Kutub; Ṣāliḥīyah and al-'Amad, *Miftāḥ al-rāḥah*, 112–115.

26 *Mabāhij al-fikar*, MS Arabe 6745, fol. 36, BnF; Ṣāliḥīyah and al-'Amad, *Miftāḥ al-rāḥah*, 115.

27 Varisco, "Zibl and Zirā'a," 132–133.

28 *Mabāhij al-fikar*, MS Arabe 6745, fols. 34–35, BnF; Ṣāliḥīyah and al-'Amad, *Miftāḥ al-rāḥah*, 115–116; Varisco, "Zibl and Zirā'a," 139.

The third category was *tarkibāt al-sirqīn*, which represented mixtures of dung and plants with ash that had been blended with components such as urine, wine, algae, and blood. All of these fertilizers were prepared according to specific recipes to be used for groups of crops and trees that had similar qualities and needs.[29]

The discussion and application of both simple and compound fertilizers reflects the care farmers took to make use of all the resources available in their environment, possibly intending to get rid of all animal and plant waste, recycle organic materials, even harmful ones such as bath ashes, to benefit from them in every possible way. Such practices were important to maintain ecological balance and ensure a safe environment. At the same time, we may assume that they resulted in the creation of small jobs and métiers that served agriculture by, for instance, collecting and transporting different animal manures, night soil, and bath ashes, as well as preparing fertilizers.

The aforementioned fertilizers were applied to various soil types based on their respective properties. The goal was to enhance the quality of soils by balancing adverse attributes with positive ones, thereby increasing overall productivity. According to al-Ghazzī, while pigeon droppings were best used with moist soil, cow dung was good for dryer and white soils, and, sheep and goat dung was used for soft soil.[30] Distinguishing trees and other plants by their resilience to and dependance on fertilizers, he proposed three categories: first, plants that exhibited tolerance for fertilizers (e.g., almond, palm, pear, pomegranate, olive, fig, pistachio, and similar plants); second, plants that were susceptible to adverse effects of fertilizers (among them basil, jasmine, citron, and banana), in addition to those that could be completely destroyed by fertilizers (such as quince, prune, apple, and rose); and third, plants that did not need fertilizers, as they flourished naturally, grew in the wild, or were of a robust nature (such as walnut, hazelnut, carob, oak, and wild olive).[31]

3 The Preparation, Application, and Benefits of Bird Droppings

A wide variety of different birds (wild birds, water birds, migratory birds, etc.) was found in Mamluk Egypt because of its diverse environment.[32] Domestic

---

29 *Mabāhij al-fikar*, MS Arabe 6745, fols. 36–39, BnF; Ṣāliḥīyah and al-ʿAmad, *Miftāḥ al-rāḥah*, 115–118.

30 *Jāmiʿ farāʾiḍ al-milāḥa*, MS 8407, fol. 5, Al-Asad National Library.

31 *Jāmiʿ farāʾiḍ al-milāḥa*, fols. 5b, 6a.

32 Heba M. S. Abdelnaby, *Al-ṭuyūr fī al-ʿaṣr al-Mamlukī* (Cairo: Ein for Human and Social Studies, 2021), 32–39.

birds were part of the capital of farmers in Egypt: they were bred to be sold, occasionally consumed,[33] and to have their excrement used as fertilizer for soil and plants. Pigeons in particular were of great importance for their usage as food and in transporting the mail.[34] They were among the domestic birds bred in the countryside, where pigeon towers marked the agricultural landscape of Egypt.[35] Mamluk sources therefore emphasize the abundance of pigeons and their availability on Egyptian markets,[36] and tend to mention them in relation to many famous dishes.[37] Agricultural practices in Bilad al-Shām, the Mamluk territory in the Levant, were similar to those in Egypt, and the mail transport network connected the two countries, ensuring an abundance of bird droppings.

Most agricultural manuals suggest that all bird droppings are useful—except those of geese and water fowl, which are poisonous to plants.[38] As in the case of al-Waṭwāṭ, the anonymous Syrian author, and al-Ghazzī, pigeon droppings were generally considered the best animal manure.[39] Characterized by excessive heat and intense moisture with no dryness,[40] they are described as "the salvation for plants weakened by intense cold."[41] Moreover, they were known to increase the yield of fruit trees and accelerate plant growth. Just a small quantity produced similar effects to large amounts of other types of

33 Magdi Bakr, *Al-qariah al-mṣriya fi ʿaṣr Salāṭīn al-Mamālīk 648–923 AH/1250–1517 CE* (Cairo: Al-Hay'a al-Maṣriya al-ʿAāmma lil-kitāb, 1999), 185–188.

34 Abdelnaby, *Al-ṭuyūr*, 50–54.

35 Varisco, "Zibl and Zirāʿa," 134.

36 Al-Maqrīzī, *Al-mawāʿiẓ wa-al-iʿtbār bi-dhikr al-khiṭaṭ wa-al-āthār* (Cairo: Al-Hay'a al-ʿĀmma li-Quṣūr al-Thaqāfa, 1999), 3:29, 97. He mentioned that different types of birds, including pigeons, were sold on Bayn al-qaṣrīn market—once, an inspector/supervisor of markets (al-muḥtasib) found a storage containing 34,196 rotten birds, including 1,196 pigeons stored by a merchant to be sold. That such a large number of pigeons was owned by one merchant indicates their abundance on Egyptian markets.

37 Yūsuf ibn Muḥammad ibn ʿAbd al-Jawād ibn Khidr Al-Shirbīnī, *Hazz al-quḥūf fi sharḥ qaṣid Abī Shādūf* (Būlāq: Al-Matbaʿa al-Amīriya, 1308 AH [1890 CE]), 166, 183. Although this source dates to the Ottoman period, it records different dishes that have been prepared for generations in Egyptian villages and the countryside.

38 *Jāmiʿ farāʾiḍ al-milāḥa*, MS 8407, fol. 5a–b, Al-Asad National Library; MS Zirāʿah Taymūr 42, fol. 20, Dār al-Kutub. The anonymous author added that a small amount of goose droppings or water fowl droppings could destroy plants. See *Miftāḥ al-rāḥah*, MS Ahlwardt 6208, fol. 18a, Staatsbibliothek zu Berlin; Ṣāliḥīyah and al-ʿAmad, *Miftāḥ al-rāḥah*, 112.

39 *Mabāhij al-fikar*, MS Arabe 6745, fol. 33, BnF; *Miftāḥ al-rāḥah*, MS Ahlwardt 6208, fol. 17b, Staatsbibliothek zu Berlin; *Jāmiʿ farāʾiḍ al-milāḥa*, MS 8407, fol. 5b, Al-Asad National Library.

40 *Mabāhij al-fikar*, MS Arabe 6745, fol. 34, BnF; *Miftāḥ al-rāḥah*, MS Ahlwardt 6208, fol. 18b, Staatsbibliothek zu Berlin.

41 *Miftāḥ al-rāḥah*, MS Ahlwardt 6208, fol. 18b; Ṣāliḥīyah and al-ʿAmad, *Miftāḥ al-rāḥah*, 114.

manure.[42] Therefore, some older sources highlight that only an expert should apply pigeon droppings; otherwise, their excessive use could burn the plant.[43]

The preparation of manure and fertilizers was a long-term process that required a great deal of work and follow-up. Mamluk agricultural sources generally disregarded the preparation of simple/single fertilizers and focused on the preparation of compound ones—understandably, given their complexity. Still, al-Ghazzī's description of how to prepare *zibl* ("kayfiyat ʿamal al-zibl") can give us insights. He explains that *zibl* was prepared in large troughs, where it was mixed with cauliflower leaves, vine leaves, and moist river sludge. Additionally, it was sprinkled with poor-quality wine and human urine and turned continuously until it went black and its fetid stench had disappeared. Lastly, ashes were blended in thoroughly and the mixture was left to stand for a long period of time, after which it was spread flat and left to air-dry.[44]

On the other hand, another source mentions that the *zibl* of pigeon droppings was prepared by mixing one part (*kayl*) of droppings with twenty parts of dirt (*turāb*) and letting it sit for a year to turn into a good *zibl*.[45] Both the anonymous Syrian author and al-Ghazzī agreed that the manure should be composted for at least a year—the longer the better, as this would remove its musty smell and wet texture, which could attract vermin. Yet different uses required different composting times: if the manure was to be used for trees, it could be composted for less than a year, while if used for legumes, the anonymous author recommended the use of older compost. All kinds of manure benefitted from being composted for two or three years.[46]

The application of fertilizers varied according to the type and size of plants: farmers generally applied them under and around the roots of trees, while in the case of small plants, they scattered fertilizers on the surrounding soil or in irrigation streams that then carried the nutrients to the roots. In all cases, agricultural manuals instructed farmers not to apply fertilizers to branches or leaves as it could burn them.[47]

Pigeon's *zibl* in particular was beneficial for soil and plants in several ways. As a fertilizer for plants, pigeon's *zibl* was used at different stages of plant

42 *Jāmiʿ farāʾiḍ al-milāḥa*, MS 8407, fol. 5b, Al-Asad National Library.

43 See, e.g., Abu al-Khayr al-Andalusī, *Kitāb fi al-filāḥa* (Fes: Al-Matbaʿa al-Jadīda, 1357 AH [1938 CE]), 90. Abu al-Khayr was an Andalusian agricultural writer who lived in Seville in the second half of the eleventh century and was known for the aforementioned manual.

44 *Jāmiʿ farāʾiḍ al-milāḥa*, MS 8407, fol. 5b, Al-Asad National Library.

45 Abu al-Khayr al-Andalusī, *Kitāb fi al-filāḥa*, 11.

46 *Jāmiʿ farāʾiḍ al-milāḥa*, MS 8407, fol. 5b, Al-Asad National Library; Ṣāliḥīyah and al-ʿAmad, *Miftāḥ al-rāḥah*, 118.

47 *Miftāḥ al-rāḥah*, MS Ahlwardt 6208, fol. 21a, Staatsbibliothek zu Berlin.

growth to ensure a good yield. Mamluk agricultural manuals recommend adding it to the soil before planting cucumbers, gourds, and eggplants;[48] for trees such as pomegranate, oak, and olive, they suggest applying it after they have grown about 40 cm tall, as this was believed to accelerate fruiting[49] and to increase the yield of apple trees, while also intensifying the redness of fruits.[50]

4 Medical Benefits of Bird Droppings

In his concise encyclopedia of science classification, Ibn al-Akfānī,[51] the famous Mamluk physician, classified natural science as the third category of science taxonomy. He explained that natural science focused on the study of physical bodies. Therefore, this category included sciences such as botany, zoology, metallurgy, cosmology. He also enumerated ten subsciences related to those categories, including human medicine, veterinary medicine, falconry, agriculture, magic talismans, semiotics, physiognomy, chemistry, vision and dream expression, and star constellations.[52] The first four subsciences were related: just as human medicine revolved around maintaining the health of the human body, veterinary medicine and falconry revolved around the health of the animal body, and agriculture around the health of the plant. It is no wonder, then, that a mutual influence linked these four sciences and those who specialized in them. For example, in his agricultural manual *Al-filāḥa al-muntakhaba*, Ṭaybughā al-Jariklamishī not only focused on how to grow and maintain different plants and trees but also mentioned their medical benefits and their usage in treatment.[53] Al-Ghazzī, who was discussed above as an agricultural author, also wrote works on medicine: for instance, his *ʿArf al-nafḥa*

48 *Mabāhij al-fikar*, MS Arabe 6745, fols. 65–68, BnF; *Al-filāḥa al-muntakhaba*, MS Arabe 2805, fol. 78b, BnF; Ṣāliḥīyah and al-ʿAmad, *Miftāḥ al-rāḥah*, 144–146.

49 *Al-filāḥa al-muntakhaba*, MS Arabe 2805, fol. 87a, BnF; *Jāmiʿ farāʾiḍ al-milāḥa*, MS 8407, fol. 5b, Al-Asad National Library.

50 *Jāmiʿ farāʾiḍ al-milāḥa*, MS 8407, fol. 49a, Al-Asad National Library.

51 Muḥammad ibn Ibrahīm ibn Sāʿid al-Ansārī, known as Ibn al-Akfānī (d. 749 AH/1348) was among the notable physicians of the eighth/fourteenth century AH/CE. He worked in Egypt, successfully practiced medicine there, and also wrote several books on medicine and other sciences. According to Ibn Aybak al-Ṣafadī, nothing was bought for the bīmaristān (hospital) in the Qalawun complex unless approved by him, because of his wide knowledge of tools, medications, animals, and minerals. Ibn al-Akfānī, *Irshād al-qāṣid ila asnā al-maqāṣid fi anwāʿ al-ʿulūm*, introduced by ʿAbd al-Munʿim Omar and revised by Aḥmad Hilmī (Cairo: Dār al-Fikr al-ʿArabī, n.d.) 33–40.

52 Ibn al-Akfānī, *Irshād al-Qāṣid*, 168–170.

53 *Al-filāḥa al-muntakhaba*, MS Arabe 2805, fols. 33a–35b, BnF. See also note 18.

fī ḥifẓ al-ṣiḥah (Knowing the whiff in maintaining the health),[54] which is an *argūza* or poem about the benefits of air, water, food, drink, sport, and many other aspects that can preserve health.

An examination of Mamluk agricultural manuals and medical texts reveals that bird droppings in general, and pigeon droppings in particular, were used not only as fertilizers but also for treating ailments in both plants and humans. Thus, studying the use of bird droppings in treatments provides evidence of the link between agriculture and medicine, and the dialogue between the two sciences. Agriculture manuals, zoological books, and medical sources emphasize the beneficial heating effect of bird droppings in order to justify their usage.

Many Mamluk agricultural manuals document that pigeon *zibl* was used for treating certain tree diseases. Ṭaybughā al-Jariklamishī, for instance, discusses many treatments for the delayed germination of a palm tree or the incomplete germination of its fruits, as caused by cold weather—among them, the use of pigeon *zibl*.[55] This type of *zibl*, he argues, is also an essential ingredient for the treatment of palm trees suffering from a decrease in the sweetness of fruits.[56] In his manual, al-Waṭwāṭ makes similar recommendations: to treat a palm tree that was either completely sterile or bore fruit every other year, one should light a fire two feet away from the tree and fertilize it with pigeon *zibl* and night soil.[57] On the other hand, he cautions that excessive use of pigeon *zibl* and night soil on palm trees could be one of the three causes of "jaundice" (in modern terms, chlorosis)—the other two being excessive thirst and air stagnation in July and August—and he suggests treating this with a mix of water, vinegar, and semolina flour, which is poured onto the palm core and sprinkled on its fronds.[58] Al-Ghazzī's manual in turn argues that, when mixed with water and

54 *ʿArf al-nafḥa fī ḥifẓ al-ṣiḥah*, A3802, n. 34, Süleymaniye Library, Istanbul.

55 *Al-filāḥa al-muntakhaba*, MS Arabe 2805, fol. 33a, BnF. Taybughā also mentions a treatment involving lettuce leaves decomposed with pigeon droppings and night soil which are combined with vine leaves and human urine and allowed to ferment for twenty-one days. Once dried, the resulting mixture is blended with ashes and subsequently applied to the tree.

56 *Al-filāḥa al-muntakhaba*, MS Arabe 2805, fol. 35b, BnF. Taybughā also describes a treatment for palm trees in which the ashes of cow dung and pigeon droppings are combined as fertilizer, and hot water is applied to the roots and fronds.

57 *Mabāhij al-fikar*, MS Arabe 6745, fol. 118, BnF; Ṣāliḥīyah and al-ʿAmad, *Miftāḥ al-rāḥah*, 182.

58 *Mabāhij al-fikar*, fol. 120; Ṣāliḥīyah and al-ʿAmad, *Miftāḥ al-rāḥah*, 184. Al-Ghazzī used the word *Yaraqān* (يرقان), which refers to a human disease, and commented that jaundice is not a disease of the palm tree only but occurs in other trees and most plants for the same reasons mentioned in the text. Its symptoms include dryness of the plant, a yellowing of leaves, and dropping fruits before they are ripe. *Jāmiʿ farāʾiḍ al-milāḥa*, MS 8407, fol. 47a, Al-Asad National Library.

poured on the roots, pigeon *zibl* is effective in treating both fig trees and apple trees attacked by worms. This method, according to al-Ghazzī, might even prevent the attack if used early.[59]

Due to its heat and moisture, moreover, both al-Watwāt and al-Ghazzī tout pigeon *zibl* as an effective remedy for plants damaged by cold, dry weather: al-Watwāt recommends sprinkling a citron tree with warm water if weakened by the cold, and cold water if weakened by the heat, as well as fertilizing its roots with pigeon's *zibl* mixed with dirt.[60] The same method is proposed by al-Ghazzī with regard to linen plants damaged by the cold.[61]

As for medicine, bird droppings were among the *materia medica*[62] used during the Mamluk period for treating various diseases and symptoms. The properties of different droppings were determined according to the habitat of the bird and its fodder, and physicians used this information to determine their benefits for treatment. According to both the fourteenth-century Syrian scholar al-ʿUmarī (d. 749 AH/1349)[63] and the Egyptian zoological author

59 *Jāmiʿ farāʾiḍ al-milāḥa*, fols. 48b, 49a.

60 *Mabāhij al-fikar*, MS Arabe 6745, fol. 180, BnF; *Jāmiʿ farāʾiḍ al-milāḥa*, MS 8407, fol. 49b, Al-Asad National Library; Ṣāliḥīyah and al-ʿAmad, *Miftāḥ al-rāḥah*, 236.

61 *Jāmiʿ farāʾiḍ al-milāḥa*, MS 8407, fol. 63b, Al-Asad National Library.

62 *Materia medica* refers to a wide variety of elements of plant, animal, and mineral origin that were used for preparing medicine. Note that the basic theory of Islamic medicine was derived from ancient Greek humoral pathology. According to this theory, maintaining a balance and harmony between humors was the key to good health, and such an equilibrium was achievable through dietary adjustments or medication. Therefore, a clear understanding of the properties of each *materia medica* was essential for the success of the treatment. Muḥammad Kāmil Ḥussīn, *Al-mūjaz fī tārīkh al-tibb wal-ṣaydala ʿind al-ʿArab* (Cairo: ALESCO publications, n.d.), 38–43; Peter E. Pormann and Emilie Savage-Smith, *Medieval Islamic Medicine* (Washington, DC: Georgetown University Press, 2007), 43–44.

63 Al-ʿUmarī was born in Damascus in 700 AH/1301 to a family already distinguished in the Mamluk civil service. His father was head of the chancery in Damascus, and subsequently in Cairo; al-ʿUmarī started his carrier as assistant to him. Following in his father's footsteps, he was head of the chancery in Damascus from 741 AH/1340 to 743 AH/1342. Apart from his official position, he was known for his brilliance as a writer and expert on a wide variety of subjects related to politics and administration. His notable works are *Masālik al-Ābṣār fī Mamālik al-Āmṣār* and *al-Taʿrīf bil-muṣṭalaḥ al-sharīf*, in addition to several minor essays and letters. Al-Maqrīzī, *Al-Sulūk li-Maʿrifat Duwal al-Mulūk*, vols. 2–3 (Cairo: Maṭbaʿat Dār al-Kutub, 1970–1973), 792; Ibn Taghrībirdī, *Al-nujūm al-zāhirah fī mulūk Miṣr wa-al-Qāhirah* (Cairo: Al-Hay'ah al-ʿĀmmah li-Quṣūr al-Thaqāfah, 2008), 10:234–235; Ibn Taghrībirdī, *Al-Dalīl al-Shāfī ʿla al-Manhal al-Ṣāfī*, ed. Fahim Muḥammad Shaltūt (Cairo: Dār al-Kutub al-Miṣrīyah, 1998), 1:96; *Encyclopaedia of Islam New Edition*, s.v. "Ibn Faḍl Allāh al-ʿUmarī," by K. S. Salibi, 2012, http://dx.doi.org/10.1163/1573-3912_islam_SIM_3153.

al-Damīrī (d. 808 AH/1405),[64] different types of bird droppings were used for preparing topical medicines that were widely used for dermatological diseases. Among those diseases was the common skin problem melasma, that is, a condition that causes dark, discolored patches on the skin and was treated by coating it with the droppings of *zummaj* (steppe eagle), *ʿuqāb* (eagle), or *ʿuṣfūr* (sparrow), on the condition that the *ʿuṣfūr* be fed on rice only.[65] Similarly, physicians considered *zummaj* droppings useful for minor skin issues such as freckles.[66] Furthermore, both al-ʿUmarī and al-Damīrī noted that serious disorders such as pigmentation disorder or vitiligo (a genetic disorder that causes the skin, hair, or eyes to have little or no color) were treated by the droppings of *ḥamām* (pigeon) and *rakhma* (vulture), as those were effective in changing gradually the color of skin.[67] Both authors agreed that ringworm, a contiguous fungal infection causing an itchy circular rash on the skin, was usually treated by coating it with the droppings of *ḥamām* (pigeon), *naʿām* (ostrich), *khuffāsh* (bat)—which was considered a bird—or *zarzūr* (starling).[68] Even warts that grew on different parts of the body were treated with the droppings of *ṭāwūs* (peacock) or the droppings of *ʿuṣfūr* or *qubra* (lark) mixed with saliva.[69]

Physicians also treated cataracts, one of the most common ophthalmological diseases prevailing in Egypt, with bird droppings. The clouding of the eye

64 Al-Damīrī, Muḥammad ibn Mūsa ibn ʿĪsa Kamāl al-Dīn, was born in Cairo in 742 AH/1341 and died there in 808/1405. He started his career as a tailor in his native town, then decided to become a professional theologian. He was taught by famous scholars such as al-Subkī and al-Asnawī, and took up posts in several places of learning and devotion. As an author, he rose to fame both in the Eastern and Western worlds for his compendium of Arabic zoology *Ḥayāt al-ḥayawān*, with which he intended to correct misconceptions about animals. See al-Damīrī, *Ḥayāt al-ḥayawān al-kubrā*, ed. Muḥammad ʿAbd al-Qādir al-Fāḍilī (Beirut: al-Maktaba al-ʿAṣriya, 2004), 5–6; *Encyclopaedia of Islam New Edition*, s.v. "al-Damīrī," by L. Kopf, 2012, http://dx.doi.org/10.1163/1573-3912_islam_SIM_1685.

65 Al-ʿUmarī, *Masālik al-ābṣār fi mamālik al-āmṣār*, ed. Kāmil Sulimān al-Jabbourī (Beirut: Dār al-Kutub al-ʿIlmiya, 2010), 20:92, 96–97; al-Damīrī, *Ḥayāt al-ḥayawān*, 3:13.

66 Heba Mahmoud Saad Abdelnaby, "Treating with Birds: The Insights of Two Mamluk Sources about the Medical Benefits of Birds," *Mamluk Studies Review* 25 (2022): 161–197, on 171.

67 Al-ʿUmarī, *Masālik*, 20:81, 88; al-Damīrī, *Ḥayāt al-ḥayawān*, 2:461. The remedy for vitiligo, as explained by al-ʿUmarī, was to mix pigeon droppings with barley flour, water, and a little tar, and crush them to form an ointment that was applied on the affected area with a linen cloth, which was to remain in place for three days, then removed, with the procedure being repeated until results were achieved. Another approach was to use vulture droppings and mix them with wine vinegar to form an ointment with which to coat the skin patches.

68 Al-ʿUmarī, *Masālik*, 20:81, 83, 92; al-Damīrī, *Ḥayāt al-ḥayawān*, 2:377, 4:428; Abdelnaby, "Treating with Birds," 172.

69 Al-ʿUmarī, *Masālik*, 20:94–95; al-Damīrī, *Ḥayāt al-ḥayawān*, 3:114, 149, 4:290.

lens was treated with *babaghā'* (parrot) droppings, which were also considered effective against ophthalmia.[70] According to al-ʿUmarī, *khuṭṭāf* (kingfisher or swallow) droppings, too, could counteract the clouding of the eye lens; he described in particular their wonderous effects on cataracts, emphasizing that they were a tried and tested remedy.[71]

The botanist and pharmacologist Ibn al-Bayṭār (d. 646 AH/1249)[72] stated that bird droppings were used for preparing topical medicines to heal certain types of bites, burns, and ulcerations. For instance, dog bites were treated with *dīk* (cock) droppings mixed with vinegar.[73] For burns, one used pigeon droppings burned in linen cloth that were then combined with oil before being applied to the burnt area of the skin.[74] Nose ulcerations were remedied by utilizing dried and crushed droppings of *ghurnīq* (heron), which were mixed with water to form a wet suppository applied in the nose.[75]

On the other hand, bird droppings were rarely used for preparing oral medications. Chicken (*dajāj*) droppings were useful for certain gastrointestinal diseases, including diphtheria caused by eating poisonous mushrooms, which physicians treated by drying chicken droppings that they crushed and mixed with vinegar and water to be used as an emetic. The same remedy is advised as a drink for patients suffering from colic pain.[76]

Because of its excessive heat, pigeon droppings were seen as uniquely beneficial in treating a wide variety of diseases. They were generally used to remedy cold diseases whose treatments required heating, especially chronic ones such as gout, migraine, vertigo, and pain in the shoulders, back, stomach, kidneys, and joints. Pigeon droppings were considered successful in reducing swellings

---

70 Al-ʿUmarī, *Masālik*, 20:112; al-Damīrī, *Ḥayāt al-hayawān*, 1:150.

71 Al-ʿUmarī, *Masālik*, 20:83.

72 An Andalusian botanist born in Malaga at the end of the sixth/twelfth century AH/CE. He emigrated to the East and worked in Egypt in the court of Sultan al-Salih Najm al-Din Ayyub, where he wrote his famous book *Al-jāmiʿ li-mufradāt al-ādwiya wa-l-āghdhiya*, which was used during the Mamluk period as a basic guide to botany and other materials that could be used in preparing medicine, making it an essential source for writers discussing medical remedies. After the sultan had passed away, Ibn al-Bayṭār moved to Damascus, where he died. See Ḥussīn, *Al-mūjaz fi tārīkh al-tibb*, 414–418; Georges Anawati, *Tārīkh al-ṣaydalah wa-al-ʿaqāqīr fī al-ʿahd al-qadīm wa-al-ʿaṣr al-wasīṭ* (Beirut: Awrāq Sharqiya, 1996), 189–192; *Encyclopaedia of Islam New Edition*, s.v. "Ibn al-Bayṭār," by J. Vernet, 2012, http://dx.doi.org/10.1163/1573-3912_islam_SIM_3115.

73 Ibn al-Bayṭār, *Al-jāmiʿ li-mufradāt al-ādwiya wa-l-āghdhiya* (Bagdad: Al-Muthana Library, 1964), 2:89; al-ʿUmarī, *Masālik*, 20:85; Abdelnaby, "Treating with Birds," 173.

74 Al-ʿUmarī, *Masālik*, 20:81. He was quoting Ibn al-Bayṭār.

75 Al-Damīrī, *Ḥayāt al-ḥayawān*, 3:224.

76 Ibn al-Bayṭār, *Al-Jāmiʿ*, 2:89; al-ʿUmarī, *Masālik*, 20:85.

and scrofula when used in a poultice. They were dried, mixed with barley flour (or linen seeds) and water, cooked with vinegar and honey, and then bandaged on the swelling or scrofula.[77] Moreover, droppings of red pigeons, if mixed with *darṣīnī* (clove cinnamon) and drunk as liquid oral medication, were attributed the potential to dissolve kidney (?) stones—provided that the pigeons had been fed on linen seeds.[78] Patients with dysuria were advised to sit in warm water with pigeon droppings cooked into it, which proved beneficial, while patients suffering from edema used pigeon droppings mixed with vinegar to coat the affected areas of the body to reduce pain.[79]

As for folk medicine, it was well known that magical and folkloric practices, some of which were ancient beliefs inherited for generations, contributed to the medical pluralism of Islamic civilization. Practitioners of these traditions promoted the use of talismans and amulets, citing certain verses of the Qur'an, invocations, and other ancient practices believed to be beneficial for protection from various harms and to ensure health and good fortune. Plants, animals, and minerals were employed in magical or folk medicine based on the premise that everything in nature had hidden or occult properties that could be activated, and that the disease might be cured by recognizing and utilizing these properties.[80] The body parts of animals or their products or excrements were hung around the neck or bound to some part of the body to effect a remedy. They could also be used as a fumigant or as oral medication, although the latter method was quite rare. Bird droppings were no exception to the rule, and were used in folk medicine. The droppings of *bāzī* (hawk) were warmed with water and drunk to facilitate pregnancy even in infertile women.[81] The droppings of *rakhma* (vulture), by contrast, were used in fumigation to cause miscarriage.[82] Both al-ʿUmarī and al-Damīrī suggested wrapping the droppings of *ghurāb* (crow) in a piece of cloth to stop coughing,[83] as well as the hanging of the droppings of *fākhita* (collared dove) on a boy to treat his nocturnal epilepsy.[84]

---

77 Ibn al-Bayṭār, *Al-Jāmiʿ*, 2:34; al-ʿUmarī, *Masālik*, 20:81. Pigeon droppings were mixed with linen seeds and cooked with vinegar for treating scrofula. For hard swellings, they were mixed with barley flour and cooked with honey.

78 Ibn al-Bayṭār, *Al-Jāmiʿ*, 2:34; al-ʿUmarī, *Masālik*, 20:81. He did not mention what type of stones but probably meant kidney stones.

79 Ibn al-Bayṭār, *Al-Jāmiʿ*, 2:35; al-ʿUmarī, *Masālik*, 20:81; al-Damīrī, *Ḥayāt al-ḥayawān*, 2:336.

80 Pormann and Savage-Smith, *Medieval Islamic Medicine*, 148.

81 Al-Damīrī, *Ḥayāt al-ḥayawān*, 1:146.

82 Al-ʿUmarī, *Masālik*, 20:88.

83 Al-ʿUmarī, 20:99.

84 Al-ʿUmarī, 20:100; al-Damīrī, *Ḥayāt al-ḥayawān*, 4:240.

Clearly, healers used the droppings of many types of birds from different habitats and controlled their diet so that their droppings would produce specific effects for treatment purposes. The sources reveal that bird droppings were generally used for preparing dentifrices and electuaries (*maʿājīn*), suppositories (*fatāʾil*), poultices, dressings, and bandages (*ḍammādāt*), or powders and ointments applied to the eye (*akḥāl*), all of which were used as topical treatments. They were, in rare cases, also used as oral treatment (*sufūf* or *sharāb*) for kidney stones and for treating diphtheria and colic. For kidney stones, the droppings were mixed with Cinnamomum to make them palatable. However, when employed for the treatment of the other two conditions, the droppings served as an emetic, indicating their inherently unappealing taste.[85] In folk medicine, bird droppings played a role as remedies to protect the bearer from harm, help them fulfill a desire, or cure an illness.[86] For such purposes, the droppings were either drunk, fumigated, or worn.

5 Conclusion

This chapter has revealed that bird droppings were used in both agriculture and medicine during the Mamluk period. The heating effect of bird droppings was the dominant property determining their usage in treatments, with authors and practitioners aware that the droppings of domestic birds were generally weaker than those of wild birds, with pigeon droppings considered to have the strongest heating effect. Excessive heat was the primary property motivating the usage of bird droppings in agriculture as well as in medicine. Thus, they were used to treat humans suffering from cold and chronic diseases that required a warming effect, and, additionally, to drain lumps or reduce swellings. Other bird droppings were used topically—and in rare cases, orally—to treat, among others, dermatological, ophthalmological, and pulmonological diseases, as well as dog bites, burns, and ulceration. Farmers used them to increase the productivity of both soil and plants, while also exploiting the heat effect to kill tree worms, and to accelerate both the growth of some plants and the coloring of fruit. Bird droppings were also successfully applied in the treatment of several plant diseases. Pigeon droppings played an important role as they were extremely cheap, readily available, conveniently located near pharmacists, and their heating effect could be relied upon.

85 Al-ʿUmarī, *Masālik*, 20:85. This unpalatable character is implied by al-ʿUmarī's stating that the patient was forced to drink the medication: "gharr bihi al-shrāb."

86 Abdelnaby, "Treating with Birds," 182.

According to both medical and agricultural sources, some diseases occurred in both humans and plants, and they had the same name and even similar symptoms such as jaundice or warts. Bird droppings were used to treat the yellowing signs of jaundice in humans and palm trees, the warty bumps on human skin, and the bumpy swellings on palm tree trunks and roots. Bird droppings were also used to increase the likelihood of conception in infertile women and to encourage fruiting in sterile palm trees. In other words, Mamluk agrarian authors and physicians prescribed treatments involving bird droppings to both plants and humans suffering from ailments with similar symptoms. In addition, the terminology used in both medical and agricultural sources to specify the application of bird droppings confirms the affinity between medicine and agriculture in this matter. Sources from both fields use verbs such as *yudaq*, meaning "to be finely ground," *yuṭla*, meaning "to be coated on," and *yudhar*, meaning "to be sprinkled," when referring to treating plant and bodily ailments.

Moreover, bird droppings were very common in folk medicine due to their alleged apotropaic qualities and healing effects. All of these examples confirm an ongoing dialogue between medicine and agriculture, providing a context to the metaphor expressed by al-Ghazzī, who authored works on both sciences, when he analogized soil with skin, articulating the connection between the human body and soil and between medicine and agriculture. This chapter has demonstrated both the agricultural and medical importance of bird droppings for Mamluk agrarian authors and physicians and the reciprocal relationship and dialogue between the two sciences.

Acknowledgments

This article was made possible by the generous support of the Max Planck Institute for the History of Science (MPIWG), Berlin.

CHAPTER 4

Soil Treatment in Two Late Ming Farming Manuals

Soil as Body

Jörg Henning Hüsemann

1 Introduction

Since the Song dynasty (960–1279), soil fertility and its management appeared as a concern in agronomists' writings, gaining importance as agricultural production intensified to meet the demands of a growing population.[1] In a world where a functioning agriculture was deemed the foundation of an ordered state, *quannong* 勸農 (to encourage farming) became more than a mere ritual lip service. Probably in response to the many crises of the times, during the Ming (1368–1644) and Qing (1644–1911) dynasties the interest in farming and the number of farming manuals exploded, and agricultural knowledge circulation between northern and southern China increased.[2] Agronomy, as Francesca Bray has pointed out, became a "science of state," and indeed the state and its officials played an active role in the "production, dissemination and application of agronomic knowledge."[3] According to Bray, agronomists' writings codified technical information that they had gained from peasants, or they reproduced knowledge found in earlier farming manuals. In other words, agronomy was a field of knowledge bound to tradition and practice, and many of the extensive agronomic writings that form the center of Bray's studies reflect this. Using two examples from the sixteenth century, this chapter shows

1 Although dated, still useful is Ping-ti Ho, *Studies on the Population of China, 1368–1953* (Cambridge, MA: Harvard University Press, 1959).

2 For a concise overview of the various crises, see Timothy Brook, *The Troubled Empire: China in the Yuan and Ming Dynasties* (Cambridge, MA: Belknap Press of Harvard University Press, 2013). As agricultural historian Wang Da 王达 has pointed out, 329 of the 541 titles that Wang Yuhu 王毓瑚 has recorded in his *Zhongguo nongxue shulu* 中國農學書錄 (Bibliography of Chinese agronomy) were written or compiled during the Ming and Qing dynasties, thus amounting to nearly two thirds. Wang Da, "Zhongguo Ming Qing shiqi nongshu zongmu" 中国明清时期农书总目, *Zhongguo nongshi* 中国农史 20, no. 1 (2000): 102–113, on 102.

3 Francesca Bray, "Science, Technique, Technology: Passages between Matter and Knowledge in Imperial Chinese Agriculture," *British Journal for the History of Science* 41, no. 3 (2008): 319–344, on 319, 330.

 | DOI:10.1163/9789004748484_006

that not all authors of farming manuals followed well-trodden paths. Some writings demonstrate new approaches to explaining agricultural practice, approaches that might even have previously been considered incompatible with the agricultural realm.

Such new approaches were likely to be found in shorter, regional manuals, whose content was often later incorporated into encompassing and widely distributed volumes and thus became also known in other parts of the empire. Written by landowners, state officials, or scholar-recluses, these manuals mostly deal with the agriculture of a certain region of the empire and contributed to the diversification of the genre. In addition, this "agricultural regionalism" enhanced the understanding of the empire's geographical diversity and the various ecological and climatic conditions in different areas of northern and southern China.[4]

This chapter focuses on two of these regional agronomic writings from the second half of the sixteenth century: the *Nongshuo* 農說 (Explanations on agriculture) by Ma Yilong 馬一龍 (1499–1570), and the slightly later *Baodi quannong shu* 寶坻勸農書 (Manual for encouraging agriculture in Baodi) by Yuan Huang 袁黃 (1533–1606).[5] Both manuals focus on rice cultivation: Ma Yilong's work was influenced by the agriculture of his hometown Liyang 溧陽 in southern Jiangsu province, a traditional rice-growing area in the Yangzi delta;[6] Yuan Huang promoted the exploitation of the "benefits of water" (水利) and the cultivation of wet rice in Baodi county 寶坻縣, part of present-day Tianjin in northern China.[7] Despite their different geographical orientations, the writings share many similarities. Both authors were well-educated men who wrote their *nongshu* around the same time and focused on paddy field farming. This chapter shows that Ma Yilong and Yuan Huang used similar lines

4 The term "agricultural regionalism" is derived from Marta Hanson's study of medical regionalism during the Ming dynasty. There are many parallels in the development of agronomy. Marta E. Hanson, "Northern Purgatives, Southern Restoratives: Ming Medical Regionalism," *Asian Medicine* 2, no. 2 (2006): 115–170.

5 Both writings are discussed in greater detail in Wang Yuhu, *Zhongguo nongxue shulu* 中國農學書錄 (Beijing: Nongye chubanshe, 1964) 148, 159–160. The dates of Ma Yilong's life given here are found in Ni Genjin 倪根金, "*Nongshuo* zuozhe Ma Yilong shengzu niandai" 《农说》作者马一龙的生卒年代, *Zhongguo shi yanjiu* 中国史研究 3 (2001): 176.

6 Not much is known about the life of Ma Yilong. Ni Genjin and Lu Jiaming 卢家明 were able to extract some details about his life and his family background from local gazetteers. Ni Genjin and Lu Jiaming, "*Nongshuo* zuozhe Ma Yilong jiashi, shengping, zhushu chu kao" 《农说》作者马一龙家世、生平、著述初考, in *Nongye lishi lunji* 农业历史论集, ed. Zhou Zhuanji 周篆基 and Ni Genjin (Nanchang: Jiangxi renmin chubanshe, 2000), 404–417.

7 Wang, *Zhongguo nongxue*, 159. Unless otherwise noted, all translations by the author.

of argument to explain agriculture, which is why the manuals can be understood as representations of the established agronomic knowledge of the time. The fact that both writings were regularly quoted by later agronomists indicates that their content remained part of ongoing scholarly discourses and was considered valuable.[8]

Drawing on the accounts on soil in the *Nongshuo* and the *Baodi quannong shu*, this chapter argues that these manuals represent the evolution of Ming agronomy from a practice- to a science-based profession. As the foundation of agricultural work, soil is discussed in all farming manuals, either in an individual chapter or in the context of cultivation methods. The accounts on soil in the two regional agronomic writings offer particularly rich cases of how agronomists examined inherited agricultural knowledge through a local lens and developed new ideas based on their local experience. Instead of reproducing the knowledge on soil found in earlier writings or giving practical advice on how to understand soil, Ma Yilong and Yuan Huang focused on explaining the theoretical principles underlying agricultural practice through "basic concepts of nature."[9] These concepts were essential in various fields of scientific inquiry such as medicine, astronomy, mathematics, or alchemy. As Benjamin A. Elman has argued, "Ming-Qing literati believed that a rational and orderly cosmos informed the microcosmic patterns of differentiation and organization in the creation and evolution of all things in the world."[10]

The two basic concepts most relevant to understanding the accounts on soil in both agronomic writings are qi (氣) and yin and yang (陰陽). *Qi* is often translated as "air," "vapor," "breath," "pneuma," or "vital force," an energy existing in all animate and inanimate entities. In his discussion of traditional accounts on qi, British sinologist and philosopher Angus C. Graham demonstrated that it is often presented in binary oppositions, such as hot and cold, pure and impure, or heaven and earth.[11] The core manifestation of this binarism is the concept of yin and yang, two opposing but at the same time complementary forces. The pristine qi can develop either into yin qi or yang qi and is responsible

8 For example, only a few years after its completion, the *Baodi quannong shu* was quoted by Xu Guangqi 徐光啟 (1562–1633) in his *Nongshu caogao* 農書草稿. Apart from the preface, the *Nongshuo* was quoted in the *Nongzheng quanshu* 農政全書 in its entirety and had a strong influence on the *Zhiben tigang* 知本提綱 by Yang Shen 楊屾 (1687–1785).

9 The term "basic concepts of nature" is borrowed from Ho Peng Yoke, who offers a concise overview of the various concepts. Ho Peng Yoke, *Li, Qi and Shu: An Introduction to Science and Civilization in China* (Seattle: University of Washington Press, 1987), 3.

10 Benjamin A. Elman, *On Their Own Terms: Science in China, 1550–1900* (Cambridge, MA: Harvard University Press, 2005), 22.

11 Angus C. Graham, *Yin-Yang and the Nature of Correlative Thinking* (Singapore: Institute of East Asian Philosophies, 1986), 32–34. For a detailed list of binary oppositions, see 32–33.

for various processes in the universe.[12] Many farming manuals, including the *Nongshuo* and the *Baodi quannong shu*, took the concept of yin and yang as the foundation for explanations concerning the seasons of the year as well as other natural conditions relevant for the growth and ripening of plants.

This chapter proposes that both agronomists understood agriculture in terms of these traditional, universal cosmological concepts that allowed them to build bridges between agriculture and other fields of knowledge. In the case of soil, Ma Yilong and Yuan Huang used medical reasoning to explain the properties of soil and its improvement.[13] Chen Fu 陳旉 (1076–c. 1154), active under the Southern Song (1127–1279), had already advised farmers to "use fertilizers like medicine" (用糞猶藥);[14] however, he did not elaborate further, nor does his farming manual show evidence of medical concepts influencing his ideas on agriculture. Although later farming manuals frequently refer to Chen Fu's idea of "fertilizer medicine" (糞藥), it was not until the Ming dynasty that agronomists began to interpret agriculture in terms of medical concepts. Using Ma Yilong's and Yuan Huang's writings as an example, this chapter shows that both agronomists explained soil and its improvement along the lines of medical reasoning about the human body and the physicians' treatment of their patients.

2 Soil as Body

In Chinese thought, heaven is commonly conceived of as male (yang) or father, and earth as female (yin) or mother. Accordingly, the growth of plants

12 See, e.g., Ho, *Li, Qi and Shu*, 11.

13 Note that Ma Yilong even quoted from the medical classic *Huangdi neijing* 黃帝內經. Du Xinhao 杜新豪 mentions the influence of medical theories on manuring in an article, but mostly focuses on soil nutrition. Du Xinhao, "Qilun yu yidao: Songdai yijiang shiren dui shifei lilun de chanshu" 气论与医道：宋代以降士人对施肥理论的阐述, *Zhongguo nongshi* 中国农史 4 (2016): 23–30. As Heba Mahmoud Saad Abdelnaby has shown in the previous chapter, this phenomenon was not a unique feature of Chinese agronomy, but striking similarities can also be found in sources from Mamluk Egypt. The use of excrement as an ingredient in medicinal substances in China has been discussed by Catherine Despeux. See Catherine Despeux, "Chinese Medicinal Excrement: Is There a Buddhist Influence on the Use of Animal Excrement-Based Recipes in Medieval China?," *Asian Medicine* 12 (2017): 139–169. Excrement was regularly discussed by Li Shizhen 李時珍 in his *Bencao gangmu* 本草綱目 (Compendium of *materia medica*). See Li Shizhen, *Ben Cao Gang Mu*, vol. 9, *Fowls, Domestic and Wild Animals, Human Substances*, trans. and annotated by Paul U. Unschuld (Oakland: University of California Press, 2021).

14 Chen Fu 陳旉, *Chen Fu nongshu jiaozhu* 陳旉農書校註, collated and commented by Wan Guoding 萬國鼎 (Beijing: Nongye chubanshe, 1965), 34.

is understood as the result of an interaction between heaven and earth. In his *Yuantu dayan* 元圖大衍 (Great expansion of the chart of origins), Ma Yilong pointed out that "the qi of heaven descends and interacts with the earth, and the qi of the earth ascends and interacts with heaven" (天氣下交于地，地氣上交于天), and "through the interaction of father [i.e., heaven] and mother [i.e., earth], the ten thousand things are brought forth" (父母交而萬物生).[15] The earth receives the seeds and nourishes them.[16] In addition to this analogy between earth and the female body, both agronomists transposed into the agricultural context several ideas that are found in medical thought.

As Marta Hanson has pointed out, Ming-dynasty physicians had a clear conception of a geographical division of China into north and south, which they used "to systematize the correspondences they thought connected different climates, diseases, and bodily constitutions and required different therapeutic strategies in medical practice."[17] A similar differentiation can be noted in several agricultural writings since the Southern Song dynasty.[18] Despite the growing awareness of regionally specific climatic and ecological conditions, the most common system for categorizing different types of soil was still the traditional account in the chapter "Yugong" 禹貢 (Tribute of Yu) of the Zhou dynasty (1045–221 BCE) *Shangshu* 尚書 (Venerable records), generally regarded as one of the earliest writings to analyze soil.[19] According to the *Shangshu*, after the mythical emperor Yu 禹 (trad., third millennium BCE) had tamed the great flood, he differentiated the territory into nine provinces according to their soil

15 Ma Yilong 馬一龍, *Yuantu dayan* 元圖大衍, in *Shuofu xu* 說郛續, ed. Tao Ting 陶珽 [China: s.n.], Shunzhi 3 [1646], *juan* 1, 1B, Harvard-Yenching Library, Chinese Rare Books Digitization Project-Collectanea: https://curiosity.lib.harvard.edu/chinese-rare-books/catalog/49-990067838030203941, accessed May 26, 2023.

16 Sometimes the seed grain is also compared to the mother that produces offspring. See, e.g., Ma Yilong, *Nongshuo* 農說, in *Nongshuo de zhengli he yanjiu* 《农说》的整理和研究, ed. Song Zhanqing 宋湛庆 (Nanjing: Dongnan daxue chubanshe, 1990), 42–43.

17 Hanson, "Northern Purgatives," 118.

18 One of the earliest farming manuals comparing northern and southern methods is the *Wang Zhen nongshu* from the Yuan dynasty (1279–1368). See Francesca Bray, "Agricultural Illustrations: Blueprint or Icon?," in *Graphics and Text in the Production of Technical Knowledge in China: The Warp and the Weft*, ed. Francesca Bray, Vera Dorofeeva-Lichtmann, and Georges Métailié, Sinica Leidensia 79 (Leiden: Brill, 2007), 521–567, on 538–539. Yuan Huang also repeatedly refers to different regional practices.

19 See, e.g., Lin Putian 林蒲田, *Zhongguo gudai turang fenlei he tudi liyong* 中国古代土壤分类和土地利用 (Beijing: Kexue chubanshe, 1996), 29–40; Zeng Xiongsheng 曾雄生, *Zhongguo nongxue shi (xiuding ben)* 中国农学史(修订本) (Fuzhou: Fujian renmin chubanshe, 2012), 66–71.

and the quality of their fields.[20] For these nine provinces, ten different types of soil are described and distinguished by color, texture, and moisture. Yuan Huang even included a "Song about the color and nature of the nine provinces' soil" (九州土色性歌) in his study of the geography of the "Yugong" chapter.[21] Despite his appreciation of the "Yugong" chapter, in the *Baodi quannong shu* Yuan Huang criticized Yu's very rough differentiation of the territory (然禹亦辨其大概耳). Based on his experience, he concluded that not only did "the veins of the soil all differ" (土脈各異) in a single province, but also in a small county like Baodi, various types of soil (i.e., soils with different veins) could be found.[22] Giving a detailed account of the different soils in Baodi, he emphasized that, "for all of these [regions], the qi of the earth is not uniform" (此皆地氣之不齊者也).[23] Yuan Huang differentiated the properties of soil with a set of binary oppositions:

> Moreover, the territorial advantages are not the same: there are strong and weak soils, light and heavy soils, compact and loose soils, dry and wet soils, immature and mature soils, cold and warm soils, fat and lean soils; you must examine what suits each of them and till accordingly.
>
> 又地利不同，有強土、有弱土，有輕土、有重土，有緊土、有緩土，有燥土、有濕土，有生土、有熟土，有寒土、有暖土，有肥土、有瘠土，皆須相其宜而耕之.[24]

20 For a translation of this *Shangshu* passage, see Bernhard Karlgren, "The Book of Documents," *Bulletin of the Museum of Far Eastern Antiquities* 22 (1950): 1–81, on 12–18. For a thorough discussion of the soils in the "Yugong" along with partial translations, see Joseph Needham and Gwei-Djen Lu, "Chinese Geo-botany *in Statu Nascendi*," *Journal d'agriculture traditionnelle et de botanique appliquée* 28, nos. 3–4 (1981): 199–230.

21 Yuan Huang 袁黃, *Yugong tushuo* 禹貢圖說, in *Yuan Liaofan wenji* 袁了凡文集, vol. 7 (Beijing: Xianzhuang shuju, 2006), 849B (67B).

22 Yuan Huang, *Baodi quannong shu* 宝坻劝农书, in Yuan Huang, *Baodi quannong shu; Quyang shuili; Shanju suoyan* 宝坻劝农书、渠阳水利、山居琐言, collated and commented by Zheng Shousen 郑守森, Kuang Qingkai 况清楷, and Zhai Qianxiang 翟乾祥 (Beijing: Zhongguo nongye chubanshe, 2000), 4. *Tu mai* 土脈, the channels or meridians of the soil, are a concept that had apparently been adopted from Chinese medicine. Like the human body, the soil contained veins through which qi energy flowed. Accounts on *tu mai* are not frequently found in sources written prior to the Song dynasty.

23 Yuan, *Baodi quannong shu*, 4–5. According to Yuan Huang, the land in the northwestern part of Baodi has white, loose soil; the southeastern soil is black and loamy. Moreover, depending on the territory's altitude or its distance from the sea, further types of soils are mentioned.

24 Yuan, *Baodi quannong shu*, 5.

These properties might again be associated with the fundamental opposition between yin and yang; indeed, many of the soil properties mentioned by Yuan Huang are discussed elsewhere in different contexts (see Table 4.1).[25]

TABLE 4.1 Soil properties arranged according to yin and yang

yang 陽	**yin 陰**
strong (*qiang* 強)	weak (*ruo* 弱)
light (*qing* 輕)	heavy (*zhong* 重)
loose (*huan* 緩)	compact (*jin* 緊)
dry (*zao* 燥)	wet (*shi* 濕)
immature (*sheng* 生)	mature (*shu* 熟)
warm (*nuan* 暖)	cold (*han* 寒)
fat (*fei* 肥)	lean (*ji* 瘠)

It is important to note that all of the terms used to characterize soil properties were also used by authors of medical writings to describe the constitution of the human body.

According to Paul U. Unschuld, "the health of a human organism depended on balanced exposure to the various climatic factors, such as wind, cold, and heat, and on unimpeded and proper flow of these and other so-called qi through vessels in the body," and the soil was subject to environmental influences as well.[26] Ma Yilong made this very clear in his discussion of the relationship between soil and the increase and decrease of yin and yang throughout a year.

25 These terms are found, for example, in the writings of Song-dynasty thinker Zhu Xi 朱熹 (1130–1200); see Yung Sik Kim, *The Natural Philosophy of Chu Hsi (1130–1200)* (Philadelphia: American Philosophical Society, 2000), 44–46. When only one of the properties is associated with either yang or yin, the author grouped the second property accordingly, as, e.g., in the case of *fei* 肥 and *ji* 瘠.

26 Paul U. Unschuld, *Huang Di nei jing su wen: Nature, Knowledge, Imagery in an Ancient Chinese Medical Text* (Berkeley: University of California Press, 2003), 96. In Chinese medical thought, the human body, like the soil in farming manuals, interacts with heaven (yang) and earth (yin), and both are subject to the same processes and forces in nature. Geoffrey Lloyd and Nathan Sivin, *The Way and the Word: Science and Medicine in Early China and Greece* (New Haven: Yale University Press, 2002), 220–221. Lloyd and Sivin (233) point out that, "in medicine, … the conceptual structure of health, disease, and therapy linked the body to the cosmos and the state."

FIGURE 4.1 Yearly cycle of yin and yang depicted by arrangement of twelve *Yijing* 易經 hexagrams
SOURCE: LI GUANGDI 李光地 (1642–1718), *YUELING JIYAO* 月令輯要, THE CHINESE UNIVERSITY OF HONG KONG LIBRARY, BEI SHAN TANG LEGACY, CHINESE RARE BOOK DIGITAL COLLECTION, HTTPS://REPOSITORY.LIB.CUHK.EDU.HK/EN/ITEM/CUHK-963253 (CC BY-NC-ND 4.0)

The yearly cycle of yin and yang (Figure 4.1) shows that yang (white) comes into existence in the eleventh month,[27] increases monthly until it is balanced with yin (black) in the first month (spring equinox), and reaches its zenith in the fourth month (summer solstice). In the fifth month, yin comes into being, increases until it is balanced with yang in the seventh month (autumnal equinox), and reaches its zenith in the tenth month (winter solstice), before the cycle starts anew.

Ma Yilong pointed out that yin is dominant during winter, causing cold that freezes and seals the surface of the soil. This seal allows yang to accumulate

27 Note that the months of one year are calculated according to the lunisolar calendar. The first month therefore roughly corresponds to February in the Gregorian calendar. Arabic numerals added.

and grow stronger underneath the surface without any leakage. Yang slowly rises up until, in spring, as soon as the yang qi of the sun is strong enough to thaw the seal of the frozen earth, the yang qi in the soil begins to leak and cover the surface, allowing the plants to grow. Ideally, while yang is leaking out, an equal amount of yin will flow back into the soil (陽洩一分於外，陰入一分於中) so that, according to the annual cycle of yin and yang, the equilibrium between the two forces is restored.[28]

A similar concept is applied in medical accounts that describe the relationship between temperature and the pores of the human body. In the warm southern climate, they are open and the yang qi is depleted through leakage. In the cold north, the pores are tightly closed and the yang qi is complete and full.[29] Thus, the topsoil and human skin react to environmental conditions in a similar way.

These examples demonstrate that Ma Yilong and Yuan Huang drew many parallels between agricultural and medical concepts. The connection that they established between soil and the human body also influenced their reasoning about proper soil treatment.

3 Treating the Soil

In traditional Chinese medicine, a basic principle underlying the treatment of illnesses caused by the disharmony of qi is the need to restore the balance of qi in the patient's body. For this purpose, physicians used different therapies: invasive ones such as acupuncture, bloodletting, or scraping, as well as substance and heat therapies.[30] An illness was cured either by physical means (the physicians' skilled hands) or by the effects of medical substances or heat on the human body. The *Nongshuo* proposes a comparable differentiation of farmers' methods to improve their soil:

> There are two [kinds of] procedures by which to nourish [the soil]: by using human strength: watering, hoeing, weeding, cleaning, and raking; by using [other] things' strength: [using] mud, dung, ashes, oilcake, stalks, and grasses.

28 Ma, *Nongshuo*, 12.
29 Hanson, "Northern Purgatives," 135. Note that the pores of the skin neither open nor close.
30 See, e.g., Unschuld, *Huang Di nei jing*, 265–318.

所滋之事有二；以人力者，灌溉鋤耘塗盪也；以物力者，泥糞灰〈米凡〉稿卉也.[31]

In modern terms, human strength transforms the physical properties of soil, such as its texture, density, and humidity, while utilizing the strength of other things improves its fertility. Although Ma Yilong differentiated between these two approaches, they were both needed to provide a suitable, that is, a balanced, foundation for plant growth. An imbalance, hence an excess of one property associated either with yin or with yang, will result in a soil that is unfavorable for cultivation due to its being too cold or too warm, too dry or too wet, too compact or too loose (see Table 4.2). The agronomists' practical advice for restoring or maintaining a balanced soil is based on the principle of choosing tillage methods or fertilizers whose nature is opposite to the unfavorable soil property in question. In medical terms, this corresponds to the "normal treatment" (正治) of illnesses that antagonizes the disease.[32]

Yuan Huang explained the methods that farmers should use to treat disadvantageous soil properties in a lengthy account summarized in the following table:

TABLE 4.2 Treatment of soil properties according to the *Baodi quannong shu*

Soil type	Soil property	Desired effect	Methods
hard and solid land (堅硬強地) black loamy soil (黑壚土)	strong 強	weaken 弱	– Plow in spring – Smash the clods
light soil (輕土) weak soil (弱土)	weak 弱	strengthen 強	– Plow several times – Plow after rainfall – Let buffaloes or sheep trample on it
compact soil (緊土)	compact 緊	loosen 緩	– Deep plowing – Thorough harrowing (several times) – Enrich with ashes – Mix with sand

31 Ma, *Nongshuo*, 34.

32 Volker Scheid, Dan Bensky, Andrew Ellis, and Randall Barolet, *Chinese Herbal Medicine: Formulas & Strategies*, 2nd ed. (Seattle: Eastland Press, 2009), xxiv.

TABLE 4.2 Treatment of soil properties according to the *Baodi quannong shu* (*cont.*)

Soil type	Soil property	Desired effect	Methods
loose soil (緩土)	loose 緩	compact 緊	– Roll with stone roller – Tamp with heavy tamper – Enrich it with river mud
dry soil (燥土)	dry 燥	moisten 濕	– Plow after rain – Store water for irrigation – Prevent snow to be blown away
cold soil (寒土)	cold 寒	warm 暖	– Burn weeds and roots – Use lime
immature soil (生土)	immature 生	mature 熟	– Remove all tendrils – Plow and harrow (several times)
mature soil (熟土)	mature 熟	revive 生	– Field rotation (fallowing) – Crop rotation
(overly) fat and rich soil (肥沃之土)	fat 肥	leaner 瘠	– Weaken with immature soil
(overly) hard and barren soil (磽确之土)	lean 瘠	fatten 肥	– Nurture with manurial soil

As the table shows, for each excess of a particular soil property, Yuan Huang mentioned a suitable treatment that would strengthen the antagonistic property and thus create a balanced soil that offers the best conditions for vigorous plant growth. Apart from making the soil fat and lean, all of the above-mentioned soil properties can be treated with human- or animal-powered tilling methods that improve the physical properties of soil.

According to Yuan Huang, utilization of fertilizers is based on the same principles and is comparable to the treatment of illnesses or disharmonies with medical substances. Quoting Chen Fu, he argued that fertilizers should be used like medicine, and that famers had to avoid any mistakes in managing "cold and warmth, openness and obstruction" (寒溫通塞).[33] In the writings of both agronomists, fertilizers served different purposes: depending on the type

33 These are common terms used in medical writings, referring to the yin (cold) and yang (warm) nature of diseases and cures as well as to open and obstructed veins of qi.

of fertilizer and the time of manuring, they could warm the soil, strengthen it, change its nature, or nourish seedlings.[34]

Yuan Huang explained that there were two possible times when the fields could be manured, each with its own effect: applied before sowing, the fertilizers "transform the soil" (化土); applied after sprouting, they "nourish the seedling" (滋苗). In his eyes, "applying fertilizers should, in general, cause the transformation of the soil and not merely nourish seedlings" (大都用糞者，要使化土，不徒滋苗), since late manuring has only short-term benefits. While "sprouts and branches" (苗枝) grow in abundance, the fruits remain "scarce" (不繁).[35]

Agronomists understood clearly the different effects of fertilizers on soil and were well aware that constant cultivation would deplete a field's strength, resulting in decreased yields.[36] Ma Yilong wrote:

> The growth of a cereal seedling depends on the soil. When the strength of the soil is exhausted, it perishes, but by fertilizing the soil, you can remedy the depletion of the soil's strength.
>
> 禾苗資土以生，土力乏則衰，沃之所以助土力之乏.[37]

However, agronomists also saw limits to the strengthening effect of fertilizers. Any excessive use of manure would harm the soil instead of improving it. Just as physicians carefully calculated the dose of a medical substance in a treatment, so did farmers have to consider the amount of fertilizer to apply to their fields. According to Yuan Huang,

---

34 Chinese farmers used a variety of substances as fertilizers, including human waste (night soil, urine, hair), animal waste (dung, horns and hooves, bones), plants (rotten plant material, ashes, plant material remaining after the expression of oil), and mud from waters and minerals (e.g., lime). Some of these materials were used as single fertilizers, others were mixed to form compound fertilizers. For an overview, see Cao Longgong 曹隆恭, *Feiliao shihua (xiuding ben)* 肥料史话(修订本) (Beijing: Nongye chubanshe, 1984).

35 Yuan, *Baodi quannong shu*, 28.

36 The Song/Yuan-dynasty *Zhong yi biyong* 種藝必用 (Necessary skills for planting), for example, quotes the words of an experienced old farmer (*lao nong* 老農): "If land is tilled for a long time, it gets poor. Today, an ear of grain contains one-third fewer grains than thirty years ago" (地久耕則耗. 三十年前禾一穗若干粒, 今減十分之三). Wu Yi 吳懌 and Zhang Fu 張福, *Zhong yi biyong* 種藝必用, collated and commented by Hu Daojing 胡道靜 (Beijing: Nongye chubanshe, 1963), 17.

37 Ma, *Nongshuo*, 34.

> Fertilizing is not about applying large amounts but about using what is suitable. If you fertilize carelessly and go against what is suitable, you will, however, harm the seed.
>
> 糞不在多，在用得其宜耳·糞苟失其宜，反害稼矣.[38]

Suitability was important not only when it came to the right amount of fertilizer, but also for the choice of the right type of fertilizers. While peasants probably manured with whatever substance was available to them, agronomists clearly pointed out that each type of soil required a specific type of fertilizer. Yuan Huang discussed this variety by referring to a passage in the *Zhouli* 周禮 (*Rites of Zhou*) that explains the "manuring of seeds" (糞種), namely, the soaking of seeds with a lye cooked from the bones of animals, chosen according to the soil properties.[39] Yuan Huang adapted this passage and used it to explain

TABLE 4.3 Soil types and appropriate manure

Soil type	**Soil characteristics**	**Animal**
xing gang 騂剛	red and hard	cow
chi ti 赤緹	red and loose, said to be barren	sheep
fen rang 墳壤	veins of qi rise high, soft, yang soil	Père David's deer
jie ze 渴澤	dried marshland, yin soil	sika deer
xian xi 鹹潟	salty and often moist	badger
bo rang 勃壤	loose and often dry	fox
zhi lu 埴壚	loamy and black	pig
qing biao 輕㶛	light and loose	dog

SOURCE: TABLE BASED ON YUAN HUANG, *BAODI QUANNONG SHU*, 26

38 Yuan, *Baodi quannong shu*, 26.

39 The *Zhouli*, for the most part, is considered a Han-dynasty (206 BCE–220 CE) writing that describes an ideal government under the Zhou dynasty. Jin Chunfeng 金春峯, *Zhouguan zhi chengshu ji qi fanying de wenhua yu shidai xinkao* 周官之成書及其反映的文化與時代新考 (Taipei: Dongda tushu gongsi, 1993), 223–243. For a collection of *Zhouli* passages dealing with agriculture, see Xia Weiying 夏纬瑛, *"Zhouli" shu zhong youguan nongye tiaowen de jieshi* 《周礼》书中有关农业条文的解释 (Beijing: Nongye chubanshe, 1979).

the manuring of different soils with the excrement of the animals mentioned in the *Zhouli*.[40]

The same principles underlying Yuan Huang's explanation of tillage are relevant to his comments on fertilizers, since animal excrement must be of a nature opposite to the prevailing soil property. For example, Yuan Huang described cow dung as "balanced and loose" (和緩), thus suitable for treating hard soil, and sheep's feces as "dry and dense" (燥密), hence suitable for improving loose soil. As the sika deer possesses a lot of yang, its excreta were used for yin soil. The Père David's deer, by contrast, possesses a lot of yin, so its excreta were used to treat yang soil.

Yuan Huang referred to various traditional sources as a useful base for understanding the general principles of agriculture, rather than recommending their practical advice:

> If you understand their [i.e., the ancients'] ideas and further develop them, then you will apply fertilizers according to the soil and each will have what is appropriate.
>
> 得其意而推之，則隨土用糞，各有攸當也.[41]

Thus, although bound to natural conditions, peasants still were able to create an agroecosystem that ensured vigorous plant growth through careful timing and application of suitable methods to treat their soil.[42] Just as physicians had to diagnose their patients' symptoms before treating an illness, peasants were supposed to examine climate and soil prior to tillage. Or as Ma Yilong put it,

> If you adjust to what suits the heavenly seasons, the veins of the soil, and the nature of things, so that none of them is amiss, then with half the effort you will gain twice as much.
>
> 合天時、地脈、物性之宜，而無所差失，則事半而功倍矣.[43]

40 In addition to eight different animals, the *Zhouli* also mentions hemp, which Yuan Huang did not discuss.

41 Yuan, *Baodi quannong shu*, 26.

42 Note that a few years after Ma Yilong and Yuan Huang, Xu Guangqi promoted similar ideas in his agricultural writings. See Guo Wentao 郭文韬, "Shilun Xu Guangqi zai nongxue shang de zhongyao gongxian" 试论徐光启在农学上的重要贡献, *Zhongguo nongshi* 中国农史 3 (1983): 19–23.

43 Ma, *Nongshuo*, 7.

4 Conclusion

In their writings, Ma Yilong and Yuan Huang offered new ways of understanding soil beyond the established practice of codifying technical knowledge and reproducing traditional accounts, choosing instead to explain the general principles governing successful tillage. Although traditional approaches to soil fertility remained an important source for them, they took liberties in using, reinterpreting, and supplementing them. Yuan Huang argued that the ways of the ancients could not be carried out anymore. Still, they offered valid ideas that readers could develop further—just as he and Ma Yilong did. By relying on traditional concepts of nature that were fundamental in various fields of scientific inquiry, they were able to establish connections between agricultural practices and medical ideas about the human body: agricultural soil, they observed, reacted to environmental and climatic conditions in ways comparable to the latter. Therefore, dissatisfactory conditions of agricultural soil could be interpreted as diseases to be prevented or treated following the same methods applied by physicians: namely, careful analysis followed by the selection and application of remedies that were considered to maintain or restore a balance of yin and yang qi through antagonistic effects—in the case of soil, those remedies were the correct fertilizer and suitable tilling methods.

Ma Yilong and Yuan Huang may have owed much of their knowledge of soil types and cultivation techniques to local conditions in Jiangsu and Tianjin, but from their specific examples, they were able to deduce what they considered general principles of successful tillage. Thus, the knowledge they generated was neither bound to their time nor limited to Liyang and Baodi agriculture, which is reflected in its being highly influential to later agronomists' discourses on soil.

Acknowledgments

I would like to thank the following people and institutions for giving me the opportunity to present earlier drafts of this chapter: in particular, Dagmar Schäfer for her support, critical reading, and detailed feedback; Roel Sterckx, who discussed it in the Department AAK (Artifacts, Action, Knowledge) Colloquium at the Max Planck Institute for the History of Science on September 21, 2021; and Marta Hanson and Volker Scheid, who provided valuable advice on the medical side of this chapter. Moreover, I would like to thank Friederike Assandri, Melanie Luise Glienke, and Gina Partridge-Grzimek for their critical reading and helpful comments on earlier versions.

PART 3

Governing the Soil: Taxonomy, Expertise, and the State

CHAPTER 5

Understanding Soil in Ilkhanid and Post-Ilkhanid *filāḥa* Books (1300–1600)

Himmet Taşkömür

The literature of the Ilkhanid period (1256–1353) is characterized by an efflorescence of new writings on agriculture. Far from being mere copies or collections of old theories, these works on farming (*filāḥa* manuals) written during the period immediately following the Mongol conquests in the thirteenth century demonstrate considerable innovation in their approach to agriculture, the conceptualization of the soil, and soil taxonomies.[1] Nevertheless, they owed much to their predecessors as well as the continued interest in the earlier works that had been systematically collected by the royal libraries. Most of the extant early *filāḥa* manuals have survived mainly through the tremendous efforts of Mamluk and Ottoman agronomists, copyists, librarians, and state initiatives.

Agronomy, conceived as both a practice and a domain of knowledge, was understood by the medieval and early modern Islamic writers of *filāḥa* manuals to rest upon certain methodological and epistemic premises. Knowledge related to agriculture was therefore grouped under the umbrella of the natural sciences (*ʿilm-i ṭabīʿī* or *furūʿ al-ṭabīʿiyya*).[2] Similarly, agriculture and peasants also feature prominently in ethical and political-philosophical texts (*akhlāq*) written during the Ilkhanid and post-Ilkhanid periods in Iran and the Ottoman Empire. These texts frequently underscored the foundational role of agriculture in stabilizing and sustaining polities and societies. Medieval and early modern encyclopedias had a separate section on agronomy (*ʿilm-i filāḥa*), further attesting to its intellectual significance. In his *Akhlāq-i Nāṣirī* (*Nasirean Ethics*), the renowned scholar Naṣīr al-dīn al-Ṭūsī (d. 1274) notably integrated agronomy (*filāḥa*; also termed *ʿilm-i baghbānī* or *kishāverzī*) into the domain of

1 For a discussion on the renewed interest and their presence in the Ottoman Royal library in the fifteenth century, see Aleksandar Shopov, "'Books on Agriculture (*al-filāḥa*) Pertaining to Medical Science' and Ottoman Agricultural Science and Practice around 1500," in *Treasures of Knowledge: An Inventory of the Ottoman Palace Library* (1502/3–1503/4) (Leiden: Brill, 2019), 1:557–568.

2 Ibn al-Akfānī Muḥammad ibn Ibrāhīm [d. 1348], *Irshād al-qāsid ilā asnā al-maqāsid*, ed. Ṭāhir ibn Ṣāliḥ al-Jazāʾirī (Beirut: s. n., 1904), 103–104.

natural philosophy, classifying it explicitly as a natural science—a framework that subsequently became widely adopted (*ʿilm-i ṭabīʿī*).[3]

Al-Ṭūsī, who worked closely with Hülāgu (r. 1256–1265; d. 1265), the first Ilkhanid ruler, envisioned the social and political stability (*mizān-ı iʿtidal*) of any polity as resting on the equilibrium of four social classes: the military, the men of the pen (bureaucrats), the people of commerce, and the men of the soil (peasants).[4] Written at a time marked by Mongol conquests and increased connections within Eurasia, al-Ṭūsī's emphasis on agriculture and soil as one of the four pillars of the social order had considerable bearing on political and economic developments in the following centuries. His work continued to impact the sciences (logic, trigonometry, astronomy, ethics, etc.) in the late medieval period and was well known in the early modern Islamic empires (Ottoman, Safavid, and Mughal). Jalāl al-dīn al-Dawwānī (d. 1502), a highly influential Āq-qūyūnlū scholar and philosopher, argued that the earth is capable of improvement (*arz qābil-i ʿimārat-ast*).[5] In a striking reference to the natural world, the sixteenth-century Ottoman political thinker and scholar Kinalizāde ʿAli Çelebi (d. 1572) wrote that farmers are like soil (*türāb gibidir*): "They toil on black soil [*hāk-i siyeh*] day and night, and society receives most of the benefits; ... despite their continuous humiliation by people walking on the soil, which is their face, they [the farmers] still get up and continue to work."[6]

The Mongol Empire and its subsequent offshoots brought novel additions to the Islamicate apparatus of state. The Mongols and, later, the Turkic dynasties incorporated Turco-Mongol political ideas into Perso-Islamic statecraft, bringing novel bureaucratic practices along with new scientific knowledge that

3 The two manuscript copies of the *Nasirean Ethics* that I have consulted referred to the science of agronomy in different ways. The first uses the Arabic terms *filāḥa* and *kishāvarzī* for the science of agriculture. Naṣir al-dīn Ṭūsī, *Akhlāq-i Nāṣirī*, Istanbul University, Persian Manuscripts, MS 758, fol. 5b. The second copy of the same book from Safavid Iran uses the term *ʿilm-i baghbānī* for the science of gardening or horticulture. *Akhlāq-i Nāṣirī*, Istanbul University Rare Books Library, Persian Manuscripts, MS 479, fol. 6b.

4 Al-Tūsī follows a quadripartite division of society that was well established among earlier medieval Islamic thinkers. See, e.g., Abū al-Ḥasan Muḥammad ibn Yūsuf ʿĀmirī, *al-Saʿādah wa-al-isʿād fī al-sīrah al-insānīyah* [On seeking and causing happiness], ed. Mojtaba Mīnovī (Wiesbaden: Steiner Verlag, 1957), 307–308.

5 Jalāl al-dīn al-Dawwānī, *Lawāmiʿ al-ishrāq fī makārim al-akhlāq* [known as *Akhlāq-i Jalālī*], ed. Abd Allah Masʿūdī Arānī (Tehran: Intishārāt-i Ettalaʿat, 1391), 258–260. Unless otherwise noted, all translations are by the author.

6 Kinalızāde, *Ali Çelebi, Ahlāk-ı Alāʾī*, ed. Mustafa Koç (Istanbul: Klasik Yayınları, 2007), 485. On al-Ṭūsī, see *Encyclopaedia of Islam*, s.v. "al-Ṭūsī, Naṣīr al-Dīn," by Hans Daiber and Jamel Ragep, 2012, https://doi.org/10.1163/1573-3912_islam_COM_1264.

incorporated Chinese science into the Islamic world of the time.[7] Rashīd al-Dīn Hamadānī (1247–1318), the writer at the center of this chapter, was a scholar, physician, and vizier of Ghāzān Khān (r. 1295–1304), the first Ilkhanid ruler to convert to Islam.[8] Under Ilkhanid patronage, Rashīd al-Dīn initiated an exceptionally long-lasting intellectual, political, and economic project aimed at synthesizing Chinese knowledge with the received wisdom of the Islamic world.

A prolific writer, Rashīd al-Dīn wrote many works, among them *Jāmiʿal-tawārīkh* (Compendium of chronicles), which is considered the first world history. One of his lesser known works is our focus here: *Āṡār va aḥyāʾ* (Monumental traces and revival; known as *Kitāb al-āthār wa-l-aḥyāʾ* in Arabic), a book on agriculture. Highlighting hitherto neglected aspects of the book, this chapter demonstrates that a significant number of the economic elite, particularly landowners, appear to have recognized the importance of experiment and the mobilization of local and regional knowledge about soil; and that there are indications of a (possibly idealized) demonstrative sympathy toward peasant farmers, all of which made a lasting impression on subsequent political philosophies.[9] This larger project was initiated during what some have described as the "reforms" of Ghāzān Khān.[10] Following the end of Ilkhanid rule in 1353, the Ilkhanid agricultural policies of *ʿimārat* (improvement of the land), shaped agricultural developments in successive political entities across western and southern Asia (including the empires of the Timurids, Ottomans, and Mughals).[11]

---

7 Halil İnalcık, "Turkish and Iranian Political Theories and Traditions in *Kutadgu Bilig*," in *The Middle East and the Balkans under the Ottoman Empire*, ed. H. İnalcık (Bloomington, IN: Indiana University Turkish Studies, 1993) 1–19.

8 See Dorothea Krawulsky, *The Mongol Īlkhāns and their Vizier Rashīd al-Dīn* (Frankfurt am Main: Lang, 2011). On the Ilkhanids and the Ghāzān era, see Bertold Spuler, *Die Mongolen in Iran* (Berlin: Akademie Verlag, 1955), 91–127, 314–322.

9 On the topic of manure in *Āṡār va aḥyāʾ*, see Riaz Tony Howey, "'Rotten and Useful': Compos(t)ing Knowledge in Mongol Iran," *Journal of Material Culture* 30, no. 1 (2025), https://doi.org/10.1177/13591835251318164.

10 For a critical analysis of the reforms of Ghāzān Khān, see Osman G. Özgüdenli, *Gâzân Han ve Reformları (1295–1304)* (Istanbul: Kaknüs Yayınları, 2009), 298–302. Ghāzān Khān's far-reaching reforms and their legacy echoed in the mid-sixteenth-century Ottoman Empire. Inferring from Ghāzān's conversion to Islam, Lütfi Paşa, the Ottoman statesman and grand vizier of the Sultan Süleyman, ranks Ghāzān Khān among the archetypal religious reformers (*mujaddid*), as foretold in the reports relayed from Prophet Muhammad. Lütfi Paşa, *Tevarih-i Al-i Osman*, (Istanbul: Matbaʿa-i ʿĀmire, 1341), 11.

11 Soviet historian Ilia Pavlovich Petrushevsky summed up the complex situation of this period as a tension between "a centralized feudal state together with a ramifying

Experimental approaches to soil and the active promotion of the dissemination of local knowledge about these practices were central to the Ilkhanid intellectual and economic project. Indeed, Rashīd al-Dīn comments extensively on the economic and agrarian improvements that were made during the rule of Ghāzān Khān. The Ilkhanid agricultural reforms implemented during Rashīd al-Dīn's time were far reaching and encompassed state finances and the central treasury. For example, in order to alleviate the financial burden of the peasants, it was prohibited to charge high interest on loans. Access to affordable credit was intended to encourage people to "engage in agriculture and commerce and to busy themselves with beneficial crafts" (bishtar-i mardum bā zirāʿat va be tijārat va pīshahā-yi nāfiʿ mashġūl shodand). Other reforms included revitalizing the land, building dams for irrigation, encouraging the resettlement of peasants in areas where population density was not sufficient to cultivate, setting clearly defined policies regarding fallow land, as well as creating a bureaucracy to keep track of such lands.[12]

Al-Dawwānī presents two legal views as to the standing of agriculture among other occupations. While Imām Idrīs al-Shāfiʿī (d. 820), the founder of the Shāfiʿī school of jurisprudence, saw trade as the best option and ranked it between agriculture and other crafts, the political theorist and scholar al-Mawardī (d. 1058), a member of the Shāfiʿī school, saw agriculture as the best way to make a living and considered it one of the essential and most necessary (*ẓarūrī*) of crafts.[13]

bureaucratic apparatus and that of feudal disintegration together with a system of military fiefs. ... In the thirteenth century, the struggle between these two tendencies ... became mixed up with that between two political trends in the upper strata of the Mongol conquerors of Iran and other countries of Asia and Eastern Europe." Ilya Pavlovich Petrushevsky, "Rashīd al-Dīn's Conception of the State," in "Rashīd al-Dīn Commemoration Volume (1318–1968)," ed. J. A. Boyle and K. Jahn, special issue, *Central Asiatic Journal* 14, no. 1 (1970): 148–162. While Petrushevsky saw the Ilkhanid period as a static and consolidated political bloc, I suggest that medieval Iran under Mongol rule cannot be reduced to a situation of perpetual confrontation; it must be viewed as a reciprocal relationship. For a more recent approach to the question of state longevity in medieval Iran, see Ishayahu Landa, "Famines, State, and the Stability of Mongol Eurasia: Preliminary Remarks," *Central Asiatic Journal* 66, nos. 1–2 (2023): 115–154.

12 Rashīd al-Dīn Fażlallāh Hamadānī, *Jāmi ʿal-tawārīkh: Tārīkh-i mubārak-i Ghāzānī*, vol. 2, ed. Muḥammad Rūshan und Muṣṭafā Mūsavī (Leiden: Brill, 2019), 1335.

13 Jalāl al-Dīn Dawwānī, *Lawāmiʿ al-ishrāq fī makārim al-akhlāq*, Süleymaniye Library, Istanbul, MS Ayasofya 2906, fols. 99a–b.

1 Contact with the Chinese World and Chinese Science: A New Beginning for Persianate Agricultural Texts

The Mongol ruling class that founded the Ilkhanid dynasty brought with them Chinese knowledge about agriculture, plants, medicine, and astronomy.[14] Rashīd al-Dīn's writings on agricultural matters and his personal investment in agricultural production must be understood against this background of a direct connection with China under the Mongols. His *Āṣār va aḥyā'* was part of an attempt to combine knowledge from China, India, the Islamic sciences, and the ancient Greek sciences to support the agricultural improvements being undertaken in Iran, Iraq, and Anatolia during this period.[15] Agriculture was central to the political economy, and scholars who wrote on ethics and politics dedicated lengthy sections to the politics of agriculture. Persianate texts written in medieval Iran and the Mongol period devote a considerable amount of space to matters of land and estate management (*dihqānī*).[16] Al-Ṭūsī penned a short treatise on "The Rites and Customs of Ancient Kings […]," which demonstrates his interest in Chinese statecraft and land tenure.[17] It opens with a striking comparison between the land administration systems of the Chinese emperors and the Iranian rulers:

> The kings of the world and the essence of peace and security [*amn va amān*] requested from the teacher of humanity, Muhammad b.

14 For a discussion on Rashīd al-Dīn and China, see Karl Jahn, "Rashīd al-Dīn and Chinese Culture," *Central Asiatic Journal* 14, nos. 1–3 (1970): 134–147, http://www.jstor.org/stable/41926868.

15 For a general introduction on exchanges between Chinese agronomists and Rashīd al-Dīn on agriculture during the Mongol era and on Rashīd al-Dīn's collaboration with Bolad, head of the Office of the Grand Supervisors of Agriculture, see Thomas T. Allsen, *Culture and Conquest in Mongol Eurasia* (Cambridge: Cambridge University Press, 2001), 115–126.

16 *Dihqān* was one of the major social and political categories in ancient and medieval Iran. This is very much reflected in Firdawsi's *Shāh-nāma*, which views *dihqān* as the major source of language and mores in ancient Iranian society. One of the best-loved stories of *Shāh-nāma*, "Dāstān-i Rostam va Sohrāb," uses *dihqān* to reconstruct ancient Iranian history: "Zi-koftār-i dihqān yakī dāstān, bi-payvandam az gofta-i bāstān" (From the speech of the estate lord, I am writing an epic story of ancient times). Quoted in Muhammad Riżā Rāshid Muḥassil, "Jay-gāh-i Pahlavān va Dihqān dar Shāh-nāmah," *Payīz va Zamistān* 3 (1983): 39–50.

17 The only existing manuscript bears the title *Risāla dar rasm va āyīn-i pādishāhān-i qadīm-i Irān-zamīn az mu'allafāt-i Khāja* [Treatise on the rites and customs of ancient kings of lands of Iran]. See Şerefeddin Yaltkaya, "İlhanîler Devri İdari Teşkilatına Dair: Nasîr-ed-Dini Tûsî'nin Bir Eseri," in *Türk Hukuk ve İktisat Tarihi Mecmuası* 2 (1939): 7–16, https://isamveri.org/pdfsbv/D00133/1939_2/1939_2_YALTKAYAS.pdf.

> Muḥammad al-Ṭūsī, [to write] about the customs and manners [*rasm u rāh-i pādishāhān-i qadīm*] of the ancient kings where the lands are in prosperity and the army and the subjects [*raʿiyyat*] are at peace. … The master, upon this request, wrote on what [the practices of] the ancient kings were. He suggested that, since the kings of Khitay [China] have ancient lands [*molk-hā-yi qadīm*], every year, those kings knew the revenues of these ancient and large lands, and their rules and rites were one and remained the same. But, as for these lands [*vilāyat-hā*], that is, Iran, since the rulers [*pādishāhān*] of these lands change so quickly, the lands that they rule are small, and the laws [*rasm-hā*] are different.[18]

He further remarked that all rulers' income is derived from four classes of persons, the second class being that of the farmers (*ahl-i zirāʿat*). The revenue collected from the farmer class is assessed according to their land and the condition of the soil. The farmers are either wealthy (*tavāngar*) or poor (*darvīsh*), and the areas where they practice agriculture are either vineyards (*bāġ*) or areas in which the soil is well watered (*āb-i zamīn nīgū*). The farmers whose soil is good pay one-tenth, while those whose soil is bad pay one-twentieth, and the poor pay nothing. These are the ancient practices. He details other tax-related regulations based on the Genghisid *yasa*, the Mongolian ethical and legal code introduced during this time. Rashīd al-Dīn's work was composed against a background of intensified political and cultural contact that took place during the century-long Mongol rule in the regions of Khorasan, Iran, and Anatolia. He boasted in his writings that:

> Most of the new inventions are novel things [*akthar mukhtaraʿāt mustajiddāt-ast*], and there is pleasure in every new thing [*li-kulli jadīd ladhdhatun*], and every novelty contains many benefits [*favāid-i har mukhtaraʿ bisyār tavānad buvad*].[19]

18 Naṣīr al-Dīn al-Ṭūsī, *Risāla dar rasm va āyīn-i pādishāhān-i qadīm-i Irān-zamīn*, Beyazıt State Library, Istanbul, Veliyüddin Efendi, MS V2542M, fols. 168b–169a. This small treatise is a rich source for understanding how Islamic land law and Mongolian *yasa* were implemented; the reference to *yasa* is explicit. For a discussion on Genghis Khan's *yasa*, see R. Y. Pochekaev, "Chinggis Khan's Great Yasa in the Mongol Empire and Chinggisid States of the 13th–14th centuries: Legal Code or Ideal 'Law and Order,'" *Golden Horde Review* 4, no. 4 (2016): 724–733, https://doi.org/10.22378/2313-6197.2016-4-4.724-733.

19 Rashīd al-Dīn Fażlallāh Hamadānī, *Tansūq-nāma-i Ilkhānī dar funūn-i ʿulūm-i khatāʾī*, Süleymaniye Library, MS Ayasofya 3596, fol. 2b.

Rashīd al-Dīn explained that the benefits of novel things may be difficult to recognize using traditional knowledge, for "there is no end to the sciences and subtle meanings" (*ʿulūm va maʿānī nā-mutanāhī-ast*), and one cannot conceive of unlimited things from a limited viewpoint.[20] Even though very few of his translations from the Chinese have survived, his goal can be seen in the context of his other works, particularly *Āṣār va aḥyāʾ*. For him, in science and learning, even when one had learned everything that could be known (in one's own land), this knowledge was not complete. His humanistic attitude thrived on the diversity of people and regions. Rashīd al-Dīn remarks several times that the books of Khiṭāy, Chīn, Machīn, and their neighboring lands had not reached Iran. Of the few books that did reach Iran—from the western Islamic world (*maghrib*), the books of the Greeks and Europeans (Franks), and the books from India—most remained untranslated.[21] He actively sought out "books and useful knowledge" (*kutub va favāʾid*) that had either not previously been available in these domains (Iran) (*dar-īn mulk pīsh az-īn futūḥat*) or that had not yet been translated because of a lack of knowledge of the (Chinese) language. He likened his translation initiative from Chinese into Arabic and Persian to the medieval Abbasid translation project.[22]

The only surviving manuscript of the *Tansuq-nāma*, Rashīd al-Dīn's translation of a Chinese medical text, was cataloged as part of the library of the Ottoman sultans Mehmed II and his successor Bayezid II, whose seal it bears.[23]

---

20 *Tansūq-nāma*, fols. 3a–b.

21 On the historical geography of Iranian and Central Asian cities, see Vladimir Minorsky, trans., *Sharaf al-Zamān Ṭāhir Marvazi on China, the Turks and India* (London: Royal Asiatic Society, 1942), 13–29, https://www.cambridge.org/core/journals/bulletin-of-the-school-of-oriental-and-african-studies/article/sharaf-alzaman-tahir-marvazi-on-china-the-turks-and-india-arabic-text-circa-1120-with-an-english-translation-and-commentary-by-v-minorsky-james-g-forlong-fund-vol-xxii-pp-170-53-london-royal-asiatic-society-1942/FED9A1EEF3C744031400EF5CA8A9A6F8.

22 This is a brief summary of his comments in the introduction. I am currently preparing an annotated translation of the introduction, including a historical analysis. For a Turkish translation of the introduction, see Süheyl Ünver and Abdülbaki Gölpınarlı, trans., *Tansuknamei Ilhan Der Fünunu Ulumu Hatai Mukaddimesi* (Istanbul: Milli Mecmua Basımevi, 1939), 3–18.

23 It was copied during the lifetime of Rashīd al-Dīn by the copier Muhammad b. Ahmad b. Mahmud Qiwam al-Kirmani in Tabriz, 20 Shaban, 713 AH (December 10, 1313 CE) and endowed to the library of Ayasofya (Hagia Sophia) by its founder, Ottoman sultan Mahmud I. The high quality and extensive illustrations of Chinese medicine in this deluxe manuscript suggest that it may have been the product of Rashīd al-Dīn's scriptoria. An ex libris medallion in the *Tansuq-nāma* and one in the Arabic version of the *Jāmiʿ al-tawārīkh*, copied in 714 AH (1314 CE), have the same design. See the reproductions in Sheila S. Blair, *A Compendium of Chronicles Rashid al-Din's Illustrated History of the World*

Although, since the early twentieth century, there have been a few studies on some aspects of the manuscript and a partial translation of the introduction into Turkish, no study has situated this work within Rashīd al-Dīn's larger translation project. The relevant section is worth quoting, as it allows us to understand his larger goal, of which *Āsār va aḥyāʾ* was only a part:

> There is no doubt that if [Caliph] Harūn al-Rashīd [d. 809], the legendary paragon of the house of Abbasid, had not made the effort and asked for the translations of the books of the Greeks, all of the wisdom and philosophies and practical knowledge [*favāʾid*] that are prevalent in these lands that came from Greek philosophy would not have been widespread in these lands. Many talented people would not have benefited from them, and all of this knowledge would be incomplete or lost, and the talented and intelligent people of our time who benefitted and continue to benefit from these works, and who have come up with new knowledge [*istinbāt*], would not have benefited from them, and these books would have been lost.[24]

1.1 *Soil in Agricultural Manuals Following the Mongol Conquests*

The Pax Mongolica opened up new opportunities for the diverse mobile populations of Inner Asia, Iran, Anatolia, and beyond, and changed attitudes in the region. Rashīd al-Dīn's agricultural work was written within this context of state rebuilding and experimentation. Adopting the medieval Islamic agrarian system (of *iqṭāʿ*), the Mongols allocated rural domains (*takhṣīṣ*) to their vassals and the military elite.[25]

One of Rashīd al-Dīn's practical projects was the construction of the Raʿb-i Rashīdī (Rashīdī Quarter), a large charitable estate in Tabriz, Iran, that

---

(Oxford: Oxford University Press, 1991), The Nasser D. Khalili Collection of Islamic Art 27, fols. 281b–282a.

24 *Tansūq-nāma*, Süleymaniye Library, MS Ayasofya 3596, 3b–4b. For an overview of modern scholarship, see Persis Berlekamp, "The Limits of Artistic Exchange in Fourteenth-Century Tabriz: The Paradox of Rashid al-Din's Book on Chinese Medicine," *Muqarnas* 27 (2010): 212–214.

25 Allsen, *Culture and Conquest*, 45. Allsen mentions land allocation, but does not draw parallels with either the pre-Mongolian practice of *iqṭāʿ* or the land-based local feudal lordship of *dihqānī*, which was a well-established practice in medieval Iran. See also *Encyclopaedia Iranica Online*, s.v. "Dihqān," by Isabel Miller, 2017, https://doi.org/10.1163/1875-9831_isla_COM_036009. For the *iqṭāʿ* system during the Mongol era, see Reuven Amitai, "Turco-Mongolian Nomads and the Iqtā System in the Islamic Middle East (ca. 1000–1400 AD)," in *Nomads in the Sedentary World*, ed. by Anatoly M. Khazanov and André Wink (London: Routledge, 2001), 152–171.

included cultivated lands and a town. This was intended to be an experimental space in which he could test his empire-scale agricultural project. In a letter to his son Jalāl al-Dīn, who was a governor in Anatolia, he wrote that he had ordered the construction of a garden near the Rashīdi Quarter: "a garden of such magnificence that even people with extensive knowledge [*ashab-i fitnat*] were surprised by its beautiful techniques [*husn-u sanayi*] and unique innovations." He likened it to the paradise described in the Qur'an, a corner as lovely as the Khawarnaq, a beautiful castle in Arabia said to be built by Bayram Gūr or one of the Lakhmyd Kings.[26] In the surviving charter document of the quarter, "Waqf-nāma-i, Rabʿ-i Rashīdī," Rashīd al-Dīn refers to the establishment of scriptoria that would be responsible for reproducing his works in Arabic and Persian to be sent to the major centers of the Islamic world of the time.[27] Among the books to be copied and disseminated was *Āṣār va aḥyā*ʾ, which was designed to be produced in two versions, Persian and Arabic. Early copies of the work do not survive in either language, but three extant Persian copies from the seventeenth and eighteenth centuries suggest that it was circulated in Iran.[28] Rashīd al-Dīn's plan for the dissemination of his new book and new sciences across the larger Islamic world of his time can be interpreted as a major attempt to synthetize the sciences of China along with Islamic and Greek traditions in the manner of similar well-studied efforts during the Abbasid period. An extant partial copy of the agricultural text of Rashīd al-Dīn in Persian contains Indian, Chinese, Turkish, and Mongolian words and terms that establish his knowledge of the various agricultural practices from across Asia.[29] The table of contents for *Āṣār va aḥyā*ʾ is given in *Latāiʾf al-haqāiq*, a book on theology authored by Rashīd al-Dīn. According to this, the second chapter of

26 *Encyclopaedia Iranica Online*, s.v. "Ḵawarnaq," by Renate Würsch, 2020, last modified April 23, 2013, https://doi.org/10.1163/2330-4804_EIRO_COM_11172.

27 For a detailed study of this charter, see Birgitt Hoffmann, *Waqf im mongolischen Iran: Rašīduddīns Sorge um Nachruhm und Seelenheil* (Stuttgart: Franz Steiner Verlag, 2000). The original charter was published in a facsimile edition: Rashīd al-Dīn Fażlallāh Hamadānī, *Waqf-nāma-i Rabʿ-i Rashīdī*, ed. Mojtaba Minovī and Īraj Afshār (Tehran: Intishārāt-i anjuman-i āthār-i millī, 1971), 342.

28 One needs to be cautious, however, as to the dissemination of the manuscripts throughout the larger Islamic world due to factors such as the end of dynasties and the dissolution of libraries.

29 See the editor's introduction to Rashīd al-Dīn Fażlallāh Hamadānī, *Āṣār va aḥyā*ʾ*: Matn-i Fārsī dar bārah-i fann-i kishāvarzī*, ed. Manoochehr Sotoodeh and Iraj Afshār (Tehran: McGill University Library, 1989), 1–80, on 38–39 (page numbers spelled out in Persian, not referring to the numerals used for the edited text).

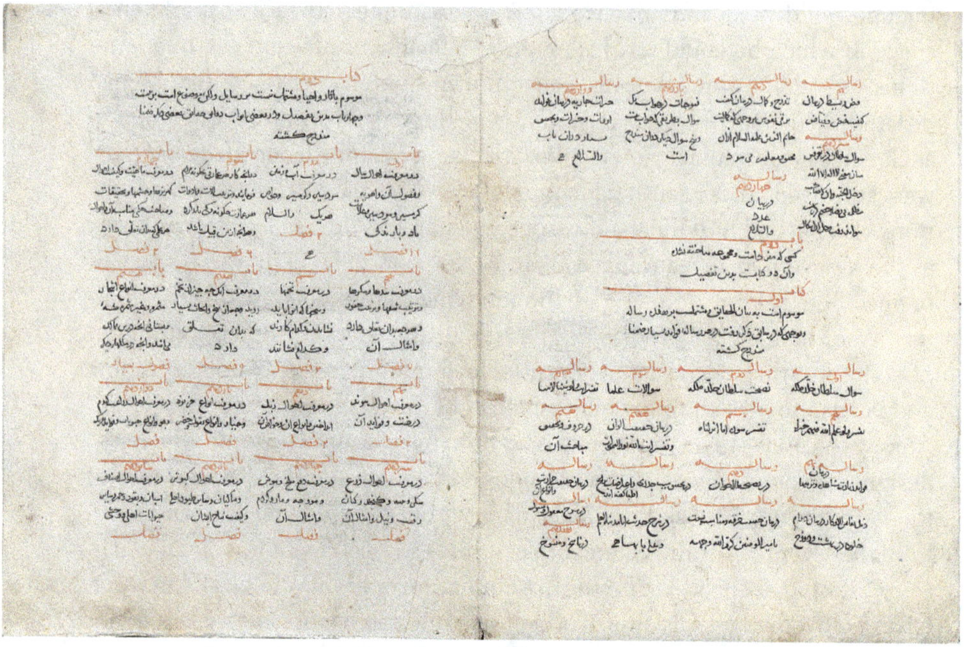

FIGURE 5.1 The table of contents of *Āṣār va aḥyā'* indicating that chapter 2 is dedicated to the subjects of water and soil
SOURCE: LATĀI'F AL-HAQĀIQ, SÜLEYMANIYE LIBRARY, MS AYASOFYA 3833, FOL. 4A–B

the book is divided into two sections exploring the properties of water and soil in cold and warm regions.[30]

2 Moral and Political Economy of Soil

In his writings, Rashīd al-Dīn went beyond the political taxonomy of the social, political, and economic classes. In his understanding, soil and its qualities provide an excellent model for an ideal governor. In his broader policy outlines, he eloquently elaborates on soil quality and its relationship with rulership. He remarks that a good governor should have the qualities of the soil, first of which is the bequest (*tark*), which does not happen unless air percolates into its holes (*furja*); but once air enters, the soil gives back in manifold ways everything it has received. The soil is self-sustaining and constantly dispenses goodness (*istigna* and *mustagni*); it is safe and reliable (*mu'temen*), as it carries divine

30 "Bāb-ı dovvum dar-ma'rifat-i āb va zamīn-i sard-sīr va garm-sīr va khavāss-ı har-yak." *Latāi'f al-haqāiq*, Süleymaniye Library, MS Ayasofya 3833, fols. 4a–b.

trust (*al-amanah*), even though they declined to take on the responsibility of the divine offer.[31] The soil is purity (*pākī*) and is used to purify oneself when performing the obligatory prayers five times a day. Soil submits and preserves that which is entrusted to it. It is a dwelling place for human beings, pasture-land for animals, mine for metals (precious or otherwise), a domain for plants (*khalwat-khaneh*), and the fountainhead of the sweet paradisial water of life (*Manbaʿ-i salsabil-i hayat*). Furthermore, the soil serves as a metaphor for moral quality and rulership. In relation to the levels of spiritual life, the soil is like a contented soul (*khāk rā mathabat nafs al-muṭmaʾinna*).[32] The fourth caliph of Islam, Imam Ali, was given the sobriquet Abu Turab, Father of Soil. This rich array of Islamic references in relation to soil gave Rashīd al-Dīn's political ecology an ethical foundation.

Just as soil served as the very core of his political ecology, the agrarian people were the bedrock upon which this system of balance rests. This agrarian social category subsumed both those who worked the land and those who owned the land—thus subscribing to the well-known idea of the circle of justice (*daira-i ʿadalat*), according to which ruled and ruling are intimately and perpetually linked to each other through various forms of dependency. That is, there is no sultanate without an army; the army can only be sustained by wealth; wealth can only be generated by subjects; and subjects can only generate wealth (i.e., pay taxes) if they are supported by justice.[33]

Agricultural terms also served Rashīd al-Dīn as metaphorical tropes, which can be seen in his letters to his sons. Giving advice to his son Jalal al-Dīn, he wrote that subjects who are people of agriculture (*arbab-ı hirāset*) need protecting (*riʿāyat kon*) and to be given the means to sustain themselves and the necessary tools to make their livelihood. Wealth and productivity (*tathmīr-i riyāʿ*) are incumbent upon their help and support. In another letter, this time to his son Hāja Saʿd al-Dīn, he counsels the latter on how to learn good deeds from the trees of knowledge (*ashjār-i ʿilm*), softness from the flowers of clemency and gentleness, and how to collect the fruits of knowledge (*athmār-ʿilm*), as together these benefits would create a paradise-like city and a world that mirrors the virtuous city of al-Farabī (d. 950), a well-known medieval philosopher.

31 "Truly, We did offer *Al-Amanah* (the trust or moral responsibility or honesty and all the duties which Allah has ordained) to the heavens and the earth, and the mountains, but they declined to bear it and were afraid of it (i.e. afraid of Allah's Torment)." Qur'an 33:72, trans. Mohsin Khan, https://noblequran.com/surah-al-ahzab/.

32 On muddy soil (*khalāb*), see Adem Uzun, "Lugat-i Halimi (Inceleme-Metin)" (PhD diss., Atatürk University, Turkey, 2005), 136.

33 Rashīd al-Dīn Fażlallāh Hamadānī, *Sawānih al-afkār-i Rashīdī*, ed. Muhammad Taqī Dānishpajoh (Tehran, 1358 AH / 1980 CE), 112.

Hence, Rashīd al-Dīn eliminated some of the taxes levied on the people of Tustar/Shustar—one of the towns of Khuzistān province—so that the people could pursue agriculture (*mardum ba-zira'at mayl konand*).[34]

Similar concerns emerged in the Mamluk Sultanate, where Ibn Khaldūn (d. 1406) noted in his *Muqaddimah*, published in 1377, that, in the cases of Andalusia and North Africa, there was a direct correlation between soil quality and market prices of food stuffs sold in the cities. He elaborates that:

> The Christians pushed the Muslims back to the sea coast and the rugged territory there, where (the soil) is poor for the cultivation of grain and little suited for (the growth of) vegetables. They themselves took possession of the fine soil and the good land. Thus, (the Muslims) had to treat the fields and tracts of land, in order to improve the plants and agriculture there. This treatment required expensive labor (products) and materials, such as fertilizer and other things that had to be procured. Thus, their agricultural activities required considerable expenditure. They calculated this expenditure in fixing their prices, and thus Spain has become an especially expensive region, ever since the Christians forced (the Muslims) to withdraw to the Muslim-held coastal regions, for the reason mentioned.[35]

He went on to explain that, while high prices in a city could be attributed to increased demand for food and luxury items, higher wages for city dwellers, taxes, and levies, another factor driving up prices was the cost of agricultural labor. This was especially the case in places where the soil quality was poor and not suited to the cultivation of grains and vegetables, thus requiring extensive treatment of soil through fertilizers and labor. In contrast, he points out, the Berber countries did not need to bring anything from outside in order to cultivate agriculture.[36] Rashīd al-Dīn is concerned about agricultural productivity and its relationship to soil in ways similar to Ibn Khaldūn. For example, he notes that, in order to cultivate pomegranates, one should plant them on high ground and in soft sandy soil.[37]

34 Rashīd al-Dīn, *Sawānih*, 113.

35 Ibn Khaldūn, *The Muqaddimah: An Introduction to History*, trans. Franz Rosenthal, 2nd ed. (Princeton, NJ: Princeton University Press, 2015), 278, parentheses in the original.

36 Ibn Khaldūn, 279.

37 Rashīd al-Dīn, *Āṣār va aḥyā'*, 10.

2.1 *The Moral Universe of Agriculture*

Agricultural writers between 1300 and 1600 moralized and politicized agriculture and the soil, imagining the earth in its totality as a site for sowing (*mazraʿa*) whose yields would be harvested in this world and the next. Furthermore, the expression of moral issues through the metaphors of agricultural terms, particularly soil qualities, was at the core of religious practices, as these play a significant role in Islamic prayer rituals. Since soil was considered pure and purifying (*ṭāhir* and *muṭahhir*), legal manuals discussed how soil lost its purifying qualities. Soil was color-coded as to its use and quality, with black soil in the top position. Yellow soil was used for the casting of metals, and various types of clay (*kil*) were used for cleaning when soap was not available.

For Rashīd al-Dīn, agricultural terms often served as metaphors to describe the learned and moral political universe. His letters in particular are full of figurative language. In a description of Mongol agricultural experiments that took place before the rule of Ghāzān Khān, his contemporary, he explained that: "All sorts of fruit bearing trees were imported from every country and planted in the gardens and orchards, and most have done well."[38] He also minutely describes processes of wall-building using available clay and mud, which could be used to protect vegetable gardens as well as to build housing for people and animals.

2.1.1 Agronomic Experiments and New Agricultural Settlements (*Mustajiddah*)

Rashīd al-Dīn's agricultural experimentations were not limited to Tabriz and its environs. Both Ghāzān Khān and Rashīd al-Dīn owned large estates in various parts of Iran and Anatolia (see Map 5.1). The novelty of the recently established agricultural sites were reflected in their name, *mustajiddah* (newly formed), which clearly indicated their difference from established settlements.[39] Two

38 Wheeler M. Thackston, trans. and ed., *Classical Writings of the Medieval Islamic World: Persian Histories of the Mongol Dynasties* (London: I. B. Tauris, 2012), 2:440. On Saljukid patronage in Anatolia, see Patricia Blessing, *Rebuilding Anatolia after the Mongol Conquest: Islamic Architecture in the Lands of Rūm, 1240–1330*, Birmingham Byzantine and Ottoman Studies (Farnham: Ashgate Publishing, 2014).

39 Notes relevant to Anatolia in Rashīd al-Dīn's letter were analyzed in a pioneering article by Zeki Velidi Togan, one of the major historians of medieval Anatolia, Iran, and Central Asia. He can be credited with drawing attention to the schemas for the settlements and exploring the social and economic life of Anatolia under Mongol rule. Zeki Velidi Togan, "Reşideddin'in Mektuplarında Anadolunun İktisadi ve Medeni Hayatına Dair Kayıtlar," *İstanbul Üniversitesi İktisat Fakültesi Mecmuası* 15, nos. 1–4 (1953), 33–50. See also Zeki Velidi Togan, "Moğollar Devrine Anadolu'nun İktisadi Vaziyeti," in *Türk Hukuk ve İktisat Tarihi Mecmuası* (Istanbul: Evkaı Matbaası, 1931), 1:1–42. The Mongol administrative

were created: one near Diyār-i Bakr, today Diyarbakir in southeast Turkey; and the other in Diyār-i Rabīʿah, the most fertile part of the easternmost and largest region in al-Jazīrah (Upper Mesopotamia).[40] Rashīd al-Dīn named the former after himself: the Mustajiddah-i Rashīdī (New Rashīdi Settlement). In a letter addressed to the officials and notables of the city of Diyār-i Bakr, he asked for the creation of fourteen fortified villages (*qarya-i muḥavvaṭa*), which were to be populated by resettling inhabitants from around Diyār-i Bakr. In addition, he arranged for these people to be provided with seeds (*tokhm*), draft animals (*ʿawāmil*), fodder (*taqāwī*), and food supplies (*muʾākalah*) so that they could happily dedicate themselves to agriculture (*zirāʿat*) and the development and cultivation (*ʿimārat*) of the area. In a unique surviving settlement plan, he indicated where each of the villages was to be located around the banks of the Zab River near Musul. He also ordered water canals to be built to irrigate the agricultural settlement named after Ghāzān Khān, Mustajiddah-i Ghāzānī, which was created in the city of Malatya, in southeastern Anatolia near the Euphrates. His son Khwaja Jalal al-Dīn who served as a governor of the Rūm, the Medieval Anatolia, received from his father, Rashīd al-Dīn, a plan of the irrigation channels and newly established villages.[41]

2.2 *Prosperity of the Domains: Improving the Soil and Plants and Creating* ʿimārat

Rashīd al-Dīn's writings on agriculture and soil are both political and deeply ethical. He remarks as a scholar and stateman that the welfare of the population is what makes rulers successful and provides them with eternal fame. Rulers were encouraged to promote the improvement and cultivation (*ʿimārat*) of their domains.[42]

---

division of the region continued well up to the Ottoman era. See Mehmet Sait Sütçü, "XIII–XVI: Asırlarda Diyār-ı Bekr Bölgesi'nin İdārī Taksimatı," *Genel Türk Tarihi Dergisi* 10 (2023): 512, 514–516.

40 *Encyclopaedia of Islam New Edition*, s.vv. "Diyār Bakir," by M. Canard, C. Cahen, Mükrimin H. Yinanç, and J. Sourdel-Thomine, 2012, https://doi.org/10.1163/1573-3912_islam_COM_0175, and "Diyār Rabʿīa," by Canard and Cahen, 2012, https://doi.org/10.1163/1573-3912_islam_COM_0173.

41 Rashid al-Dīn Fażlallāh, *Sawānih al-afkār-i Rashīdī*, Majlis Library, Tehran, MS 2188, fol. 53a.

42 The two concepts contained in the term *ʿimārat* are difficult to translate, as this term acquired multiple meanings and materialized in various ways. For example, during the Ottoman Empire, it was the umbrella term for their charitable institutions, which provided food and temporary lodging for the poor or travelers. Similarly, *ābādān* (prosperity) was understood differently across the Persianate world. For *ʿimāret* in the Ottoman usage see, Zeynep Tarım Ertuğ, "Imaret," *Türkiye Diyanet Vakfı İslam Ansiklopedisi* (Istanbul:

This concept of improvement, cultivation, and prosperity (*ʿimārat u ābādānī*) was a guiding principle of Rashīd al-Dīn's political and agronomic projects. According to him, the prosperity and development of the arts and sciences in a region depended on several factors, one of which was population. He argued that, if a city has a large population, there are likely to be more people with discerning abilities who can engage in learning and scientific pursuits. This is evident in cities like Tabriz and Baghdad, but it is rare to find many learned people in smaller cities. Rashīd al-Dīn claimed that his argument was based on his own experience and observation (*tajruba*).[43]

2.2.1 *ʿImārat* as an Agrarian Concept

One of the meanings of the word *ʿimārat* in Arabic is "to build." However, this word soon acquired the status of a concept in itself, indicating the monumental scale of building activity (mosques, hospitals, and the like).[44] In the Ilkhanid and post-Ilkhanid periods, the term was used to describe populating an area for the purposes of agriculture. In Rashīd al-Dīn's writings, the term *ʿimārat* is used in several ways: In *Āṣār va aḥyāʾ* and other works, Rashīd al-Dīn frequently discussed the importance of improving soil and plants to yield the best results in terms of both quantity and quality and referred to the method of cultivation as *kayfiyat-i zirāʿat* and the method of improvement as *kayfiyet-i ʿimārat*, always coupling the two terms. The region and specific location of a town or village and the climate and air contribute to the productivity the soil. For him, each region (*har vilayat*) or locality (*har mawżʿī*) is different in terms of water, air, and soil (*zamīn*); likewise, these qualities have a bearing on the shape (*shakl*), color (*rang*), and taste (*ṭaʿm*) of the region or locality's produce.[45]

Rashīd al-Dīn conceptualized the agricultural reforms of Ghāzān Khān, the Ilkhanid emperor, in two significant concepts that were followed by subsequent Islamic dynasties. The first was the concept of *ʿimārat dustī* (support for cultivation and agricultural reform), and the second was *taḥrīṣ* (encouraging the people to practice agriculture).[46] When terming the agricultural reforms

Türkiye Diyanet Vakfı, 2002), 20:219–220. Nina Ergin, Christoph Neumann, and Amy Singer, eds., *Feeding People Feeding Power Imarets in the Ottoman Empire* (İstanbul: Eren Yayınları, 2007).

43 *Tansūq-nāma*, Süleymaniye Library, MS Ayasofya 4695, fol. 6b.

44 On Ilkhanid art and architectural patronage, see the 119 monuments listed in Donald Newton Wilber, *The Architecture of Islamic Iran: The Il Khānid Period* (Princeton, NJ: Princeton University Press, 1955).

45 Rashīd al-Dīn, *Āṣār va aḥyāʾ*, 198.

46 Rashīd al-Dīn Fażl Allāh ibn ʿImād al-Dawlah Abū al-Khayr, *Kitāb-i taʾrīkh-i mubārak-i Ghāzānī: Dāstān-i Ghāzān Khān* (London: Luzac E. J. W. Gibb memorial, 1940), 207.

ʿimārat (cultivating), *ʿimārat* is used to describe an overarching concept of opening uncultivated lands to agriculture, establishing new settlements, and encouraging the population to adopt new agricultural methods, hence increasing revenues generated by an agricultural surplus.[47] *ʿImārat* was therefore taken as a model of agrarian empire building by the later Turco-Iranian dynasties, most notably the Ottoman Empire, during which the idea of *ʿimāret* (*ʿimārat*) led to the construction of the unprecedented institution of *ʿimāret-ḫāne* (*ʿimārat-ḫāne*), the soup-kitchen.[48]

2.2.2 Manure Application and Soil Improvement

Repurposing pastureland and reviving neglected land required investment in the soil. Manure application serves one of the most common soil improvements. Rashīd al-Dīn dedicated a specific chapter of *Āṣār va aḥyā* to manure and its application. He refers to the knowledge of recognizing different types and properties (*khawaṣ*) and benefits (*favāiʾd*) of manure as well as techniques of turning it. In this chapter, his knowledge is generally in line with medieval Islamic works on agronomy; however, his own work is distinguished by the attention he pays to the specific application of different kinds of manure (*zibl*) to specific plants ("Gūyīm zibl anvāʿ ast va har- zibl be-irtifāʿī maḫsūṣ mī bāshad").[49] For example, he explained that cow manure was best for grapes, whereas sheep manure would be too potent. Rashīd al-Dīn remarked that there is no fertilizer better than well-rotted compost (*ḫāk-i kohan*).[50] He said that the more it aged, the better, as aged manure was brackish and briny in

---

47 The word *ʿimārat* is also used in the sense of the planting and improvement of plants.

48 The frequent use of the verbs *ʿīmārat kardan* (in Persian) and *ʿimāret etmek* and *şenlendirmek* (in Turkish) have acquired meanings that differ from the original "to build." For the Ottoman institution of *ʿimāret-hāne*, see Barkan, Ömer L. "Osmanlı İmparatorluğunda İmaret Sitelerinin Kuruluş ve İşleyiş Tarzına Ait Çalışmalar," *İstanbul Üniversitesi İktisat Fakültesi Mecmuası* 23, nos. 1–2 (2015): 297–341.

49 For a discussion of *zibl* in Arabic agronomy works, see Daniel Varisco, "*Zibl* and *Zirāʿa*: Coming to Terms with Manure in Arab Agriculture," in *Manure Matters: Historical, Archaeological, and Ethnographic Perspectives*, ed. Richard Jones (New York: Routledge, 2012), 129–144.

50 Egyptian agronomists of the medieval period considered that the soil used for construction could also be used as fertilizer. The Arabic term for fertilizer is *sibākh* or *sibāq*. The major sources for the soil typology are the taxation manuals and the histories written in Medieval Egypt. Al-Makhzūmī and al-Maqrīzī list *sibāq* as a type of soil. On *sibāq*, see Anthony T. Quickel and Gregory Williams, "In Search of *Sibākh*: Digging up Egypt from Antiquity to Present," *Journal of Islamic Archeology* 3, no. 1 (2016): 89–108. For the partial edition of al-Makhzūmī's work and analysis of the text see Claude Cahen, *Makhzūmiyyāt: Études sur l'histoire économique et financière de l'Égypte médiévale* (Leiden: Brill, 1977).

taste and far superior to animal droppings. If fresh dung (*sargīn*) was applied as fertilizer, it would not be productive (*tavallud na-konad*).[51]

3   *Ās̱ār va aḥyā*: A Novel Approach to Writing about Agriculture

The title and the methods with which Rashīd al-Dīn approaches agricultural matters demonstrate his conceptual creativity and ambition. In *Ās̱ār va aḥyā*, there are two major concepts on which his larger agricultural project rests. The terms *ās̱ār* (monumental traces to be left behind) and *aḥyā*, or *iḥyā* (revivification of the dead lands [*mavāt*]) were Qur'anic references.[52] The Qur'anic allusions in his title are quite telling and fit very well with the reforms he initiated as vizier in the Ilkhanid realm. Although Rashīd al-Dīn calls Ghāzān Khān the Sultan of Islam, the honorific mimics that of the Islamic Caliphate, which had been destroyed by ancestors of the very same Ghāzān Khān in 1256 in Baghdad. Here, in referencing the Qur'anic verses, Rashīd al-Dīn likens his restorative project to God's restoration of the world after its destruction.

It is not yet known whether Rashīd al-Dīn utilized any Arabic or Persian vernacular agronomic texts in writing *Ās̱ār va aḥyā*. However, he was clearly cognizant of existing *filāḥa* texts and saw himself as writing for a community of learned elite agronomists who shared his concerns and were familiar with the *filāḥa* books. In the section on grafting, he explicitly invites comparison (*muvāzana*) with other authorities to demonstrate the original aspects of his book. He boasts of his novel technique of grafting and claims that no past agronomist had written as much on this topic. He asks his readers to use their virtuous and intellectual minds (*nazd-i uqalā va kāmilān*) to inspect and correct or amend (*ilḥāq*) anything they found unsuitable:

> By way of introduction, one should explain the grafting of trees, its origin, the different ways it can be done, its purpose, and the benefits that come from it. Indeed, not everyone has fully explored the depths of this topic, and it is not written in agronomy books in a manner that is suitable to be seen by [the] intelligent and virtuous. If someone has talked about it and

---

51   Rashīd al-Dīn, *Ās̱ār va aḥyā*, 175.
52   "Behold, then, [O man,] these signs of God's grace [*athāri raḥmat Allāh*]—how He gives life to the earth after it has been lifeless [*kayfa yuḥyī la-arḍa baʻda mawtihā*]! Verily, this selfsame [God] is indeed the One that can bring the dead back to life: for He has the power to will anything." Qur'an 30:50, trans. Muhammad Asad (modified), https://muhammad asad.com.

> written about it, this humble writer has yet to encounter a complete and acceptable explanation of the method's truth and subtleties that would satisfy virtuous and intellectual minds. If compared with what this humble person has written on this subject, they will see the difference, and if, in their illuminating opinion, the suggestions of this humble person are acceptable, it is an honor for this humble one; or, if they know of a better [method], out of their kindness and nobility they should correct [what I have written] and add to it.[53]

Rashīd al-Dīn remarks further that:

> Techniques and sciences are numerous, and one may benefit from [the knowledge of] those who give good advice. Just as there is a difference between a trained and non-trained horse in strength, plants and trees that have been improved [*ʿimārat*] are better than those that grow on their own [*khod-rūy*].[54]

Drawing on his own experiments (*tajruba*), he criticizes previous writers of agronomy and counters their suggestions:

> A group of people who wrote *filāḥa* books, and individuals who have knowledge of this discipline claimed [*taqrīr*] that grafting a tree whose fruit has seeds [*ostokhān*] with a tree that has none will not yield. But on the contrary, it is known through experience and experiment [*tajruba*] that the willow tree [*bīd*], which has no fruit, will bear extremely sweet fruit [if] peach branches are grafted on it [*shaftālū*].[55]

He outlined his method in such a manner that grafting and its methods appeared rational (*bi-ṭarīq maʿqūl*), sensible (*maḥsūs*), and established by indisputable proofs (*mubarhan*), so that the intellectuals (*ʿuqalā*) and the elect ones (*kāmilan*) could get the maximum benefit from his book. At the same time, it ought to be clearly understandable to architects (*miʿmārān*), master craftsmen (*ser-kārān*), and farmers (*fallāḥān*).[56] Written in a very strong personal voice, Rashīd al-Dīn's agronomic book(s) contain many references to himself as author (*muʾallif-i kitāb*), that is, to the original writer and compiler,

53 Rashīd al-Dīn, *Āṣār va aḥyā'*, 107.
54 Rashīd al-Dīn, 113–114.
55 Rashīd al-Dīn, 123.
56 Rashīd al-Dīn, 109.

as the term *mu'allif* suggests. On other occasions, he uses self-deprecating expressions, referring to himself as humble (*īn fakīr*, a performative phrase often employed to indicate humility). Such phrases do not appear in his diplomatic and stately communications; his use of action verbs, such as "I did/have done" (*karda-im*), or the imperative form, communicates very directly a language of power and command.

He does not make any reference to the authority of the ancients. In fact, the whole text takes contemporary practices to be more reliable than ancient ones. In one section, he says that anyone who reads this "humble" text may notice major differences compared to ancient books. Without indicating which of the previous agronomy manuals he has in mind, he implies that they have been read, with the result that their presence lends the text an authoritative voice. Two major concepts are important here. The term "ancients" probably refers to ancient Greek agronomy texts, while "moderns" refers to agronomic works written in Arabic and Persian.[57]

> That which was written in the agronomy books of the ancients [*mutaqaddimān*] and moderns [*muta'akkhirān*] is to be studied carefully if it occurs to them that what this humble author has written and explained has to be compared [*muwāzana*], so that they will understand the issues at hand [*kayfiyyat*] and see the difference.[58]

His invitation to his readers to engage critically as they read and to experiment and make corrections as they see fit is a recurring theme indicating active agronomic experiments and discourse on the topic.

Knowledge about soil, plants, and cultivation was acquired and generated through two distinct methods which he calls experience and experiment (*tajruba*). The concept of *tajruba*, and its general category, the collection of experiments and experiences (*mujarrabāt*), is one of the epistemic terms used

57 One such ancient Greek work on agriculture is by Anatolius. The Greek original is no longer available; the only surviving Arabic text was copied in 1332 for the private library of Mamluk Amīr Sarākir (c. 1400), who served as the deputy of Damascus/Sham (Nāi'b al-Shām). In the sixteenth century the copy was in Iran and was endowed by Shah Abbas II to Mashhad Library. The copy postdates the death of Rashīd al-Dīn, yet it was made during the Ilkhanid period. On Anatolius, see Fuat Sezgin, *Geschichte des arabischen Schrifttums*, vol. 4, *Alchimie-Chemie, Botanik-Agrikultur* (Leiden: Brill, 1996), 314–315. Central Library of Astan Quds Razavi, Mashhad, MS 5762, fol. 181a. For the comparative study and detailed description of the table of contents, see Carlo Scardino, *Edition antiker landwirschaftlicher Werke in arabischer Sprache*, vol. 1, *Prolegomena* (De Gruyter: Berlin, 2015), 59–218.

58 Rashīd al-Dīn, *Āṣār va aḥyā'*, 225.

in Islamic learning to describe the acquisition of knowledge. It is one of the ways for an individual to access certain knowledge (*yaqīn*). The concept is mainly discussed in works on logic, yet medical and other practical sciences also rely on *tajruba*.[59] The medieval Islamic doctor, philosopher, and scholar of ethics Ibn Miskawayh (d. 1030) titled his book *Tajārib al-umam* (The experience of nations). Sayyid Sharif al-Jurjānī (d. 1414), a Timurid era scholar, defines *mujarrabāt* as "that which leads the intellect to the certainty of judgment on the repeatability of the observation time and again" (hiya mā yaḥtāju al-ʿaql fīhi fī jazm al-ḥukm ilā taqarrur al-mushādati marratan baʿda ukhrā).[60]

In *Āsār va aḥyā'*, *tajruba* is a central concept, and Rashīd al-Dīn refers to his own experiments in a very strong first-person voice. Yet *tajruba* is not only used as an umbrella term, it also indicates nuanced epistemic practices. *Mushāhada*, observation, is yet another method he employs. Observation alone, however, does not create general knowledge, and does not hold the same level of epistemic value as *tajruba* although it may be used for a specific case. While he was discussing the mulberry, for example, he records a *mushāhada* he made in the city of Yazd:

> If they want to prepare soil for planting a mulberry orchard [*tūtistān*], they should cut fresh branches with a saw into four pieces, dig over the ground with a shovel to give a fine, well-prepared soil like for a courtyard (*qaz*), and apply manure [*zibl*], scattering it here and there, and water it, then all of [the trees] will flourish. ... The goal is for the mulberry to grow big and have lots of leaves. ... They obtain silk from it when they follow these instructions. ... I observed such growth in Yazd, where water is scarce, arable soil is very hard [*ʿazīz*] and the air is dry [*tang*] so I looked for a reason [*tatabbuʿ*]. In that particular year, they planted the cuttings according to the described method; many of the trees had silkworms without many leaves, and in that year, the silkworms did not yield well; one must prepare the soil and the [cut] branches according to the above instructions for it to bear leaves, and all of the soil nutrients [*khār va*

59 For a discussion of *tajrubah* and *mujarrabāt* in medieval Islamic medicine, see Cristina Álvarez Millán, "The Case History in Medieval Islamic Medical Literature: *Tajārib* and *Mujarrabāt* as Source," *Medical History* 54, no. 2 (2010): 195–214.

60 ʿAli b. Muḥammad Sayyid Sharif al-Jurjānī, *Kitāb al-taʿrīfāt*, entry on "al-mujarrabāt" (Beirut: Maktabat Lubnān, 1985), 213. In their introduction to *Āsār va aḥyā'*, the editors give a list of places in which Rashīd al-Dīn uses the word *tajruba*. Rashīd al-Dīn, *Āsār va aḥyā'*, 35–38.

> *muʾnat*] must be present in full. If they do not cultivate the soil in the manner described, they will not harvest many leaves.[61]

For Rashīd al-Dīn all plants and trees are improvable. The soil is the starting point for all attempts to do so. Yazd is located in the very dry central region of Iran. He uses it as an example to show that, despite regional and climatic difficulties, one can increase the production of silk if one knows how to improve the soil. Silk was a global commodity during this period, and Rashīd al-Dīn's extensive discussion of regional practices related to soil improvement was very much connected to the political and economic ambitions of the Ilkhanid elite to which he belonged.

Verified knowledge is central to Rashīd al-Dīn's approach. He employs the concept of *tajrubah* to describe two major methods of knowledge production: at the individual and the collective level. First, he uses *tajruba* to indicate his own empirical knowledge based on his own experiments. In the context of the collective, *tajruba* refers to regional agricultural knowledge and practices based on local experience, be it in specific cities in Iran and other regions, or larger domains such as Anatolia.

The emphasis on *tajruba* remained central in the *filāḥa* works of the post-Ilkhanid era. In *Irshād al-zirāʿah*, the Timurid author Abū Naṣrī-i Haravī indicates that he gathered information from the people of experience (*mardum-i ṣaḥib-tajruba*) who were famous among the cultivators (*muzāriʿān*).[62] Similarly, in the sixteenth century, the anonymous author of the first known agronomy work in Ottoman Turkish, *Revnaḳ-ı Būstān* (Splendor of the garden) writes that he compiled his work relying upon the "opinions of the scholars who wrote *filāḥa* works, and people of experience [*ehl-i tecrübe*]."[63] Rashīd al-Dīn understood that soil had different characteristics in each region and needed to be known and studied in order to improve local agriculture. He emphasized that the study and knowledge of soil was obtained through individual experience and long-term observation.

---

61 Rashīd al-Dīn, *Āṣār va aḥyāʾ*, 30–33. I have translated only the relevant sections of the given pages.

62 Qāsim ibn Yūsuf Abū Naṣrī-i Haravī, *Irshād al-zirāʿah*, ed. Muḥammad Mushīrī (Tehran: Dānishgāh-i Tahrān, 1967), 45.

63 *Revnaḳ-ı Būstān*, University of Istanbul, Turkish Manuscripts, 2720, 2a. For the political and economic context of *Revnaḳ-ı Būstān*, see Aleksandar Shopov, "The Vernacularization of Sixteenth-Century Ottoman Agricultural Science in its Economic Context," in *Living with Nature and Things: Contribution to a New Social History of the Middle Islamic Periods*, ed. B. J. Walker and A. al-Ghouz (Göttingen: V&R Unipress), 639–683.

3.1 *Soil Assessments in Ilkhanid and Seljukid Encyclopedias*

Rashīd al-Dīn introduced new concepts in discussing agricultural improvement (*ʿimārat*). His writings on agriculture and the attention that he places on experience when he discusses soil was related to his undertaking and investments in agricultural production in regions across Iraq, Iran, and Anatolia. This direct involvement in agricultural investments was quickly taken up by the Ilkhanid ruling elites in the aforementioned regions. The authors of Persian encyclopedias from the Ilkhanid period onward elevated the place of agricultural knowledge. Agriculture was treated as an independent body of knowledge. Soil and soil qualities were established in separate chapters on agriculture. They all considered agriculture to be part of the natural sciences. This categorization is particularly significant in that it enabled medieval agronomists to be open to various possibilities, and experimentation remained very central in assessing soil qualities.[64]

One example is Shams al-dīn Muhammad bin Maḥmūd al-Āmulī (d. c. 1352), a contemporary of Rashīd al-Dīn's.[65] In his book *Nafāʾis al-funūn fī ʿārayis al-ʿuyūn* (Delectable sciences in adorning the eyes), written in the first decades of the fourteenth century, al-Āmulī expounds further on the issue of soil; not only did he dedicate a separate section to agriculture, he also wrote on the issues related to soil assessments in agricultural practice.[66] These and similar assessment methods were incorporated into subsequent books on agriculture written in Persian and Ottoman Turkish in the fifteenth and sixteenth centuries.[67]

---

64 For more on soil assessment techniques, see Nicolas Roth in this volume.

65 He was a professor of a school (*madrasa*) in the city of Soltaniyeh which was established by the Ilkhanid emperor Üljeytu. Mustafa Çağrıcı, "Āmülī, Şemseddin," *Türkiye Diyanet Vakfı İslam Ansiklopedisi* (Istanbul: Türkiye Diyanet Vakfı, 1991), 3:100.

66 *Shams al-Dīn Muhammad b. Mahmūd Āmulī*, ed. Hāj Mīrzā Abū al-Hasan Shaʿrānī, 3rd ed. (Tehran, 1379), 352–356.

67 Yahya b. Ali Malkaravī (known as Nevʿī), *Netāyicü'l fünūn ve mehāsinü'l-mütūn*, Istanbul University Rare Books Library, Turkish Manuscripts, MS 588, fols. 72a–74a. Copies of the work abound. For a study and English translation of the work, see Gisela Procházka-Eisl and Hülya Çelik, *Texts on Popular Learning in Early Modern Ottoman Times*, vol. 2, *"The Yield of the Disciplines and the Merits of the Texts": Nevʿī Efendi's Encyclopaedia Netāyic el-Fünūn*, in collaboration with Adnan Kadrić (Cambridge, MA: Harvard NELC, 2015), 139–141. For a discussion of the importance of this text, see the introduction by the editors, 7–12, and the sources of the encyclopedia, 13–32. Naṣrī-i Haravī, "Irshād al-zirāʿah," 54–58; ʿAbd al-ʿĀlī ibn Muhammad ibn Husayn Birjandī (known as Fazil Birjandī), *Maʿrifat-i Falāhat: Davāzdah bāb-i Kishāvarzī* (Tehran: Markaz-i Pejohish-i Mīrās-i Maktūb, 2008), 15–18.

Soil became of central interest in the political entities that emerged following the Mongol conquest and the early modern Islamic empires. Rather than a subdiscipline of medicinal sciences, agriculture was elevated into a body of knowledge on its own and understood within the larger contexts of moral and legal subjects and statecraft, both rural and urban. This shift in the disciplinary classification of agricultural science led to two concurrent developments in late medieval and early modern Islamicate empires. First, the *filāḥa* authors began considering soil the material foundation of all human existence and endeavor. Second, experience (*tajruba* or *tecrübe* in Turkish) became the new episteme in obtaining practical knowledge pertaining to agriculture and statecraft. Finally, for agronomists, the soil provided the sole foundation for the moral life.

This epistemic shift is visible in the encyclopedias of the period. Soil and the assessment of soil constitute the core of the entries in the sections on the science of agronomy. Shams al-dīn al-Āmuli, a late thirteenth century scholar who penned one of the voluminous encyclopedias in Persian, dedicated a section to the science of agronomy, which he considered a discipline related to natural philosophy (*ʿilm-i ḥikmat*), calling it the best of all techniques (*ṣināʿāt*) that God had bestowed on humanity (*bihtarīn ḥiraf-i ṭabīʿiyya-ast, va avval ṣīnāʿatī ki īzad taʿālā ādam ra talīm dād*).[68]

For al-Āmulī, the basis of agriculture is the knowledge of the agricultural ground/soil (*zamīn*) and the time of sowing. He reports on the various techniques used to assess the quality of soil, including tasting. He suggests several methods for the application of bird droppings and animal dung, which was very much the standard practice of the time. His entry on the science of agronomy seems to have enjoyed considerable longevity, along with *Yawāqīt al-ʿulūm* (Rubies of the sciences) and Fakhrad al-Dīn al-Rāzī's *Jāmīʿ al-ʿulūm* (Summa of the sciences). These three encyclopedias formed the basis of the successive encyclopedias in the early modern period.[69]

68 Shams al-dīn Muḥammad b. Maḥmūd Āmulī, *Nafāi's al-funūn fī ʿarāi's al-ʿūyūn* (Tehran: Kitāb-furūshī-i Islāmiyya, 1379), 3:352–356.

69 Late medieval and early modern encyclopedias devoted an independent section to agronomy (*ʿilm al-filāḥa*). Nevʿī Efendi (d. 1599), in his widely circulated encyclopedia, argues that "the *filāḥa* is the principle and basis of all industries and sciences" (aṣl-i küll-ü ṣanāyiʿ). *Netāyicü'l fünūn*, Istanbul University Rare Books Library, Turkish Manuscripts, MS 588, fol. 72a. Similarly, Aḥmed ibn Muṣṭafa Taşköprülüzāde (d. 1561), a significant scholar and the compiler of the *Miftāḥ al-saʿādah* (Key to happiness), holds that agronomy (*filāḥa*) "is essential to human sustenance on earth, which is why it was called the science of survival [*al-baqāʾ*]." Taşköprülüzade, *Miftāh al-saʿāda wa misbāh al-siyyāda fī*

All of the practical philosophers emphasized the significance of agriculture and advised rulers to advocate agriculture in their domains. Timurid-era (1370–1507) scholar Husayn Vaiz-i Kashifi[70] argued that one of the methods for a ruler to demonstrate compassion toward his subjects was to encourage them to practice agriculture and cultivation (*be-zira'at va 'imāret tahrīṣ konad*), to help them to build water canals (*kāriz-hā*), and to bring the streams of water to their fields (*jūybār-hā*).[71] For Kāshifī, such signs of compassion on the part of the ruling elite, sultans, lesser potentates (*amīr*), and their state functionaries would win the hearts of their subjects (*ra'iyyet*), who are predominantly rural folk, and this was important for the stability and prosperity of the country.

Soil and soil types were significant components of lexicalized vocabulary. Contemporary dictionaries in Arabic, Persian, and Ottoman Turkish always listed various types of soil. These lexicalized items provided a much richer and localized vocabulary, which indicated local practices and understandings of the soil. For example, Halimi Çelebi, a sixteenth-century lexicographer of a bilingual Persian-Ottoman dictionary lists all types of soil, including *qaymūliyā*, of which he indicates that there are two types, white and reddish, imported from al-Andalus and Armenia.[72] Similarly, Halimi lists the word *gil-gūy*, meaning a type of black mud (*karabalçık*) that is formed through the constant trampling of sheep.[73]

4 Conclusion

This paper has delineated agricultural concerns related to soil that emerged during the Ilkhanid period after the Mongol invasion of Iran and Anatolia. During this period, the concept of *'imārat* (urban and agricultural development) became the basis of political legitimacy. The outcome of good governance is a domain where agriculture is productive, people are prosperous, and rulers are respected due to their administering of justice and the distribution of wealth

---

mawḍū'āt al-'ulūm, ed. Kāmil Kāmil Bakrī and 'Abd al-Wahhāb al-Nūr (Cairo: Dār al-kutub al-haditha, 1968), 1:332.

70 On Husayn Vā'iẓ-i Kāshifī, see Maria E. Subtelny, "A Late Medieval Persian *Summa* on Ethics: Kashifi's *Akhlāq-i Muḥsinī*," *Iranian Studies* 36, no. 4 (2003): 601–614.

71 Husayn Vaiz-i Kāshifī, *Jawāhir al-akhlāq* (Akhlāq-i Muḥsinī), Bibliothèque nationale de France, Suppl. Pers 100, fols. 42b–44a. See also Subtelny, "Late Medieval Persian Summa." On the Timurid era, see Beatrice Forbes Manz, *Power, Politics and Religion in Timurid Iran* (Cambridge: Cambridge University Press, 2007).

72 Uzun, "Lugat-i Halīmī," 206.

73 Uzun, 244.

across social classes where needed. Soil improvement may have been only one of the many concerns in the realization of this political project, but soil is the basis of all agricultural activities, and understanding, improving, and developing its cultivation is fundamental to achieving agrarian economic success. Within this context, Rashīd al-Dīn emerged as a key figure, whose works aimed to synthesize Arabic-Islamic knowledge with Chinese science in the field of agriculture. In addition, he and other members of the Ilkhanid elite invested in agriculture as agrarian entrepreneurs creating new experimental agricultural cities and towns as well as establishing many charitable foundations across the empire.

I have shown that Rashīd al-Dīn identifies his work as a departure from the established agronomic books written in Arabic between the ninth and thirteenth centuries. He engaged in empire-wide knowledge-gathering regarding practices related to the importance of soil in encouraging agricultural productivity. This was a shared interest throughout the fourteenth century in the Mamluk Empire and Islamic Andalus. This compiled information was skillfully integrated into new agronomic knowledge and his *Āsār va aḥyāʾ*. Here, soil plays a crucial role in his instructions for planting trees such as pomegranate and mulberry—the latter essential for the production of silk, a highly lucrative global commodity. He collected regional methods of soil improvement and provided examples in which he applied these methods in his own trials in Tabriz. For Rashīd al-Dīn, such trials were a major method of knowledge production that facilitated new possibilities of developing further agronomic knowledge. In *Āsār va aḥyāʾ*, he invited his reader to join his intellectual project.

Acknowledgments

I would like to thank Aleksandar Shopov, who read the drafts of this paper and offered valuable comments on various aspects. During my stay at the Max Planck Institute for the History of Science, Berlin, I benefited tremendously from discussions in the working group "Agriculture and the Making of Sciences," particularly those with Justin Niermeier-Dohoney, Chun Xu, Riaz Howey, Heba Mahmoud Saad Abdelnaby, Bethany J. Walker—and Dagmar Schäfer, who facilitated my visit to the Institute and without whom this project would not have been conceived. I would also like to thank Tekin Şahiner, Hilal Kurutan, and Osman Etoğlu, my big brothers (*abi*) and people of my soil (*toprağım*), who helped me greatly in adjusting to life in Berlin.

CHAPTER 6

Building Innovative Soil Knowledge to Improve Iberian Agriculture (16–18th Century)

Alberto González Remuiñán and Dulce Freire

1 Introduction

Studying the soil as an independent analytical object has changed significantly since the end of the nineteenth century.[1] While many people may have wondered about the different characteristics of soil as a growing medium ever since humankind first selected and cultivated seeds, the historical documents that are available only allow us to trace this process in more recent times. Nowadays, many types of soil are well known thanks to work done in different fields of science such as geology, chemistry, edaphology, pedology, biology, and agronomy, yet this empirical knowledge was obtained earlier, in a wide range of territories and under very different circumstances. This chapter broadens the historical perspective of current soil studies by drawing on agrarian instruction books published in the Iberian Peninsula between the sixteenth and eighteenth centuries.[2] The privileged geographical position of the Iberian Peninsula at this time cannot be overstressed. As the central hub of its overseas territories and colonies reaching from the East to West Indies, it received privileged access to knowledge arriving from these new lands.[3] Yet, despite the

1 Regarding the study of these issues, it is important to emphasize the role of Vasily Dokuchaev. Francisco Díaz-Fierros Viqueira, *La ciencia del suelo: Historia, concepto y método* (Santiago de Compostela: Servizo de Publicacións e Intercambio Científico, 2011), 20.

2 The books analyzed here include agrarian manuals, chorographies, and economic essays. Although they belong to different genres, all these books offer some useful suggestions for improving agriculture, and this content component is the focus of our analysis. Thus, for the purposes of this chapter, we refer to these books as agrarian instruction books.

3 The ReSEED project team's research has identified the dynamics of the dissemination of new knowledge and crops that arrived in the Iberian Peninsula from the fifteenth century onwards, see Reseed, accessed August 21, 2025, https://reseed.uc.pt. Recent publications include: Inês Gomes and Dulce Freire, "Seeds of Knowledge: Paving the Way to Integrated Historical and Conservation Science Research," *Journal of Environmental Studies and Sciences* 13 (2023): 376–388, https://doi.org/10.1007/s13412-023-00826-9; Carlos M. Faísca, Dulce Freire, and Cláudia M. Viana, "Changing Rice Geographies: A Long-Term Perspective of Portuguese Regional Production (1820–2018)," *Historia Agraria* 91 (2023): 1–31, https://doi.org/10.26882

influx of new information, early modern Iberian agricultural writers had to negotiate staunchly held beliefs stemming from the prevailing Biblical interpretation of soil as a divinely created pristine substance, which was therefore immutable. In their writings, they noted their own views on soil that incorporated factors such as human labor, new interpretations of climate, as well as empirical and experimental evidence considered relevant to fertility. With the approval of the Catholic authorities, they wrote in the vernacular, targeting a trained readership who could use this knowledge to improve agriculture and the economy.[4]

Some scholars have argued that, during the Middle Ages, the dominance of Christian doctrine in the Western world meant that writers ignored the geological knowledge espoused by Greek, Roman, Arabic, and other classical scholars.[5] However, this generalized assumption of stagnation during this period in the history of soil is contradicted by regional data. In fact, recent publications have underscored that the study of soil remained an active topic in the Middle Ages.[6] In Europe, a more systematic analysis of soil began during the Renaissance, as part of the study of natural resources.[7]

After reviewing the contributions of the classics, some current historians of the Iberian Peninsula skip to theories from the eighteenth century and later, without considering anything that was written about soil in the intervening centuries.[8] As noted in this volume's introduction, it is as if soil science was an exclusive domain of historians of science studying the last two centuries of the modern period. But our analysis of source material helps

/histagrar.091e07f; Inês Gomes, Alberto González Remuiñán, and Dulce Freire, "Exotic, Traditional and Hybrid Landscapes: The Subtle History of the Iberian Peninsula Maize between 'Tradition' and 'Modernity,'" *Plants, People, Planet* 6, no. 5 (2024): 1047–1059, https://nph.onlinelibrary.wiley.com/doi/full/10.1002/ppp3.10458; Carlos Manuel Faísca, and Dulce Freire, "Not All Wheats Are the Same: Selection and Improvement of Wheat in Portugal since Early Modern Times (16th to 20th Centuries)," *Investigaciones de Historia Económica* 21 (2025): 59–86, https://doi.org/10.33231/j.ihe.2025.01.03.

4 On long-term Iberian agricultural dynamics, see, e.g., Dulce Freire and Pedro Lains, *An Agrarian History of Portugal, 1000–2000: Economic Development on the European Frontier* (Leiden: Brill, 2017); Ramón Garrabou Segura and Ricardo Robledo Hernández, *Sombras del progreso: Las huellas de la historia agraria* (Barcelona: Editorial Crítica, 2010).

5 Eric C. Brevik and Alfred E. Hartemink, "Early Soil Knowledge and the Birth and Development of Soil Science," *Catena* 83, no. 1 (2010): 23–33.

6 Eric C. Brevik, "A Brief History of Soil Science," in *Encyclopedia of Life Support Systems*, vol. 6, *Encyclopedia of Land Use, Land Cover and Soil Sciences: Soils and Soil Sciences (Part 1)*, ed. Willy H. Verheye (Oxford: UNESCO–EOLSS, 2009), 40–63.

7 Brevik and Hartemink, "Early Soil Knowledge," 26.

8 Díaz-Fierros Viqueira, *La ciencia del suelo*; Roque Ortiz Silla, "Síntesis de la evolución del conocimiento en Edafología," *Eubacteria* 34 (2015): 51–64.

fill this historiographical gap, as we discuss authors who wrote between the sixteenth and eighteenth centuries. In this chapter, we argue that, in order to grasp the early modern understanding of soil as a natural resource, it is necessary to analyze how these writers adopted and altered the existing material on soil they had inherited from previous centuries. We also assess how successful the writing of this period was in creating new options for increasing agricultural production in different European regions, including the Iberian Peninsula.

Our research for this chapter is based on several agrarian instruction books published in Portugal and Spain between the sixteenth and eighteenth centuries—for part of this period, the two territories were united in the Iberian Union (1580–1640) due to dynastic issues until the outbreak of the Portuguese Restoration War. Despite their differences, these books share several common traits. Like other books published throughout Europe during this period, these were written in vernacular languages—Portuguese and Spanish—rather than Latin, which was still the language of international communication among the erudite elite. These books were intended to inform and educate local people on various subjects related to the geographical characteristics of different regions, stressing how natural resources like soil could be exploited to increase income. So it is important to address literacy in order to estimate who could access the information in these texts. Since most of the population was nonliterate, the reading audience was limited to the cultural elite, who may have shared the knowledge with their settlers and tenants. But it is noteworthy that these authors sought effective modes of communication to try to explain how the land should be worked, urging that the labor invested in the land, rather than the land itself, was a critical factor in agricultural productivity. The apparent lack of effort—that is, a lack of labor—led some of these authors to interpret agriculture as an activity in crisis. Although these agrarian instruction books had diverse objectives and made different theoretical conjectures, their authors provided concrete data and practical advice regarding the territories they investigated. This innovative knowledge was mainly a result of personal experience and observation. The writers of these books presented general information on soil and identified existing elements that were essential to ensure productive agriculture throughout these territories, even if they could not explain why this was so.

Since the fifteenth century, Portugal and Spain had been protagonists in an intensification of maritime connections that provided access to different territories around the globe and opened a gateway for new plants to enter the old world. These new plants were recognized as commodities ripe for commercialization in what Bartolomé Yun-Casalilla has called a dialogue of cross-cultural consumption.[9] New information obtained from overseas biospheres influenced not only what was already known about other species but also existing knowledge of the soil, acting as an incentive and a challenge: new data generated new ideas that called into question previously held theories. Thus, as members of the literate elite, authors of Iberian agrarian instruction books drove the discussion raised by information and objects from the Americas that had until recently been unknown to Europeans. Through the analysis of soil, different authors produced new knowledge that greatly influenced the commercial and state activities of the Iberian empires. These novelties, which were admired and imitated by other interested European states, fostered the development of empirical observation, contributed to the erosion of the medieval epistemological framework, and, eventually, laid the groundwork for modern science.[10]

The maritime world was recognized as a pioneering field in the development of European science, but by no means the only one.[11] The works analyzed in this chapter are further evidence of a scientific transition in which authors used direct experience and observation to question the authority of classical writers, whose knowledge was dependent on textual interpretation. The use of direct, hands-on experience by elite agrarian writers was emblematic of the broader development of experimental science in Europe.[12] As more recent historians have noted, Edgar Zilsel's classic sociological thesis—namely, that the collapse of social barriers between artisans and scholars contributed to the rise of science—extends to the Iberian Atlantic as well as the practice of agriculture.[13] The authors collected here symbolize a convergence of the roles

9 Bartolomé Yun-Casalilla, "The Spanish Empire, Globalization, and Cross-Cultural Consumption in a World Context, c. 1400–c. 1750," in *Global Goods and the Spanish Empire, 1492–1824: Circulation, Resistance and Diversity*, ed. Bethany Aram and Bartolomé Yun-Casalilla (London: Palgrave Macmillan, 2014), 277–306, https://doi.org/10.1057/9781137324054_15.

10 Antonio Barrera-Osorio, *Experiencing Nature: The Spanish American Empire and the Early Scientific Revolution* (Austin: University of Texas Press, 2006); Jorge Cañizares-Esguerra, *Nature, Empire, and Nation: Explorations of the History of Science in the Iberian World* (Stanford: Stanford University, 2006).

11 Henrique Leitão and Antonio Sánchez, "Zilsel's Thesis, Maritime Culture, and Iberian Science in Early Modern Europe," *Journal of the History of Ideas* 78, no. 2 (2017): 191–210.

12 This issue of direct experience and experimentation also plays a major role in the chapters by Justin Niermeier-Dohoney and Himmet Taşkömür, this volume.

13 Leitão and Sánchez, "Zilsel's Thesis"; Edgar Zilsel, "The Sociological Roots of Science," *American Journal of Sociology* 47, no. 4 (1942): 544–562.

of artisan—or agricultural specialist in this case—and scholar, which broke social barriers of knowledge. These writers gained knowledge either from their travels or through their status as landowners. While either option allowed them to collect data from people of many different social backgrounds, the latter lent them credibility as empiricists who worked closely with the land. However, knowledge of soil advanced in particular ways within the field of "Iberian science," as the contemporary Catholic interpretation of soil was that it was an incorruptible, divine creation. Sometimes the writers of these volumes adhered to traditions derived from Catholic religious beliefs, and at other times they progressively disregarded classical textual practices and sought to implement empirical methods.

At the same time, these authors wrote in kingdoms where a strict control over publications was imposed. The revisions to which these and other books were generally subjected clearly demonstrate the control exercised by the Catholic Church. Thus, among the first pages of these books, there are notes signed by an ecclesiastic, assuring the reader that the contents do not contradict Catholic canons and recommending the grant of a publishing license. The presence of such an imprimatur could hint at intellectual stagnation or at least the absence of disruptive statements, considering that the contents of these texts had been revised and approved by the proper Catholic organizations. But it is important to stress that this censorship rarely hindered the circulation of books, since there were various strategies to gain access to them.[14] And the very publication and circulation of these writings is evidence that their contents did not directly conflict with the views of the conservative authorities. Nevertheless, we argue that the Church's strict control over the circulation of new ideas led to a discourse adaptation, a mediation between classical knowledge (including some notions from Islamic authors), the Christian doctrine of the divine creation of the universe, and more recent empirical evidence. As we will see, an analysis of the agrarian instructions written by authors since the sixteenth century shows how original explanations appear next to material that demonstrates an integration of classical natural philosophy with Christian doctrine. Without contradicting the established tenets, some of these authors restructured this rigid knowledge, adding new variables and their own experience of the land in a way related to the study of agriculture.

Some of the books published between the second half of the sixteenth and the last years of the seventeenth centuries—including several in this

14 Hannah Marcus, *Forbidden Knowledge: Medicine, Science, and Censorship in Early Modern Italy* (Chicago: University of Chicago Press, 2020).

study—fit within the arbitrismo. This was a Hispanic political and economic movement that sought, through the publication of essays, to offer solutions to problems that affected the monarchy, including the lack of available credit, stagnant industry, and unproductive agriculture.[15] Aware of being immersed in what has come to be called a "Global Crisis" by some historians—marked in Spain by the loss of imperial territories, constant wars, and even climate change—these authors tackled the problem from different perspectives, proposing solutions to restore the past glory of the Hispanic Monarchy.[16] They addressed, in particular, local and regional authorities, whom they understood to be the only agents capable of implementing the necessary measures.[17] The movement also influenced Portuguese authors analyzed in this study, especially between 1580 and 1640, when the Iberian crowns were united, facilitating the circulation of people and ideas.[18] In some ways, the arbitrismo movement resembled mercantilism, as both were founded on the idea that the current economic situation was in decline, although this view has been questioned.[19] So arbitristas tried to find solutions to new challenges in the Iberian context. To some extent, they did this in a patriotic manner, taking advantage of the knowledge from new overseas dominions that was converging in this area. Although their approach was often theoretical, the arbitristas' equation of agricultural improvement and economic advancement led to the improvement of new information about soils, which allows us to observe how knowledge evolved within the aforementioned scientific framework.

In this study, we chose eight books: five published in Spanish and three in Portuguese, all shown in chronological order in Table 6.1.

---

15 José Ignacio Fortea, "Economía, arbitrismo y política en la Monarquía hispánica a fines del siglo XVI," *Manuscrits: Revista d'història moderna* 16 (1998): 155–176.

16 Geoffrey Parker, "Crisis and Catastrophe: The Global Crisis of the Seventeenth Century Reconsidered," *American Historical Review* 113, no. 4 (2008): 1053–1079, https://doi.org/10.1086/ahr.113.4.1053.

17 Consolación Baranda Leturio, "Diálogo y arbitrismo: De *Los diálogos de la fertilidad y abundancia de España* al *Despertador que trata de la gran fertilidad que españa solía tener*, o 'Cómo se desmonta un diálogo,'" *E-Spania* 29 (2018), https://doi.org/10.4000/e-spania.27360.

18 Margarida Sobral Neto, "Conflict and Decline, 1620–1703," in *An Agrarian History of Portugal, 1000–2000: Economic Development on the European Frontier*, ed. Dulce Freire and Pedro Lains, Library of Economic History (Leiden: Brill, 2016), 101–131; Graça Almeida Borges, "¿Un imperio ibérico integrado? El arbitrismo y el imperio ultramarino portugués (1580–1640)," *Obradoiro de Historia Moderna* 23 (2014): 71–102.

19 Enrique Ujaldón, "Arbitrismo y mercantilismo en la España de Saavedra Fajardo," *Res Publica: Revista de Historia de las Ideas Políticas* 19 (2008): 299–312.

TABLE 6.1 List of Spanish and Portuguese books analyzed in this chapter

Author	Original title	English title	Year	Country
Gabriel Alonso de Herrera	*Obra de agricultura**	Treatise on agriculture	1513	Spain
Juan Valverde de Arrieta	*Despertador, que trata de la gran fertilidad, riquezas, baratos, armas y caballos que España solía tener*†	A wake-up call that takes into account the great fertility, wealth, arms, and horses that Spain used to have	1581	Spain
Duarte Nunes de Leão	*Descripção do Reino de Portugal*	Description of the Kingdom of Portugal	1610‡	Portugal
Lope de Deza	*Gobierno político de agricultura*	Political government of agriculture	1618	Spain
Fray Miquel Agustí	*Libro de los secretos de agricultura*§	Book of agricultural secrets	1626	Spain
Francisco de Gilabert	*Agricultura práctica con la cual puede uno llegar a ser perfecto agricultor*	Practical agriculture with which one can become a perfect farmer	1626	Spain
Manuel Severim de Faria	*Notícias de Portugal*	News from Portugal	1655	Portugal
João António Garrido	*Livro de agricultura*	Book of agriculture	1749	Portugal

Notes:

* Title variants appear across reeditions, including *Libro de agricultura*, and, the most popular, *Agricultura general* (both titles here in their abridged form).

† An enlarged version of a text originally published in 1578. The original title was *Diálogos de la fertilidad y abundancia de España* (Dialogues of the fertility and plenty of Spain) in its abridged form.

‡ This text was written earlier, but it was published posthumously in 1610.

§ This text is an expanded translation of an earlier Catalan version published in 1617.

Garrido's *Livro de agricultura* unified previous Iberian agrarian knowledge and offered multiple examples located in both peninsular kingdoms. We consider that Garrido closed the era ushered in by Herrera, in terms of the way his book is organized and the information he provides. He collected the peninsular agrarian knowledge of the previous two centuries in what we argue is a transitional text that mediates between the authority of classic authors and the emergence of scientific methodology. Beginning in the second half of the eighteenth century, agricultural and agrarian instructional textbooks began to change, abandoning untested theories and other information that did not have some kind of experimental foundation.

Comparing these books with their different styles and objectives allows us to observe how authors created and disseminated visions about Iberian soils between the sixteenth and eighteenth centuries. The rest of this study is divided into two parts, with a detailed analysis of soil as given by each author, first focusing on Spain and then on Portugal. We have also included notes about the reeditions of some texts in order to demonstrate their dissemination, reception, and the scope of their messages. In this context, we consider that multiple editions represent popularity and a wider readership, although obtaining data to support this is not always possible.

2 Exploring Spanish Soils

Gabriel Alonso de Herrera, a priest with agronomic training, published *Obra de agricultura* in 1513. He was commissioned by Cardinal Cisneros, the archbishop of Toledo, Grand Inquisitor, regent of Spain, and one of the most important figures in early sixteenth-century Castile. This task led Herrera to travel through the Iberian Peninsula looking for agronomy texts, simultaneously collecting peasant cultivation practices that he described in his book. As he underlined, "I'm not the first inventor of this Art of Agriculture … but the first to try to put the rules of this art in Spanish."[20] It has to be stressed that his precocious book, even by European standards, had a global and integrative vision of agriculture. Herrera criticized those who pointed out that knowledge of soil from other regions could not be applied to Spain, claiming that the invalidity of this argument "can be proved with the authority of the masters themselves," referring to classical texts.[21] Herrera drew on the teachings of Greek, Roman, and Islamic writers. As the text shows, Herrera directly referenced Ibn Wafid, under the Latinized name Abencenif. In fact, his knowledge of some of the Moorish classics is well known.[22] Talking about soil, he highlighted that "its entity and quality never change … neither losing fertility after many harvests nor losing

20 Gabriel Alonso de Herrera, *Obra de agricultura* […] (Alcalá de Henares: Arnao Guillen de Brocar, 1513), prologue. All translations by the authors.

21 Herrera, prologue.

22 Juan Estevan Arellano, *Ancient Agriculture: Roots and Application of Sustainable Farming* (Layton, UT: Gibbs Smith, 2006), 12–13.

strength with age."[23] If soil did not lose reproductive capacity, he argued, then a failure to properly work the soil was the unique cause of poor crop yields. According to Herrera, it had to be like this because God had created soil to be "perpetually fecund, giving the strength and vigor of perpetual youth."[24] This interpretation was an irrefutable argument of faith that was in accordance with the scholastic belief in the incorruptibility of divine creation.

Despite these religious beliefs about its immutability, Herrera identified great differences among different types of soil. First, he generally defined soil as either cold or dry, but argued that its nature and fertility could be altered due to certain geographic variables: agricultural land could be located in a valley, on a plain, or a mountain, and, within this last category, on a slope or a peak. Herrera considered lowlands better, because they had "more substance," which they received from the areas above. This helps to explain the existence of different qualities. As Herrera notes, soils "are either fertile, thick, outstanding, or they are completely sterile and quite bad, or they are in a medium range without being too bad or too good."[25] Finally, temperature and moisture also contributed to soil quality, which allowed Herrera to speak about hot, cold, or mild lands, depending on "the influx of the hot, cold, or mild air," suggesting that soil status and environmental factors were connected.[26] By "air" here, Herrera likely meant "wind," since, during the early sixteenth century, the concept of climate remained a local rather than regional or global phenomenon, more tied to geography and cartography than it was to proper meteorology.[27] Notably, for Herrera, the elements that most directly intervened in soil fertility were clearly local in nature.

Some of the soil traits could be observed directly through individual experience, but Herrera also listed several rules to identify the types of soil suitable for agriculture, like a series of rudimentary physicochemical principles. The color of the land was not a "sufficient witness for entire and reliable knowledge of it."[28] However, this was not the case for touch, at least if the farmer

---

23 Gabriel Alonso de Herrera, *Libro de agricultura* [...] (Alcalá de Henares: Casa de Joan de Brocar, 1539), prologue. In early versions, this quote is shorter.

24 Herrera, prologue.

25 Herrera, 3.

26 Gabriel Alonso de Herrera, *Libro de agricultura* [...] (Logroño: Casa de Miguel de Eguia, 1528), 1. This last reference to air is not found in earlier versions.

27 Franz Mauelshagen, "Climate as a Scientific Paradigm: Early History of Climatology to 1800," in *The Palgrave Handbook of Climate History*, ed. Sam White, Christian Pfister, and Franz Mauelshagen (London: Palgrave Macmillan, 2018), 565–588.

28 Herrera, *Obra de agricultura*, 4.

followed a simple formula: "take a small clod of dirt, wet it with saliva or water and put it between the fingers. If it sticks and makes a dough, it [the soil] is good and thick. But if it is rough and sandy, it is not."[29] Finally, taste was relevant too, not of the dirt itself, but of the water it was in contact with. Herrera explained two different valid methods: one related to the water that was raised directly from the ground, while the other could be used if there were no water courses or aquifers, by taking a clod of soil, dissolving it in pure water, and filtering it.[30] A sweet-tasting water signified high-quality land, while a foul-tasting water signified poor-quality land.[31] Furthermore, good soils were better at retaining heat and moisture, and wild flora growing vigorously indicated the most agriculturally productive land. Certain plant species were associated with particular kinds of soils, but Herrera did not elaborate theoretically why this was the case. Some of these statements can be found in Ibn al-ʿAwwām, who, like Herrera, cited Columella.[32] Given that not all soils were equal, Herrera wrote about the possibility of improving those that were unproductive by applying fertilization, irrigation, or introducing new fallowing techniques. The best option, however, was to "provide each soil with what is good and fit for it,"[33] a core idea that, he argued, was not respected in Spain, resulting in a less productive agriculture.[34]

Herrera's *Obra de agricultura* clearly made a deep impression on subsequent authors, considering that at least thirty-five editions have been published since 1513.[35] Even in the early days, the book reached well beyond its homeland, with five editions translated into Italian in the sixteenth century and the first Portuguese edition in 1841, though the English translation had to wait until the dawn of the twenty-first century.[36] Given its relevance, after its third centenary, some scholars of the Royal Botanic Garden of Madrid published a new

29 Herrera, 4.
30 Herrera, 4.
31 Herrera, 4.
32 Yaḥyà ibn Muḥammad Ibn al-ʿAwwām, *Kitāb al-filāḥa: Libro de agricultura*, trans. Antonio Banqueri (Madrid: Imprenta Real, 1802), 1:48–49.
33 Herrera, *Obra de agricultura*, 5.
34 Gabriel Alonso de Herrera, *Libro de agricultura* [...] (Alcalá de Henares: Casa de Joan de Brocar, 1539), prologue.
35 On the Ministerio de Agricultura, Pesca y Alimentación's "Plataforma de conocimiento para el medio rural y pesquero," see the entry "v Centenario del 'Libro de Agricultura,'" subsection "Ediciones," accessed August 4, 2025, https://www.mapa.gob.es/es/ministerio/publicaciones-archivo-biblioteca/publicaciones/ediciones.
36 Arellano, *Ancient Agriculture*.

expanded edition based on the original text.[37] The early editions also provide examples of changes in the precision of Herrera's thinking. He added new details to each new edition in order to clarify some of the concepts explained in less detail in the 1513 original. Even taking a conservative estimate for his unknown date of death, at least six editions were published during his lifetime between 1513 and 1539.[38] In fact, Herrera's *Obra de agricultura* was so popular that many subsequent authors reflexively deferred to his expertise and merely reproduced his principles in their works. In some ways, this led to a crisis of innovation in the study of soil in the Iberian Peninsula over the next century. However, many of the authors we discuss in this chapter advanced soil sciences during this time by introducing nuanced interpretations of Herrera's work.

Juan de Valverde Arrieta, an obscure treatise writer and arbitrista, is perhaps the author closest to Herrera in terms of both his arguments and the breadth of the dissemination of his work, as shown by his *Despertador*, published in 1581. A first print run of two hundred copies was sent to all of the cities that were represented in the courts of Castile by the civil servants who operated on behalf of their interests, the procurators. However, its diffusion achieved even higher levels after it was included among the appendices of the 1598 reedition of the Herrera book and in later reeditions.[39] In his text, Valverde upheld the principle that soil did not become exhausted, which he reinforced by way of two allusions already mentioned: an ever-flourishing nature and the classical writers. The conjunction of both arguments led to the identification of the same problem Herrera had noted, that agricultural labor was a divine punishment, because "God cursed the soil, so it is necessary to work on it until you sweat."[40] However, Valverde also identified the crux of the problem: the decline of agricultural productivity in Spain was due to the replacing of oxen in farming with mules, as the oxen had enabled deeper plowing. He also suggested a practical solution, which was to return to traditional farming practices and reinstate the oxen.

Broadening his vision, he recognized that soil's capacity to produce depended on three variables: soil's own virtue, the influences of heaven, and good human industry.[41] Unlike Herrera's precise language concerning soil

37 Antonio Sandalio de Arias Costa et al., eds., *Agricultura General de Gabriel Alonso de Herrera* (Madrid: Imprenta Real, 1818).

38 Mariano Quirós García, "El *Libro de Agricultura* de Gabriel Alonso de Herrera: Un texto en busca de edición," *Criticón* 123 (2015): 105–131, https://doi.org/10.13039/501100003329.

39 Baranda Leturio, "Diálogo y arbitrismo."

40 Valverde Arrieta, *Despertador que trata de la gran fertilidad, riquezas, baratos, armas y caballos, que España solia tener* (Madrid: Don Josef de Urrutia, 1581), 42–44.

41 Arrieta, *Despertador*, 85.

fertility in the *Obra de agricultura*, Valverde gave more rudimentary advice. He suggested that, to obtain the maximum benefit, farmers should consider both intrinsic qualities and extrinsic environmental factors concerning soil's fertility, and combine this knowledge with the practical peasant knowledge that also considered the above variables. Lastly, soil virtues changed according to rules of "settling and position," a point that Valverde took directly from Herrera, without further explanation. He recognized Herrera by citing him among other "famous farmers" from the ancient world like Varro or Theophrastus. Valverde followed these citations with a very simple classification of three soil types, depending on their suitability for agriculture: fertile, sterile, and an intermediate category. In this particular text, references to classical authors can be understood in the intellectual context of the arbitristas' texts: as these essays were written in an attempt to convince the regional authorities who received them, strong arguments—drawn from the classics—were necessary to persuade and assure the reader of the author's competence in the subject.[42]

In 1618, Lope de Deza, arbitrista and lawyer, wished to contribute to finding a solution to the agrarian crisis experienced by some Iberian Peninsula regions at the time. His book *Gobierno político de agricultura* is divided into three parts that deal with the need for and importance of agriculture, the causes of its decline, and suggestions to alleviate the situation. In addition to numerous references to the political and social aspects of agriculture, the text offers interesting reflections about natural resources, including soil. According to Deza, understanding the structure of soils and their agricultural uses depends on understanding soil's divine nature and placing it within the proper context of Aristotelian physics and cosmology. In his opinion, the "earth is the center of this universal globe. ... For this noble element, Agriculture serves her in order to receive the rest of the elemental and celestial influences."[43]

Influenced by Aristotle and the Greek cosmogonic notion of *archē*,[44] he argued that soil "is one of the four elements or principles that make up all sublunar corporeal things, for which Agriculture was made."[45] In Aristotle's logical system, agriculture represented the "final cause" of the substance of soil, and this interpretation was confirmed by scriptural precedents. Soil could not become exhausted, scarce, or diminished, as the "Creator infused it to

42 Baranda Leturio, "Diálogo y arbitrismo."

43 Lope de Deza, *Gobierno político de agricultura* (Madrid: Viúda de Alonso Martín de Balboa, 1618), 1–2.

44 Adam Drozdek, *Greek Philosophers as Theologians: The Divine* Arche (New York: Routledge, 2007).

45 Deza, *Gobierno político de agricultura*, 19.

multiply, remake, increase itself, without any possibility of consumption or alteration."[46] Thus, although agriculture was a tool used by humans to take advantage of soil for their own benefit, the religious principle determined that its fertility was inexhaustible, a view that clearly followed the arguments of previous authors. According to this last line of reasoning, "in Spain, those who claim that sterility is inevitable, due to natural causes, such as the weariness and weakness of the soil, are deluding themselves," because if that were the case, meadows and forests would also show signs of fatigue.[47] Deza's argument was a nearly exact copy of what Valverde had written. In a similar way, referring to what Herrera had already said, Deza stated that "we can see that where they labor the soil properly, manure it and water it, it is always in vigorous youth."[48]

Even as agrarian authors from Herrera to Deza adhered to the Biblical position of the incorruptibility of soil and deferred to the authority of classical authors, they nevertheless began to suggest that human labor and novel agricultural techniques could positively contribute to soil fertility. Despite the introduction of these empirical and labor-oriented challenges to traditional soil knowledge, more than a century after Herrera's *Obra de agricultura*, Biblical ideas about the incorruptible, divine nature of soil and the authority of classical writers still maintained a stranglehold over Iberian soil visions. However, two books published in 1626 offered some challenges to these prevailing views, suggesting that, after all, both interpretations and practices were changing: one by Fray Miquel Agustí, prior of the Order of Knights of the Hospital of Saint John of Jerusalem; the other by Francisco de Gilabert, a court noble who had written several political and economic treatises. Yet novel interpretations did not mean disruption, as several ideas already explained by Herrera persisted in both writers' works. Apart from Herrera's prestige, these ideas surely endured because they continued to be so practical and convenient.

Despite its conservatism, Agustí's *Libro de los secretos de agricultura* presented more innovations than any other agricultural work published in the previous century due, above all, to a systematization of soil knowledge. Focused on providing agrarian instructions, the book was remarkably successful, with at least eight Spanish editions in the first half of the eighteenth century, but without reissues in the original Catalan.[49] The first part lacked

46 Deza, 19.

47 Deza, 19.

48 Deza, 19–20.

49 Luis Pablo Núñez, "Ediciones e historia textual del 'Libro de los secretos de agricultura' de Miguel Agustín," *Boletín de la Real Academia de Buenas Letras de Barcelona* 51 (2007): 199–224.

novelty, as it presented well-known quotes from classical authors supporting, or even reinforcing, Biblical or mythical interpretations of the nature of soil. However, in the second part, Agustí provided a fresh perspective on the study of soils, reduced the number of references to authorities, and gave space to more practical, experiential, and dynamic knowledge.

In his discussion of soils, Agustí shows that there had been a shift in the understanding of some concepts. Firstly, he demonstrates a rudimentary understanding of climatic variations: while Herrera and Valverde's interpretations were based on the idea of air as local wind influencing the soil according to criteria of geographical position and orientation of the cultivated plots, Agustí's reasoning has a more regional range, considering in "various provinces and kingdoms ... a variety of complexions according to the air and climate of the sky."[50] But the most outstanding aspect is his analysis of the soil. He approached soil traits according to a novel binary taxonomy that contained obvious references to their physical characteristics: strong/soft, dry/moist, stony/gravelly, thick/thin, and sandy/coarse.[51] Here, Agustí classified soil according to criteria obtained by observing the size of its particles or its water content, which indicates a more detailed and empirical approach than that of his predecessors. This type of classification can be understood as a background to other analyses that allowed an increasingly accurate approach to the physicochemical nature of soil in the following centuries, demonstrating that "past conceptions of soil are both the foundations and building stones for the conceptions prevailing now."[52] This aspect, together with his overcoming of the strictly local scope in soil observation, allows us to place Agustí as a transitional author, as there is no doubt that he shows some significant differences from previous writers.

Gilabert, whose book *Agricultura práctica* was also published in 1626, is probably the most original of the seventeenth-century authors cited here. As with Agustí, one of his key interventions was a new understanding of climate. He pointed out some notions that, although obvious, were not present in previous authors, whom he criticized. For example, the existence of different seasons was evident, so it was logical that

> not everywhere will a month have suitability for the same farm, and it is wrong to give it to them with the sole distinction of the land being hot

50 Miquel Agustí, *Libro de los secretos de agricultura, casa de campo y pastoril* [...] (Perpignan: Casa de Luys Roure, librero, 1626), 223.

51 Agustí, 223.

52 Roy W. Simonson, "Concept of Soil," in *Advances in Agronomy*, ed. A. G. Norman (New York: Academic Press, 1968), 20:1–47, https://doi.org/10.1016/S0065-2113(08)60853-6.

> or cold ... and thus [previous authors] deceive those who follow them, because they do not warn of this diversity of climates.[53]

Gilabert points out here—going beyond a local logic—that seasons and climates, when talking about agriculture and soils, introduce other variables that add diversity in the different regions. However, his understanding of climate did not adhere to our current definition. He was referring to a contemporary version—"that the experience of the moderns has shown us," in his own words—of the Ptolemaic notion of climate zones organized in bands parallel to the equator.[54] In this view, there were forty-six total climates, each of which became warmer the closer it came to the equator. According to Gilabert, bands equidistant to the equator on either side possessed similar climates, though this was not a universal characteristic and could vary. Travels to new lands had progressively eroded the foundations of Aristotelian natural philosophy that recognized an uninhabitable, torrid zone.[55] If it was not entirely clear in the case of Agustí, in Gilabert's text we see evident changes in the analysis of the environment. He now applied this idea to his thinking about soil, which was crucial to understanding a global—not local—idea of soil's variety.

Gilabert assured his readers that, rather than relying on the authority of classical or Biblical authors, he came to his conclusions empirically, claiming: "I do not follow the opinions of authors, but trust my own experience."[56] As a landowner, he validated his knowledge through many years of experience and dealings with his tenants as evidence to support his arguments.[57] Nevertheless, he did not completely abandon previous authors, who provided him with a series of logical concepts. In any case, despite the unquestionable existence of similar ideas taken from other writers, particularly Herrera, Gilabert presented the arguments of his work with a better structure and in a clearer way. This is particularly evident in the sections of the book where he discusses the influence of the sun on the soil and its relation to the geographical position and orientation of farmland.

First of all, Gilabert wrote about the "particular qualities of the territory," referring to the fact that, regardless of whether it was in a cold, temperate, or hot climate, terrain could be hot or cold "because of the nature of land itself,

---

53 Francisco de Gilabert, *Agricultura práctica con la cual puede uno llegar a ser perfecto Agricultor* [...] (Barcelona: Sebastián de Cormellas, 1626), prologue.

54 Gilabert, 5.

55 Craig Martin, "Experience of the New World and Aristotelian Revisions of the Earth's Climates during the Renaissance," *History of Meteorology* 3 (2006): 1–15.

56 Gilabert, *Agricultura práctica*, prologue.

57 The resemblance to Rashid ad-Dīn in Himmet Taşkömür's chapter, this volume, is obvious.

or its circumstances," which resembled Herrera's position.[58] By its nature, according to Gilabert, soil was hot if it was also dry and thick, and cold if it was moist and light. Depending on the circumstance, soil would be hot "because it was located facing the midday sun, or in a deep place, or sheltered from cold winds," and it would be cold "because it was located in a dismal place and dealing with cold winds."[59] Finally, just as Herrera had written, Gilabert asserted that soils found in valleys were the most useful to agriculture, with easy access to water and the fertile silt that flowed to the valley floor.[60]

The author continued to classify soils based on their internal, physical composition using a cold-hot system and a dry-moist binary taxonomy similar to Agustí. However, he introduced variables that could disrupt these dichotomies and change the nature of soils. Using the concept of "substance" (*sustancia*), meaning a soil's material density, he established the existence of strong/soft lands, while the presence of water also led him to classify irrigated/nonirrigated soils. According to Gilabert, all of these could be improved through specific techniques, including irrigation, fertilization, and manuring. Gilabert's innovation here lay in the fact that these characteristics, connected with other variables like climate or air influx, resulted in a wide variability of scenarios that prevented a generalized application of agricultural patterns to all soils. This is why he criticized previous scholars—mainly classical authors as well as some contemporaries—who had encouraged the dissemination of excessively general labor calendars among farmers. These texts described soils only as a container for seeds and did not consider other wider factors that affect agrarian rhythms.

These fresh observations, spearheaded by early seventeenth-century authors like Agustí and Gilabert, created new natural frameworks that contributed to an awareness of the diversity of soils and the external, climatological factors that could change them. But we are also facing a phenomenon already noted in the historiography of soil science: the possible concurrence of different conceptions of soil. That is, the existence of variations that lead to the same individual using one conception in some circumstances and another one in others.[61] This apparent paradox helps to explain the encounter between these visions of a developing science and other, more conservative, religious views. As shown in the following, the Portuguese case is similar.

---

58 Gilabert, *Agricultura práctica*, 17.
59 Gilabert, 17.
60 Gilabert, 26.
61 Simonson, "Concept of Soil."

3 Knowing the Portuguese Land

Though fewer in number than their Spanish language counterparts, Portuguese authors also contributed significantly to the study of soil in early modern Iberia. The political situation and intellectual influences—the two kingdoms were united for sixty years, ruled by the Spanish Habsburgs—may help to explain the scarcity of specific books expressing Portuguese views about agriculture and soils. Instead, there are copies of the original Spanish books, and some of the texts cited in the previous section circulated widely in Portugal.[62] However, one Portuguese book written during the Iberian Union and two others written after its dissolution stand out as examples of changes in the study of soil: Duarte Nunes de Leão's *Descripção do Reino de Portugal* (1610), Manuel Severim de Faria's *Notícias de Portugal* (1655), and João António Garrido's *Livro de agricultura* (1749). In terms of dissemination, according to the National Library of Portugal, multiple editions of the latter two were published, including two eighteenth-century editions of the *Notícias de Portugal* and six reissues of the *Livro de agricultura*.

The *Descripção* by the jurist and historian Duarte Nunes de Leão, a chorography written before 1610, was the final legacy of an already renowned author, underlining the excellence of Portugal at a time when the kingdom was ruled by Spanish kings. Although this political situation may have contributed to some overvaluation of his own homeland, the author tried to link his descriptions to direct and indirect regional observations. Attempting to describe the realm of Portugal, the author wrote a chapter addressing soil fertility, albeit with poorly developed and very general reasoning. In his opinion, Portuguese soil fertility was robust enough to make possible the growth of any crop. Nevertheless, he stressed that, in recent times, part of this prosperity had disappeared due to the lack of respect for old agrarian laws, something that had led to the reversion of many fertile lands into scrubland. He reinforced his ideas with arguments that idealized the Portuguese countryside, once again with references to classical authors. Leão's treatment of soil was rather inconsequential since he assumed that the divinely appointed fertile substrate guaranteed crop growth. While this approach may seem a bit shallow, it can also be understood as an example of how the process of debating the merits of soil treatment changed over these two centuries. Considering the political

62 Ana Duarte Rodrigues, "Gardening Knowledge through the Circulation of Agricultural Treatises in Portugal from the Sixteenth to Eighteenth Centuries," in *Gardens, Knowledge and the Sciences in the Early Modern Period*, ed. Hubertus Fischer, Volker R. Remmert, and Joachim Wolschke-Bulmahn (Basel: Birkhäuser, 2016), 305–317.

context, Leão's flattery of his homeland is novel in his use of the classics for a patriotic goal. Thus, Leão proudly recognized that numerous water resources contributed to the productivity of Portuguese land, a well-known aspect since ancient times, as the Latin geographer Strabo "called Lusitania a happy land because of what it produces."[63] Both Athenaeus of Naucratis and Polybius are also cited for the same purpose. This demonstration of patriotism supported with authoritative references is also evident in the presentation of Portuguese historical figures, such as King Denis, known as "the Farmer King."[64]

Manuel Severim de Faria, a Catholic precentor who wrote on agriculture, also cultivated other disciplines, including theology, history, and geography.[65] In his *Notícias de Portugal*, published in the year of his death, he analyzed very different issues influenced by the arbitrista style, from descriptions of landscape and society to advice on how to face challenges in the Portuguese overseas territories. Part of its recognition lies in its comprehensive approach to the problems affecting the Portuguese economy.[66] But especially useful were the pages about how to enlarge the wealth of the Portuguese kingdom, how to increase the population and its economic activities, including agriculture, and how to use soil as a tool to this end. Faria argued that cultivation was one of the ways to guarantee the livelihood of people as well as the wealth, power, and happiness of every kingdom. He focused on ensuring the presence of a sufficient number of inhabitants for a proper care and maintenance of soil, theorizing that agricultural labor secured the productive capacities of soil. Optimistically, Faria essentially concluded that no soil was infertile. Rather, it was necessary to pair each kind of soil with the most suitable crop. Even though his reasoning was similar to Herrera's idea of accommodation, Faria recognized the presence of new plants from America in Portuguese territory.[67] Notably, Faria's advice was practical and, although the author was a priest, we suggest that *Notícias de Portugal* lacked religious references because the

63 Duarte Nunes de Leão, *Descripção do Reino de Portugal* [...] (Lisbon: Jorge Rodríguez, 1610), 66.

64 Leão, 63.

65 Liam Matthew Brockey, "An Imperial Republic: Manuel Severim de Faria Surveys the Globe, 1608–1655," in *Portuguese Humanism and the Republic of Letters*, ed. Maria Berbara and Karl A. E. Enenkel (Leiden: Brill, 2012), 265–285.

66 Inês Amorim, "Manuel Severim de Faria: Uma releitura dos remédios para a falta de gante—1655," *História: Revista da Faculdade de Letras da Universidade do Porto* 5 (1988): 151–172, on 157.

67 Manuel Severim de Faria, *Noticias de Portugal* [...] (Lisbon: Officina Craesbeeckiana, 1655), 2–3.

development of agriculture depended essentially on the work of those who care for the soil.

Finally, we examine João António Garrido's *Livro de agricultura* of 1749. Garrido was a Spanish teacher who traveled through the Iberian Peninsula. In 1749, he was already a naturalized Portuguese citizen and taught at the royal court in Lisbon, writing the first Portuguese agrarian manual entirely focused on agriculture as a broad subject.[68] He began by recognizing that, "although many authors have written about every art and science, I see very few in this kingdom that address the very useful art of Agriculture," a subject that is "the oldest, the noblest, and the most useful for mankind."[69] Crossing direct experience with the observations of others, he wrote about how to raise farm animals and how to grow plants (including flowers, cereals, vegetables, and fruit trees) in gardens, fields, and orchards. On some issues, Garrido followed Herrera's interpretation of soil. For example, he wrote about the convenience of sowing wheat on soil that is preferably more humid than dry, and he stated that the land in the valleys has "more substance than that on a mountain."[70] But what made Garrido stand out in didactic terms was the way he expressed his own experience, giving specific examples collected in his travels through Castile, Extremadura, and Portugal. In this way, the *Livro de agricultura* took an integrative approach to the entire Iberian Peninsula that was not present in the work of any of the other authors analyzed here. Garrido also included advice about the preparation of soil and seed, suitable soils, and cultivation skills for new overseas crops like tomatoes, maize, and peppers.[71]

Some of the ideas expressed by Garrido about the soil are well-known precepts. These include both the divinity of the land and its production capacity—though Garrido provided nuance to the human role: "so, although abundance originates in God, it comes from a good industry and human diligence."[72] It should be noted, however, that references to this divine condition are rare in his book, which underlines the importance of labor and, especially, the correct application of manure, something that was probably the result of more accurate knowledge about the effects of fertilization. Garrido stresses the significance of this idea when addressing the preparation of soil for different types of farm crops.

68 Carmen Soares, ed., *BiblioAlimentaria: Alimentação, saúde e sociabilidade à mesa no acervo bibliográfico da Universidade de Coimbra* (Coimbra: Imprensa da Universidade de Coimbra, 2018).

69 João António Garrido, *Livro de agricultura* (Lisbon: Officina Alvarense, 1749), prologue.

70 Garrido, 93.

71 Garrido, 25, 29–30.

72 Garrido, prologue.

For Garrido, the starting point for understanding soil was the well-known classification of land as cold, mild, or hot. However, he introduced new characteristics that caused soil to be classified into these groups: "cold land is the one that does not produce or bring up any fruit. … Mild land grows all types of fruits, vines, and olive trees, except lemon and orange. Hot land is the one that grows all types of fruits and lots of lemons and oranges."[73] On the one hand, this concept, although similar to Herrera's nomenclature, had different content, since it referred to the productive character of soil fertility, taking as its scale the possibility that citrus trees could only grow in a given land. On the other hand, while he identified three types of land, other characteristics must be considered, because "there are soils that produce seeds well, and others [that produce seeds] badly, just as clay soil produces good cantaloupes, and sandy soil watermelons."[74] It was up to the good farmer to "observe and be sure about attributes of the land," so experience played a key role. Knowledge about these attributes had to be complemented with knowledge of the winds that contributed to temperature and humidity. Garrido named four types of winds that come from the cardinal points, all different from each other and with their own particular qualities: North (cold and dry), South (hot and humid), East (hot and dry), and West (cold and humid).[75]

Our analysis of the Portuguese authors ends with Garrido integrating Iberian knowledge of soil and agriculture from the previous two centuries. Significant changes can be noted since the work of the first authors, including the Spanish ones. The writings around soil and its conception have been transformed. Expert and professionalized farming methods described by Garrido replace general references to techniques such as fertilization, irrigation, and fallowing. The work of the land, understood as divine punishment due to God's cursing the soil, was now interpreted within the context of a more flexible view of agriculture as something that could be improved with practical and experimental knowledge. Recently, some scholars have recognized the work of these authors as representing a stage during which the main interpretation of soil is as a medium for plant growth.[76] However, the great variety of nuances and the discernable trajectory toward reasoning based on observable factors is undeniable, as we can see in Garrido's work. Moreover, this evolution reveals that these authors were influenced by local and regional practices and

73 Garrido, 75.
74 Garrido, 75.
75 Garrido, 74.
76 James G. Bockheim et al., "Historical Development of Key Concepts in Pedology," *Geoderma* 124, nos. 1–2 (2005): 23–36, https://doi.org/10.1016/j.geoderma.2004.03.004.

knowledge, bearing witness to intellectual changes in the perception of nature in everyday life.[77]

4 Conclusion

Agrarian manuals contained both soil knowledge and soil imaginaries in sixteenth- and seventeenth-century Iberia. However, these texts were not the only ones to discuss soil: chorographies or arbitrista proposals, among others, feature different approaches, some of which have been discussed above. Over the centuries, these genres provided agrarian instructions that reached different regions of the Iberian Peninsula and of the wider world. What was the impact of the knowledge offered in these books? The answer is still uncertain, but we can draw some broad conclusions.

First, hardly any information exists about the range or dissemination of most of these books in the Iberian Peninsula, as this is linked to literacy levels, about which we know little. Reading skills went far beyond mere learning because reading demanded not only valuable time but also training. This obvious requisite means that these books had a specific target audience. The developing body of scientific knowledge about soil also contrasted with other peasant and classical erudite types of knowledge and was more concerned with the correct development of crops that guaranteed subsistence than with a technical understanding of how soil actually functioned.

Second, the international diffusion of these books needs consideration. On this topic, further research could investigate whether the authors' choices of writing in vernaculars jeopardized the dissemination of their innovative achievements outside the Iberian Peninsula. An analysis of reeditions and translations is not always possible, and its results are variable. We may not be aware of all the international reprints that may have been lost or preserved in private places inaccessible to researchers.

Our analysis of soil knowledge in the Iberian Peninsula during these centuries displays a panorama of disparate theoretical positions on soil entwined with historical sources. On the one hand, religious-based reasonings emerge, as well as ideas taken from ancient authors and even from cosmogonic theories. These books could easily be in conflict with the censorship authorities, but the presence of licenses approved by experts in faith issues demonstrates that there was enough space granted so that the adaptation and improvement

77 Gomes and Freire, "Seeds of Knowledge," 378.

of soil knowledge was not seen to contravene Scripture. On the other hand, the authors of these agrarian instruction books placed growing importance on individual experience and on the experience and know-how available in their social backgrounds, which they drew upon to build increasingly robust explanations. References to practical action in the form of cultivation advice (even for new plant species), soil classification based on the observation of particles, and the determination of suitable lands for new crops are all proof of this.

At the same time, these characteristics show that early modern agrarian authors deserve just as much attention as eighteenth- and nineteenth-century ones, who classified soils in a similar fashion. The lack of modern analytical tools and the tensions between their divergent arguments may have hampered the development of original propositions by early modern Iberian authors, and this could explain why they have been largely disregarded by recent scholars. However, our chapter has shown that one would be ill-advised to skip them by placing a kind of dividing line between classical antiquity and the dawn of modern times.

The Iberian Peninsula was the gateway for new crops from the east and the west, due to the colonial empires of the two peninsular crowns. Consequently, the Spanish and Portuguese were sometimes the first to try growing new plants on European soil. Thus, interest in the study of soil and the importance of experience were associated with investigations being carried out by thousands of anonymous protagonists who aimed to cultivate these new seeds. In fact, the necessity to adapt seeds to different types of soil was quickly recognized by many of the scholars and farmers who worked in the laboratories that were the Iberian fields. When these new crops were at the heart of intense discussions, and still unknown in many European regions, Iberian authors contributed to improving innovative knowledge about them and the soils in which they grew. The Iberian books discussing agrarian issues published during the sixteenth, seventeenth, and early eighteenth centuries reveal another dimension of the transition to modern science that was taking place across Europe.

Acknowledgments

This research has been carried out in the framework of the project ReSEED—Rescuing Seeds' Heritage: Engaging in a New Framework of Agriculture and Innovation since the 18th Century. It has received funding from the European Research Council (ERC) under the European Union's Horizon 2020 research and innovation program (GA no. 760090, https://doi.org/10.3030/760090), and, with Dulce Freire as Principal Investigator, is hosted by the University of Coimbra at the Centre for Interdisciplinary Studies (UIDB/00460/2025). This paper reflects only the authors' views. The European Commission and European Research Council Executive Agency are not responsible for any use that may be made of the information it contains. This chapter has benefited greatly from the discussions held within the ReSEED project. In particular, discussions with Carlos Manuel Faísca, Inês Gomes, Leonardo Aboim Pires, Caroline Delmazo, and Anabela Ramos.

CHAPTER 7

Agricultural Manuals and the Economic Taxonomy of Soils in Early Modern Poland

The Price of Soil Knowledge

Monika Kozłowska-Szyc

1 Introduction

The society of preindustrial Poland was agrarian by nature, the economy was based primarily on grain, the shortage of which often translated into various types of socioeconomic crises, such as famine, overpricing, and subsistence crises. As agricultural production largely depended on natural conditions, such as climate, soil, hydrological features, and topography, those living in the early modern period were much more dependent on the surrounding natural environment for their livelihoods than current generations. They were astute observers of their natural surroundings, not only adapting to them but also influencing them, for instance, by draining and developing marshes, creating new farmlands, and improving the existing soil conditions. It is clear that such interventions must have been based on knowledge, often passed down from generation to generation, but the nature of this knowledge and its transmission have not been examined sufficiently in the Polish case.

This chapter argues that landowning elites of the Polish-Lithuanian Commonwealth used cutting-edge knowledge about soil and novel taxonomic methods to economically improve agricultural productivity. Surveying agricultural manuals, I discuss soil terminology and taxonomy and methods used to test soil fertility as well as recommendations of how to improve its quality. Moreover, I examine how the authors debated soil knowledge across the pages of these agrarian manuals. One outcome of these debates was the emergence within the Polish-Lithuanian Commonwealth of a new soil science that the gentry could apply to their economic advantage. As the inventories and quantitative data from the ducal estates in Mazovia attest,[1] manor administrators used this knowledge to achieve higher crop yields. This chapter traces the

1 A region in central and northeastern Poland situated around the middle course of the Vistula and its basin. During the sixteenth and seventeenth centuries, Mazovia was known for its significant exports of grain and agricultural products.

DOI:10.1163/9789004748484_009

transfer of soil knowledge from ancient and medieval treatises to early modern Polish ones, and the adaptation of external innovations to local needs.

Today, one of the basic agricultural sciences is pedology, that is, the study of soils, their formation and structure; their physical, chemical, and biological properties; soil systematics; and potential applications. Although pedology as a science did not develop until the nineteenth century, people had been aware of the variety of soils and their range of fertility and infertility since ancient times. The first soil classifications in the Polish-Lithuanian Commonwealth appeared in the second half of the sixteenth century, when a four-level scale of land distribution was introduced during the state land reform (the famous Volok Reform).[2] For tax purposes, the soil was divided into four categories: good, medium, poor, and extremely poor. The new owners of the farm granges (Polish: *folwark*) established during this period needed simple technical and organizational guidelines. This created a demand for specialist literature on how to run exemplary farms, on animal husbandry, horticulture, and the exploitation of meadows and forests. Agrarian reformers published books that focused on the utilitarian aspects of nature, giving advice on topics such as the organization of work, cultivation techniques and systems, fertilization, the influence of weather on crops, the timing of fieldwork, and soil requirements of crops, to name just a few.[3]

The most notable works of this type include a Polish translation of Pietro de' Crescenzi's *Ruralia commoda* (ca. 1304–1309), first published in 1549, and several original manuals by Polish authors (see Table 7.1). The authors were members of the wealthy nobility and owners of large landed estates. They were responsible for organizing work on the farm, keeping order, taking care of the profitable sale of agricultural products, and increasing their income. The manuals they wrote were of an empirical nature, based on their own experience, practice, and observation of nature. The authors showed a respect for practical local knowledge in matters of plowing, sowing, and harvesting. For instance, the authors (especially Anzelm Gostomski and Jakub Kazimierz Haur) advised their readers to look directly at the work of peasants, real farmers working on the land, and to benefit from their experience.[4] On the other hand, farm books of the

2 Krzysztof Łożyński, "Charakterystyka gleb w powiecie grodzieńskim w XVI wieku," in *Szkice z dziejów społeczno-gospodarczych Podlasia i Grodzieńszczyzny od XV do XVI wieku*, ed. Józef Śliwiński (Olsztyn: Wydawnictwo Uniwersytetu Warmińsko-Mazurskiego, 2005), 61–69, on 64.

3 Władysław Ochmański, *Wiedza rolnicza w Polsce od XVI do połowy XVIII wieku* (Wrocław: Zakład Narodowy im. Ossolińskich, 1965), 21–31.

4 Anzelm Gostomski, *Gospodarstwo*, ed. Stefan Inglot (Wrocław: Zakład Narodowy im. Ossolińskich, 1951), 55; Jakub K. Haur, *Skład abo skarbiec znakomity sekretów oekonomiej ziemianskiej* (Kraków: Mikołaj Aleksander Schedel printing works, 1693), 12, 22.

Old Polish period (between the sixteenth and eighteenth centuries) relied on the agronomic literature of Roman antiquity.[5] However, it should be noted that this was not unique: many European agricultural manuals referenced classical Greek and Latin works, adapting the advice to suit local conditions.[6] It should also be noted that the writers looked at nature from a practical and utilitarian view point: their prose was devoid of artistic sensitivity and intellectual reflection. These manuals were didactic works; their main purpose was to provide basic knowledge about plant cultivation and breeding and to improve the agricultural competencies of landowners and administrators.[7]

TABLE 7.1 A selection of extant agricultural manuals from the sixteenth and seventeenth centuries referring to the Kingdom of Poland and the Polish-Lithuanian Commonwealth

Author	Title	Lifetime	First edition	Geographical background of the author's activities	Comments
Andrzej Trzecieski or Andrzej Glaber	*Księgi o gospodarstwie y o opatrzeniu rozmnożenia rozlicznych pożytków, każdemu stanowi potrzebne* [Books on farming and on the propagation of the various benefits each social class needs]	ca. 1530–1584 or 1500–1555	1549	–	Polish translation of Pietro de' Crescenzi's *Ruralia commoda* (ca. 1304–1309), the first agricultural book published in the Polish language

5 Polish authors mainly used the book by Pietro de' Crescenzi, who was very familiar with ancient works.

6 See Justin Niermeier-Dohoney and Aleksandar Shopov's introduction to this volume.

7 Tadeusz Bieńkowski, *Wiedza przyrodnicza w Polsce wieku XVI* (Wrocław: Zakład Narodowy im. Ossolińskich, 1985), 112–119.

TABLE 7.1 A selection of extant agricultural manuals (*cont.*)

Author	Title	Lifetime	First edition	Geographical background of the author's activities	Comments
Anzelm Gostomski	*Gospodarstwo* [The farm]	1508–1588	1588	Mazovia	The first original Polish agricultural manual
Martin Grosser	*Kurtze und gar einfeltige Anleytung zu der Landwirthschaft* [Short and simple guide to agriculture]	Second half of the sixteenth century	1590	Lower Silesia	Written in German; the only manual on peasant farming
Teodor Zawadzki	*Memoriale oeconomicum* [Memorial to farming]	d. 1637	1608	Lesser Poland	Agricultural calendar
Jan Herman	*Liefländischer Landman* [Livonian landowner]	d. 1670	1662	Livonia	Written in German; translated into Polish in 1671; popular in the Grand Duchy of Lithuania
Jakub Kazimierz Haur	*Oekonomika ziemiańska generalna* [General landowning economics]	1632–1709	1675	Lesser Poland	The most popular agrarian book in Poland, with ten editions from the seventeenth to nineteenth centuries

TABLE 7.1 A selection of extant agricultural manuals (*cont.*)

Author	Title	Lifetime	First edition	Geographical background of the author's activities	Comments
	Ziemiańska generalna oekonomika [General landowning economics]		1679		The second edition of *Oekonomika ziemiańska generalna*, expanded by the author
	Skład abo skarbiec znakomitych sekretów oekonomiej ziemiańskiej [Compendium or treasury of excellent secrets of land economy]		1693		Popular compendium abundant in all sorts of content, full of anecdotes and superstitions

The popularity of these works, especially the economic manuals, is evidenced by the numerous reprints and reissues (see table above), and their inclusion into library collections of the nobility.[8] Moreover, they were repeatedly cited in almanacs from the turn of the seventeenth and eighteenth centuries, and often were recorded in family chronicles (the so-called *silva rerum*) in the form of extensive excerpts.[9]

8 Tadeusz Witczak et al., *Bibliografia literatury polskiej: Nowy Korbut*, vol. 2, *Piśmiennictwo staropolskie: hasła osobowe A–M*, ed. Roman Pollak (Warsaw: Państwowy Instytut Wydawniczy, 1964), 249.

9 Joanna Partyka, *Rękopisy dworu szlacheckiego doby staropolskiej* (Warsaw: Wydawnictwo Naukowe Semper, 1995), 55, 59; Piotr Kowalski, *Theatrum świata wszystkiego i poćciwy gospodarz: O wizji świata pewnego siedemnastowiecznego pisarza ziemiańskiego* (Kraków: Wydawnictwo Uniwersytetu Jagiellońskiego, 2000), 65–66.

2 Crescenzi's Work and Its Polish Translation

Before the middle of the sixteenth century, no agricultural book had been published in the Polish-Lithuanian Commonwealth, nor had there been any Polish translations of agricultural works by ancient authors. The 1549 translation of Pietro de' Crescenzi's *Ruralia commoda* was the first agricultural publication and for thirty-nine years remained the only complete compendium of farm knowledge available in the Polish language.[10] It was not until 1588 that the first original Polish work covering the entire discipline of agriculture was published by Anzelm Gostomski.

The translation of Crescenzi's book was a milestone in the development of agricultural knowledge in the Polish-Lithuanian Commonwealth—above all, because it introduced the terminology of cultivation, botany, and soil science. This was undoubtedly thanks to the translator, who was familiar with farming and ecological issues in the federative union. However, their identity remains uncertain, with two hypotheses in the literature suggesting either Andrzej Trzecieski or Andrzej Glaber of Kobylin.[11] While both options are plausible, finding irrefutable evidence to favor one over the other has proved challenging and further research is required. Meanwhile, attributing the translation to Trzecieski is a practice so well established and common in Polish historical literature that this information is even reproduced in encyclopedias and dictionaries.[12]

The direct antecedent to the Polish edition may not necessarily be Crescenzi's Latin version. Modifications to the text, which consider the climatic conditions of Central Europe, suggest German-language editions of Crescenzi as the most probable source texts, though it is possible that adaptations to local conditions occurred in the Latin versions as well. Just as the identity of the translator remains uncertain, so too does the source of the translation. A comparison of

10 The second Polish edition of *Ruralia commoda* was published in 1571. It should be noted that the differences between the first and second editions are minor—the structure and names of the chapters and their content, and even the engravings and their placement are largely the same. The second edition was a reprint of the first, produced by a different printing house. The printer of the first edition was Helena Unglerowa's publishing house in Kraków, while the second edition was printed by Stanisław Szarfenberger's publishing house, also in Kraków.

11 Alodia Kawecka-Gryczowa, ed., *Drukarze dawnej Polski od XV do XVIII wieku*, vol. 1, *Małopolska*, part 1, *Wiek XV–XVI* (Wrocław: Zakład Narodowy im. Ossolińskich, 1983), 314.

12 Recent linguistic research indicates that numerous northern Polish features are present in the text, which may indicate the area from which the author came. Andrzej Trzecieski was born in Lesser Poland (south of the country), while Andrzej Glaber was born in Greater Poland (north of the country).

the tables of contents of various editions of *Ruralia commoda* reveals numerous similarities between the Polish and German ones; in particular, the 1493 Polish edition shows striking affinities to its German predecessors.[13]

Furthermore, the Polish edition was not a literal translation of *Ruralia commoda*. The author departed significantly from the original in some places and altered the content of the work on several occasions. Especially in the chapters describing plants, he ordered the information thematically rather than alphabetically. Moreover, he did not include every description of the plants and trees included in the original. He did, however, attempt to add to the existing descriptions, often incorporating knowledge of the local ecology and highlighting differences from the original (such as the inclusion of buckwheat, which was unknown in Italy in Crescenzi's day). While this indicates a liberal approach to the original, it also suggests that the translator made the changes he felt necessary to render the book comprehensible and readable for the audience.

The translation of *Ruralia commoda* largely reflects the prevailing environmental conditions in the Polish-Lithuanian Commonwealth with respect to the descriptions of terrain, climate, soil properties, and soil moisture. Most of the information on soils is contained in the second and third sections. In general, the soil could be suitable for sowing (*osiewna*) or for fallowing (*odłogowa*), fertile or infertile, wet or dry, warm or cold, loose or dense, heavy or meager.[14] Crescenzi characterized fertile soil as black, plump, subtle, and greasy. Not overly wet or muddy, it was found covered with grasses and other kinds of vegetation. In contrast, he described barren soils as dry and hard or meager and cold.[15] He also noted that the fertility of the soil was influenced by its location, suggesting that mountainous and hilly fields were less productive because they had "less substance" compared to lowland fields.[16]

Since the nature of the soil and its fertility were not invariable, Crescenzi introduced methods for testing soil quality, suggesting two main ways to check whether the soil was fertile. The first was to test its greasiness by throwing a lump of soil into water; if it disintegrated rapidly, the soil was poor or infertile. The second was to examine the salt content by dropping a lump of soil into clean—preferably river—water, waiting for it to settle at the bottom,

13 See also Anna Pankowicz, "Najstarszy polski druk z zakresu gospodarstwa wiejskiego (Piotr Krescentyn, Księgi o gospodarstwie … Kraków 1549 r.)," *Zeszyty Naukowe Akademii Rolniczej w Krakowie: Historia Rolnictwa* 150, no. 4 (1979): 47–63, on 55.

14 Pietro de' Crescenzi, *Księgi o gospodarstwie y o opatrzeniu rozmnożenia rozlicznych pożytkow, każdemu stanowi potrzebne* (Kraków: Helena Unglerowa, 1549), 88–92.

15 Crescenzi, 91–92, 138–140.

16 Crescenzi, 97–98, 141.

and subsequently tasting the water.[17] Both methods had already appeared in Roman literature (Virgil, Palladius).

Crescenzi also explored methods of correcting and improving soils with low yields, using techniques such as fertilization, land reclamation, accurate plowing, and fallowing. The timing and intensity of some agricultural work (such as plowing and harrowing), he argued, depended on the quality of the soils.[18] In order to improve soil fertility in humid areas, he recommended blending cold and wet soils with warm and dry ones. Waterlogged ground or wetlands, on the other hand, should be drained through measures such as digging channels and ditches. These were also necessary in case of flooding caused by, for instance, heavy rain or spring snowmelt.[19] This focus on drainage is one of the changes made to adapt Crescenzi's text to local conditions, as one of the problems of agriculture in Poland at that time was the excessive moisture of the land, which led to low crop yields.[20]

Crescenzi's work had a major impact on the development of agricultural knowledge in the Polish-Lithuanian Commonwealth by establishing terms for soil science in the Polish language and familiarizing readers with the opinions of Roman agronomists. For the most part, Polish agricultural writers of the next 250 years reproduced the information contained therein.[21] However, they introduced a number of innovations worthy of mention.

3 Polish Authors' Discourse on Soil

Authors of Polish agricultural manuals were acutely aware of the impact of natural conditions on the efficiency of agricultural production, both environmental (i.e., weather and climate) and the physical condition of the soil and its properties. Consequently, they recommend that the type of soil determine the type of plants to be cultivated as well as the cultivation techniques, including the frequency of fertilization and the appropriate time for the execution of

17 Crescenzi, 139.

18 Crescenzi, 104–105.

19 Crescenzi, 94–95, 97.

20 Tomasz Związek et al., "On the Economic Impact of Droughts in Central Europe: The Decade from 1531 to 1540 from the Polish Perspective," *Climate of the Past* 7, no. 18 (2022): 1556–1557.

21 Crescenzi's book was also utilized in other European countries, including Italy, France, and England, up until the mid-nineteenth century. For further details, see Mauro Ambrosoli, *The Wild and the Sown: Botany and Agriculture in Western Europe, 1350–1850* (Cambridge: Cambridge University Press, 1996), 41–95.

various agricultural tasks. Polish agricultural manuals of the time were created primarily for owners and administrators of landed estates. These works not only offered advice and guidance but also illustrated the benefits of proper cultivation of the land. They proposed that every owner and administrator of a farm grange interested in cultivating the land effectively should understand its properties.[22]

As can be seen in Table 7.1, the first original Polish agricultural manual was Gostomski's 1588 *Gospodarstwo* (The farm). Its contents focused on the broader economic impacts of agriculture because the author was an exemplary administrator rather than a farming specialist. Shortly after, in 1590, Martin Grosser, a Protestant pastor from the village of Szewce, published the *Kurtze und gar einfeltige Anleytung zu der Landwirthschaft* (Short and simple guide to agriculture), which was based on his own farming practice. Another manual, *Memoriale oeconomicum* (Memorial to farming) by Teodor Zawadzki, a lawyer and skilled writer, was published in 1616. Structured in the form of a calendar, this book was a compilation of other works rather than an original creation and clearly shows that he was inspired by Gostomski. These three agricultural manuals, published in the late sixteenth and early seventeenth centuries, are similar in many regards. Above all, they were utilitarian texts written in everyday language; the authors were not interested in soils per se, and information on this subject was included only when discussing other topics (e.g., the cultivation of particular crops, fertilization, or plowing).

It is only in literature from the second half of the seventeenth century that more attention is given to soils in particular: both Herman's *Lieffländischer Landman* (1662) and the works of Haur (Table 7.1) contain separate chapters on soils, their taxonomy, and soil fertility. Herman and Haur were experienced in administering their own estates, well educated, and familiar with foreign agricultural literature. In Haur's manuals, the influence of Crescenzi and German *Hausväterliteratur* is noticeable. He based the construction and thematic scope of his first work, *Oekonomika ziemiańska generalna* (General landowning economics, 1675), on the aforementioned Western European models. However, he took a step further in the subsequent edition (1679) by completely rewriting paragraphs from the Polish translation of *Ruralia commoda*.[23]

22 John R. McNeill and Verena Winiwarter, "Breaking the Sod: Humankind, History and Soil," *Science* 304, no. 5677 (2004): 1627–1629, on 1628; Antoni Podraza, *Jakub Kazimierz Haur pisarz rolniczy z XVII wieku* (Wrocław: Zakład Narodowy im. Ossolińskich, 1961), 122.

23 The excerpts rewritten by Haur should not be taken as a new translation of Crescenzi's original text.

Terminology provides the best evidence of the continuity between older forms of soil knowledge and early modern Polish innovations. The most common adjectives and phrases used to describe soils are presented in Table 7.2 (Polish terms are given in parentheses). They have been divided into several categories, some of which correspond to contemporary guidelines on soil classification (grain size, density and structure, color, humidity), while others are typical of preindustrial times (fertility, temperature, cultivated state of the soil). The terminology was very broad, lacking only in terms that would relate to the shape and size of the field. The words in Table 7.2 marked the first step toward a taxonomy of soils.

The first and basic criterion of the soil taxonomy was an assessment of its quality. The authors used numerous adjectives when dividing the soil into three main categories: good (fertile, greasy, choice, fecund, lush, plentiful, fruitful, etc.), bad (barren, infertile, hungry), and intermediate (meager, mediocre, poorish). In manuals by Gostomski (1588) and Zawadzki (1616), such terms rarely appear—the soil was simply good and greasy or, depending on the degree of moisture, wet or dry.[24] Many more terms are found in the work of Grosser (1590). He described healthy soil as greasy, choice, and plentiful.[25] However, it was authors in the second half of the seventeenth century (Jan Herman and J. K. Haur) who greatly expanded the terminology, using adjectives such as best, fertile, fruitful, plentiful, meager, mediocre, barren, and infertile. They also introduced intermediate categories such as "first mediocre" and "second mediocre."

Another feature of soils described in the agricultural literature of the time was their color. According to the agricultural manuals of Grosser, Herman, and Haur, black and dark gray were the colors of the most fertile soil, as opposed to white and pale. This distinction comes as no surprise, as it was humus that darkened the soil from various shades of gray to black. Herman was one of the first writers to start paying attention to soils with yellow and red coloration, which he considered intermediate.[26] It is also important to note that Polish agricultural manuals used a greater number of adjectives to describe the color of the soil than ancient literature or Crescenzi.[27] This suggests not only the

24 Gostomski, *Gospodarstwo*, 21; Teodor Zawadzki, *Theodora Zawadzkiego Memoriale oeconomicum*, ed. Józef Rostafiński (Kraków: drukarnia C. K. Uniwersytetu Jagiellońskiego, 1891), 33.

25 Grosser, *Krótkie wprowadzenie*, 242, 248, 260.

26 Jan Herman, *Jana Hermana z Neydenburku Ziemianin albo Gospodarz inflandski* (Minsk: Drukarnia Nowa J. Stefanowicza, 1823), 60.

27 Verena Winiwarter, "Prolegomena to a History of Soil Knowledge," in *Soils and Societies: Perspectives from Environmental History*, ed. John R. McNeill and Verena Winiwarter

TABLE 7.2 Soil classification and terminology used by agronomic authors

Categories	Terminology
Fertility and quality	good (*dobra, szczególnie dobra*), fertile (*urodzajna, żyzna*), the best (*najlepsza*), choice (*wyborowa*), lush (*bujna*), greasy (*tłusta, bardzo tłusta, omasna*), fecund (*płodna*), rich (*bogata*), plentiful (*obfita*), meager (*chuda*), mediocre (*mierna, słaba*), poorish (*licha*), barren (*jałowa, bezpłodna*), bad (*zła, bardzo zła*), infertile (*nieurodzajna*)
Soil composition and grain size	clayey (*gliniasta*), loamy (*ilasta, iłowata*), sandy (*piaszczysta*), stony (*kamieniasta*), rocky (*skalista*), slimy (*mulista*), salty (*słona*), rendzina-like (*rędzianna*), calcareous (*wapnista, wapienna*)
Density and structure	stout (*tęga*), hard (*twarda, zatwardziała*), dense (*gęsta*), thick (*gruba*), heavy (*ciężka*), mellow/soft (*miękka*), crumbly (*krucha*), sticky (*lepka, klejka*), plump (*pulchna*), fine (*miałka*), loose (*rzadka*), light (*lekka*), furfuraceous (*otrębiasta*)
Color	black (*czarna*), dark (*ciemna*), swarthy (*śniada*), gray (*szara, siwa*), griseous (*szarawa*), dark red/red-brown (*ciemnoczerwona*), red (*czerwona*), yellow (*żółta*), white (*biała*)
Humidity	wet (*mokra*), waterlogged (*zbyt mokra*), damp (*wilgotna, bardzo wilgotna*), of dry nature (*suchego przyrodzenia*), dry (*sucha*), burned by the sun (*spalona, oparzysta*), drizzly wet (*mokra od dżdżów*), muddy (*błotna*), marshy (*bagnista*)
Temperature and climate	cold (*zimna*), warm (*ciepła, ciepłego przyrodzenia*), free from frost (*wolna od mrozu*), frozen (*zmarzła*), open / not cold and not covered with snow (*otwarta*)
Location	even/level (*równa*), flat (*płaska*), sloping (*pochyła*), low (*niska*), of low location (*nisko położona*), high (*wysoka*), of high location (*wysoko położona*), upper (*górna*), mountainous (*górzysta*), hilly (*pagórkowata*)
Cultivated state of the soil	cultivated (*uprawna*), for sowing (*siewna*), grassy (*trawiasta*), full of couch grass (*pełna perzu*), of yore (*dawniejsza*), fallow (*ugorowa*), communally worked on (*tłoka*), new (*nowa, nowizna*), burned out (*wypalona*), manured (*zgnojona, nagnojona, ugnojona*), rotten (*przegniła*), clean (*czysta*), raw (*surowa*), horticultural (*ogrodnicza*)

SOURCE: THE AUTHOR'S OWN STUDY OF POLISH AGRICULTURAL MANUALS

development and evolution of conceptual knowledge but also the presence of a wider range of soil colors in Poland. Furthermore, this diversity enabled landowners to define soils in more detailed ways, enhancing their agricultural productivity.

An important criterion that facilitated cultivation was the classification of soils according to their structure and composition. Grosser (as well as Crescenzi) thought that heavy and clayish soils were preferable to light, sandy, and stony ones.[28] Authors from the second half of the seventeenth century held similar opinions; however, they pointed out that the soil also had to be easy to cultivate. The soil's compactness and the force with which it resisted mechanical pressure (e.g., cutting by a plow or a colter) depended on its structure and physical composition. Herman considered plump, soft, not very sticky, and not too stout soils the best, while hard and dense soils, but also too light, very loose, and furfuraceous (i.e., flaky) soils, were regarded as poorer.[29] Haur remarked that rendzina-like soils and those with an admixture of clay or loam were supposed to be better than stony, rocky, and sandy types. However, in his view, moderately greasy and plump soil was the best, as it not only gave a good yield but was also easier to cultivate.[30] According to these two authors, the ideal soil was neither too stout and hard nor too greasy.

The moisture content of soil was an important issue. Moderately moist soils were preferred, while those that were dry, muddy, or too wet were considered bad. Moreover, in agronomic literature (e.g., Grosser and Haur), more attention was paid to an excess of moisture than to dryness.[31] The authors must have noticed that excess moisture reduced the quality of the soil. Another factor to consider was the soil's temperature, which, like air temperature, had a significant impact on vegetation. Yet expressions describing the temperature can be understood in two ways: in some cases, they refer directly to atmospheric conditions (free from frost, frozen, clear); in others, especially in the books by Haur, the terms are drawn from classical authors and refer to the temperament of the soil (warm, cold). Haur's concept was rooted in humoral theory, a classical Hippocratic medical concept developed further by Galen,

(Cambridge: The White Horse Press, 2006), 203–206; Kenneth D. White, *Roman Farming (Aspects of Greek and Roman Life)* (London: Thames and Hudson, 1970).

28 Martin Grosser, *Krótkie wprowadzenie do gospodarstwa wiejskiego*, ed. Stefan Inglot, trans. Jan Piprek (Wrocław: Zakład Narodowy im. Ossolińskich, 1954), 240, 243, 245.

29 Herman, *Ziemianin albo Gospodarz*, 60, 63–64.

30 Jakub K. Haur, *Oekonomika ziemiańska generalna* (Kraków: K. Schedel printing works, 1675), 3–4; Jakub K. Haur, *Ziemiańska generalna oekonomika* (Kraków: Typis Vniversitatis, 1679), 20.

31 Grosser, *Krótkie wprowadzenie*, 242–243, 246–227; Haur, *Ziemiańska generalna*, 33.

wherein temperaments such as coldness and dryness influence an individual's complexion and overall health. Over time, physicians applied humoral theory to other physical substances, including elements such as earth. Haur sought to maintain a balance among four elements, soil, water, air, and manure, which corresponded to the classical elements of earth, water, air, and fire. An imbalance among these four elements was perceived as a cause of insufficient soil fertility, so manuring, for instance, was recommended to restore balance. Haur's use of humoral terminology suggests that he approached soil treatment akin to the way a physician would treat a patient.

Yet another element influencing the efficiency of agricultural production was the location and topographical relief of the land. Some plots were in low-lying areas, others in highlands or mountainous regions. The land to be cultivated might be level, inclined, or have steep slopes. Haur, whose estates were located in the foothills of more mountainous areas, was more likely to pay attention to these conditions. He appreciated land that was low lying and even with little or no slope. He considered upland fields comparatively less fertile, because their soil was cold, meager, arid, making them only suitable for pastureland. Roman authors and Crescenzi expressed similar opinions in their treatises. In sharp contrast, Gostomski and Herman, who farmed in low-lying areas, gave these issues scant attention.

It is worth noting that terminology related to soil fertility was consistent across Afro-Eurasia. Terms like "greasy" to describe fertile soils appeared in many vernacular languages, as did classifications of soils based on physical composition, color, and temperature, or assessments through touch, taste, and smell. These similarities can be attributed, on the one hand, to the influence of Greco-Roman writings, and on the other hand, to independent observations and agricultural practices.[32]

When describing soils, all of the authors used a variety of adjectives related to the condition of the crops, referring mainly to changes caused by human activity (fertilization, plowing), or related to the abandonment of cultivation activities. These terms, their descriptions, and valuation were a prelude to the creation of soil classifications. They first appeared in Polish agricultural manuals in the middle of the seventeenth century. The first such taxonomy can be found in the work by Jan Herman, who created an extensive classification, dividing all soils into six groups: the first two were considered good, the middle two average, and the last two the poorest. His classification is based primarily

32 See Justin Niermeier-Dohoney and Aleksandar Shopov's introduction to this volume, as well as the contributions by Alberto González Remuiñán and Dulce Freire, and Aleksandar Shopov and Himmet Taşkömür.

on variations of physical composition and grain size, color, temperature, and the soils' condition after fertilization. The two best soils (black, or dark gray, both warm, with fine sand on a clay subsoil) showed long term benefits from the use of fertilizers. Moreover, they were resistant to adverse weather conditions (mainly precipitation). Mediocre land (gray soil with white clay or gray sand, subsoil of yellow clay) could turn into good land with careful cultivation and sufficient fertilization. Herman called the worst soils (red sand or red clay alone) "the doom of all farmers." This author contributed significantly to the development of soil knowledge, as he was the first to draw special attention not only to the topsoil but also to the subsoil and its importance in terms of sustainability of fertilization and absorption of excessive water.[33]

Agricultural manuals also included attempts to define what makes soil good or suitable. To this end, Haur developed some soil testing methods based mainly on the senses of sight, smell, touch, and taste. Much of his methodology was drawn from the treatises of Virgil, Columella, Palladius, and Crescenzi.[34] Suitable soil had a pleasant smell (especially when it rained), a sweet taste, and stuck easily to the hand. Another testing method was digging a hole and then filling it with the same soil.[35] If it was overly full, the soil was greasy and plentiful.[36] Greasy soil, according to Haur, was also indicated by the presence of earth worms.[37] To determine the soil's fertility, he recommended observing the plants: a variety of flowers, herbs, and grasses, as well as smooth and straight branches, testified to fertile soil, while curvaceous and knotty ones indicated barren soil. The transcription of these methods in eighteenth-century agricultural calendars and their appearance in the first scientific farming books of the late eighteenth and early nineteenth centuries attests to their enduring popularity.[38]

---

33 Herman, *Ziemianin albo Gospodarz*, 59–60.

34 Verena Winiwarter, "Soil Scientists in Ancient Rome," in *Footprints in the Soil: People and Ideas in Soil History*, ed. Benno P. Warkentin (Amsterdam: Elsevier Science, 2006), 11–12.

35 The soil was also tested by holding a lump of earth in the hand until warm and then tasting it in the mouth. If it stung, the soil was good. This method, which also appears in Roman literature (Virgil, Palladius), served to check the content of salt in the soil (which likely refers to the content of nitrates). On taste as a method for examining soils, see Nicolas Roth's chapter, this volume; on nitrates, see Justin Niermeier-Dohoney's chapter, this volume.

36 This method was also taken from Roman literature (Virgil; Columella).

37 Haur, *Ziemiańska generalna*, 21.

38 *Kalendarz dla Królestwa Polskiego na Rok Pański 1787* (Kraków: Drukarnia Akademicka, 1786); *Kalendarz Polski y Ruski Na Rok Pański 1751 [...] Stanisława z Łazów Dunczewskiego* (Zamość: Akademia Zamojska, 1750); Michał Oczapowski, *Gospodarstwo wiejskie obejmujące w sobie wszystkie gałęzie przemysłu rolniczego teoryczno-praktycznie wyłożone*, vol. 1,

Knowledge about the quality and fertility of the soil, its composition, and structure had a significant influence on the selection of crops and their further cultivation. It was precisely this type of knowledge that elite landowners craved, because it would allow them to make their estates more economically productive. Agronomic authors explained that there were soils suitable for a particular plant or, more generally, for specific growing circumstances. Early modern manuals stated that the best soils should be used for the cultivation of wheat. According to Grosser, a wheat field should be greasy and well fertilized, while in the opinion of Herman and Haur, it had to be slightly greasy with an admixture of clay. Similar soils were also recommended for the sowing of barley.[39] Rye was much less demanding and, according to these authors, was successful on all types of soil, but it produced the best yield on light and sandy soils. Furthermore, Grosser argued that rye also grew on stony and barren lands, provided that they were well fertilized.[40] In turn, as Grosser indicated, oats could be sown on any field, but the better the soil, the greater the yield. Herman recommended sowing oats on soils that were not completely barren and somewhat manured. Similarly, Haur wrote that fields that had been fertilized a few years earlier produced the best oat harvests.[41] Agronomic writers, however, were not unanimous in defining the soil requirements of peas. While Grosser recommended greasy, black, and clayey soil, manured the previous year, Herman preferred grayish soil combined with small stones, and Haur suggested soil that was mellow, plump, and moderately fertilized.[42] On the other hand, millet required light and moist soil, as did buckwheat, which was cultivated primarily on sandy and barren soil.[43] Though authors tended to agree on certain recommendations, these differences of opinion show that additional experiments and empirical studies remained necessary before disputes could be effectively addressed and landowners offered dependable insights to enhance their yields.

Agronomia czyli nauka o gruntach, ich własnościach, wpływie na nie zewnętrznych okoliczności, oraz o klassyfikacyi ich ekonomicznej, teoryczno-praktycznie wyłożona (Warsaw: S. H. Merzbach, 1835), 140–145; Stanisław Moskal, "Michał Oczapowski jako gleboznawca i chemik rolny," *Postępy Nauk Rolniczych*, no. 1 (1989): 27–36, on 32.

39 Grosser, *Krótkie wprowadzenie*, 239, 251, 259; Herman, *Ziemianin albo Gospodarz*, 56–60; Haur, *Ziemiańska generalna*, 33, 38.

40 Grosser, *Krótkie wprowadzenie*, 243.

41 Haur, *Skład abo skarbiec*, 11; Grosser, *Krótkie wprowadzenie*, 246; Herman, *Ziemianin albo Gospodarz*, 14.

42 Grosser, *Krótkie wprowadzenie*, 256; Herman, *Ziemianin albo Gospodarz*, 12; Haur, *Skład abo skarbiec*, 37.

43 The Polish translator of Crescenzi's work was the only one to recommended sowing buckwheat on fat, plump, and clayey soils. Crescenzi, *Księgi*, 179.

Agricultural manuals also advised a specific relationship between the sowing density of individual cereals and the type of soil. Today, farmers typically follow the principle of a lower sowing rate on more fertile soils and a higher one on barren land. However, early modern Polish agricultural manuals opposed this practice—they recommended sowing more densely on good soils, and less frequently on medium- and poor-quality soils.[44] The authors of the time were also aware of the relationship between the quality of the soil and the date of sowing. Grosser and Zawadzki advised that the sowing of winter cereals should start earlier on light, sandy, and semibarren soil.[45] Herman expressed a different opinion: he believed that, in warm and light fields, sowing should take place later, while cold, heavy, and clayey fields should be cultivated earlier. He argued that, on heavy soils, cereals needed more time to develop a proper root system.[46]

Another feature of these manuals is the inclusion of plowing recommendations. Triple (sometimes even quadruple) plowing was recommended on stout, heavy, and weedy soils; for lighter ones, it was enough to plow twice. Moreover, authors of agrarian manuals drew attention to different plowing techniques depending on the quality of the soil—the furrows of lush and greasy fields should be plowed more deeply than fields with poor, meager soil.[47]

One of the problems commonly faced was the long-term loss of nutrients in the soil. Where the soil was fertile and rich, this problem could be ignored for some time, but otherwise, it required swift action. Until the nineteenth century, fertilization was the main solution to this problem. No author in Poland attempted to challenge the validity and value of fertilization; on the contrary, the authors of agricultural manuals treated fertilization as something obvious and necessary. The author of the first original Polish manual, Gostomski, stressed that the cowshed was the first and most important foundation of the farm.[48] In Poland, the most important and easily obtained source of chemical components needed to fertilize the soil was animal manure. Night soil does not appear in any Polish manuals from the sixteenth and seventeenth centuries. The most common and frequently mentioned type of manure in agronomic literature (Gostomski, Zawadzki, Herman, Haur) was that from horned cattle (cows and steers). Other examples include pig manure (Gostomski, Zawadzki, Haur), stable manure (Haur), sheep manure (Zawadzki, Haur), and manure

44 Grosser, *Krótkie wprowadzenie*, 241–242.

45 Zawadzki, *Memoriale oeconomicum*, 51.

46 Herman, *Ziemianin albo Gospodarz*, 63–64.

47 Grosser, *Krótkie wprowadzenie*, 240–241; Haur, *Skład abo skarbiec*, 23.

48 Gostomski, *Gospodarstwo*, 19.

from birds (Haur).[49] Of all the agronomic works analyzed in this article, only Haur's manuals contain attempts to evaluate the different types of fertilizers (using humoral terminology). He considered pigeon and hen droppings to be the best, but in practice the small amount available made this irrelevant. In contrast, he criticized horse manure, believing that it was temperamentally "hot" and only suitable for fertilizing gardens. He conceded that "cold" fields could be successfully fertilized with it, provided that it was plowed in as soon as possible, but the preferable fertilizers for "cold" fields were sheep, pigeon, and chicken manure.[50] The same was recommended for the cultivation of wheat. In turn, pig manure and cattle manure were to be spread on warm soil, and for sowing rye and barley.[51] Manuals by Gostomski, Zawadzki, and Haur also mention adding old thatch, weeds, straw, leaves, and reeds to the manure. However, they do not provide information on domestic waste, which—as archaeological research shows—was an important component of manure, with plant waste, animal bones, and fragments of clay pots complementing the fertilizer.[52]

Other than animal fertilizers, Haur described ash and soot as useful elements necessary for fertilizing the soil.[53] They could be used primarily to regulate acidity and were to be scattered on the soil[54]—although Herman suggested burning the fields.[55] This was supposed to bring about the same positive effect, according to the opinions of the time. The chemical compounds needed to fertilize the fields were also to be found in pond silt. Gostomski and Haur advised that, in the absence of manure, sandy ground should be fertilized with silt that had dried out through exposure to air.[56] Polish manuals from the sixteenth and seventeenth centuries make no mention of mineral fertilizers such as lime, marl, or salt. The earliest mentions of soil fertilization methods of this

49 Gostomski, 63; Zawadzki, *Memoriale oeconomicum*, 16, 30; Herman, *Ziemianin albo Gospodarz*, 31; Haur, *Skład abo skarbiec*, 29–30.

50 Both Columella and Crescenzi considered horse dung to be the worst, suitable only for fertilizing meadows.

51 Haur, *Ziemiańska generalna*, 18; Haur, *Skład abo skarbiec*, 29–30.

52 Tomasz Związek, *Krajobraz szesnastowiecznej Polski: las—ziemia—woda—ruda darniowa; Powiat kaliski i Wielkopolska w tle* (Warsaw: Instytut Historii PAN, 2022), 83–85.

53 Haur, *Skład abo skarbiec*, 30.

54 Robert S. Shiel, "Nutrient Flows in Pre-modern Agriculture in Europe," in *Soils and Societies: Perspectives from Environmental History*, ed. John R. McNeill and Verena Winiwarter (Cambridge: The White Horse Press, 2006), 216–242, on 226. See also Niermeier-Dohoney's chapter, this volume.

55 Herman, *Ziemianin albo Gospodarz*, 17.

56 Gostomski, *Gospodarstwo*, 84; Haur, *Skład abo skarbiec*, 31.

kind can be found in Polish literature dating back to the eighteenth century.[57] The limited use of lime and marl as soil additives in Poland may be attributed to their restricted availability, as they were primarily found in the southern regions of the country. Such fertilization methods have been known since ancient times, but they were more commonly practiced in other European countries, including Britain, France, and the Netherlands.[58] Nor did Polish authors writing about ways to improve the quality of the soil mention crop rotation with legumes. Grosser and Zawadzki did write about cultivating peas and broad beans in fields,[59] which seems to indicate their use as a pre-crop for winter crops; however, it was only in the second half of the eighteenth century that agricultural works consciously promoted crop rotation with legumes.[60]

How best to fertilize fields was an important issue in agricultural manuals. The early modern Polish agricultural manuals recommended using the quality of the soil to determine the technique, frequency, and duration of fertilization. They included instructions for the quick plowing, or even harrowing, of dung that was to be scattered on the field. This advice could be found in all Polish manuals, including the translation of Crescenzi's work.[61] Moreover, my analysis of agricultural manuals demonstrates that the annual fertilization considered necessary for each field was impossible due to the lack of manure. Therefore, Crescenzi, as well as Gostomski and Grosser, recommended that the largest possible area of the field be fertilized. Grosser also pointed to the triennial (every three years) application of manure in each field.[62] However, only seventeenth-century authors made the frequency of fertilization dependent on the quality of the soils. Herman pointed out that fertile and lush fields could be fertilized up to every eight years, while mediocre and meager ones could be manured every five to six years.[63] In turn, Haur noted that dung on hard soils lasted longer than on soft and light soils, so the latter required more frequent

57 Krzysztof Kluk, *O rolnictwie, zbożach, łąkach, chmielnikach, winnicach i roślinach gospodarskich*, ed. Stefan Inglot (Wrocław: Zakład Narodowy im. Ossolińskich, 1954), 134–136. Krzysztof Kluk's work was one of the earliest modern Polish agronomy manuals. It was first published in Warsaw in 1779.

58 Verena Winiwarter and Winfried E. H. Blum, "From Marl to Rock Powder: On the History of Soil Fertility Management by Rock Materials," *Journal of Plant Nutrition and Soil Science* 171, no. 3 (2008): 316–324.

59 Grosser, *Krótkie wprowadzenie*, 248, 257; Zawadzki, *Memoriale oeconomicum*, 17.

60 Kluk, *O rolnictwie*, 126–127.

61 Although in principle the advice was sound, its implementation was not feasible with the agricultural technology of the time. Furthermore, the granges lacked sufficient labor to carry out such tasks.

62 Gostomski, *Gospodarstwo*, 76; Grosser, *Krótkie wprowadzenie*, 239, 244, 248.

63 Herman, *Ziemianin albo Gospodarz*, 60.

fertilization.[64] For the authors of agrarian manuals, the connection between high yields and fertilization was obvious, although they realized that manure alone was not enough to improve soil quality.

4 The Practical Application of This Knowledge

Landowners and estate administrators made practical use of the recommendations from sixteenth- and seventeenth-century agrarian manuals in their fields. This can be verified by analyzing economic sources such as inventories and revisions of the king's estates from the area of Mazovia. These sources provide numerical data on the farms (the amount of sowing and harvesting of cereals, livestock population, and farm profitability) and describe the soil. First, the data show that administrators of these farms endeavored to adapt crop choices according to the prevailing soil conditions. On farms where the soil was described as fertile and good, we see an increase in the sowing of more demanding crops such as wheat and barley. On farms with sandy and light soil, rye is dominant.[65] Manor farm administrators used this terminology and taxonomy to argue that knowledge of the fertility, location, structure, and composition of the soil was essential to economically improve estates. They considered saturated or muddy soil and fields that were rocky, undulating or steeply sloped to be barren or unsuitable for cultivation. They classified sandy soils as poor or mediocre (depending on the amount of sand in the soil), but suitable for rye and oat cultivation.[66] These classifications mirrored the recommendations of the writers of agrarian manuals, suggesting that administrators read these works carefully and put the soil knowledge they provided to practical use.

The instructions for economic management written by various administrators and owners of large estates (collected in Baranowski et al.'s *Instrukcje gospodarcze dla dóbr magnackich i szlacheckich*) also provide evidence of the practical application of soil knowledge garnered from agrarian manuals. The instructions were of a normative character, outlining certain strategies for landowners. For example, in his instructions relating to the Krzyżanowice

64 Haur, *Skład abo skarbiec*, 31.

65 Zofia Kędzierska, ed., *Lustracje województwa rawskiego XVII wieku* (Wrocław: Zakład Narodowy im. Ossolińskich, 1965), 160–161, 164, 196, 200–202, 208–209.

66 Zofia Kędzierska, ed., *Lustracje województwa rawskiego 1564 i 1570* (Warsaw: PWN, 1959), 129; Irena Gieysztorowa and Anna Żaboklicka, eds., *Lustracja województwa mazowieckiego 1565* (Warsaw: PWN, 1968), 1:22.

estate (*Uwagi H. Kołłątaja dotyczące dóbr Krzyżanowice*), Hugo Kołłątaj stressed the importance of knowing the quality of soil and its physical composition. He argued that estate administrators needed this knowledge before they could select the right method of cultivating and fertilizing the land on their estate. Kołłątaj also wrote that it was essential to know the deeper layers of the ground, below the topsoil. In the subsequent paragraphs of his instructions, he described in detail the soils of the Krzyżanowice estate, divided them into different categories, and characterized their color, composition, structure, and grain size.[67] The specifications and guidelines thus created served one purpose—to increase the efficiency of the estate and bring even greater economic benefits.

Owners and administrators of farms were also aware of the important role played by fertilization. This is evidenced by numerous references in the instructions to low-yielding farms that cited a lack of fertilizer as the cause of the crop failure.[68] Manor farm administrators also complained about the insufficient quantities or the total absence of manures. Neither inventories nor farm instructions contain many references to the use of other types of fertilizers. The instructions relating to the Opatów estate (*Instrukcja dla pisarza w dobrach Opatów*) attest to the use of pond slime, while others only mention the need to collect ash, without any indication of its purpose.[69] The sources are unanimous, however, in their insistence that fertilization was the best method to improve the quality of the cultivated soil and increase crop yields.

5 Conclusion

When assessing the views of early modern Polish agronomists, historians must first acknowledge that the authors of agricultural manuals derived much of their knowledge from Roman sources through the intermediary of Pietro de' Crescenzi. Nevertheless, most of the terminology, recommendations, and advice was generated from local practices. What is more, knowledge of the land and compliance with the recommendations of agricultural literature,

67 *Uwagi H. Kołłątaja dotyczące dóbr Krzyżanowice (1786–1788)*, in *Instrukcje gospodarcze dla dóbr magnackich i szlacheckich z XVII–XIX wieku*, ed. Bohdam Baranowski et al. (Wrocław: Zakład Narodowy im. Ossolińskich, 1958), 1:189, 191.

68 Gieysztorowa and Żaboklicka, *Lustracja województwa mazowieckiego 1565*, 2:22; Anna Sucheni-Grabowska and Stella M. Szacherska, eds., *Lustracje województwa płockiego 1565–1789* (Warsaw: PWN, 1959), 31.

69 *Instrukcja dla pisarza w dobrach Opatów, Opatów 30 VII 1771*, in Baranowski et al., *Instrukcje gospodarcze*, 1:585.

especially concerning fertilization, could contribute to keeping the soil in good condition and prevent a reduction in yields. The soil knowledge contained in Polish agricultural manuals can be categorized into three main areas. The first focused on the terminology and classification of soils, detailing their physical characteristics. Without specialized vocabulary, it was impossible to pass this knowledge on to future generations. The second concerned more practical aspects, such as distinguishing fertile from nonfertile soils, the selection of crops, or the technique of fieldwork. The third category comprised the knowledge of how to improve the quality of the soil and fertilize it.

Polish agricultural manuals focused on practical aspects and were primarily addressed to landowners who were interested in this type of knowledge because they wanted to improve their lands for profit. These works undoubtedly originated from the economic needs of the era. This does not mean, however, that the analyzed agronomic literature had nothing "scientific" to offer. Soil science in early modern Poland emphasized knowledge that could generate profit and disregarded knowledge that could not be used for this specific purpose.

The development of science often depended on local circumstances and evolved in accordance with the needs and interests of those who were engaged in its practice or provided support. In the context of Polish agricultural manuals, the requirements of landowners resulted in the establishment of precise soil classifications and the compilation of information concerning soil quality. The accessibility of literature addressing these topics motivated property managers to use this knowledge to their economic advantage. These actions aligned with the pragmatic approach to agriculture prevalent in society at that time, where the primary objective was to enhance agricultural productivity and maximize profits.

Acknowledgments

This paper was produced as a part of project no. 2018/31/N/HS3/02187, financed by the National Science Center, Poland.

PART 4

Soil, Specialization, and Experimental Culture

CHAPTER 8

Marl and Alchemical Theories of Soil Fertility in Early Modern England

Transmuting the Soil

Justin Niermeier-Dohoney

Toward the end of the sixteenth century, a growing belief spread throughout Europe that husbandry was in crisis. There was ample reason for concern. The Little Ice Age, a widespread cooling event in the northern hemisphere, saw average temperatures drop to their lowest point between the late sixteenth and late seventeenth centuries.[1] Wetter winters, shorter growing seasons, and the failures of back-to-back harvests led to several widespread, pan-European famines between the late 1560s and the late 1590s. In Britain, poor harvests exacerbated by economic mismanagement generated successive subsistence crises in the 1620s and 1630s.[2] Possibly in response to the urgency of these shortages, husbandry manuals and agricultural improvement tracts exploded in popularity during the second half of the sixteenth century. One of the major themes of these texts concerns soil improvement to increase productivity, and many of their authors advocate direct experimental intervention to further the understanding of the chemical composition of soil in order to accomplish this.[3] By the seventeenth century, it had become a commonplace belief that improving agriculture was essential to avert famine, boost populations, and stabilize national economies.[4] This early modern notion of social and material

1 Brian Fagan, *The Little Ice Age: How Climate Made History, 1300–1850* (New York: Basic Books, 2000), 121–122. On the destabilizing effects of this event, see, e.g., Geoffrey Parker, *Global Crisis: War, Climate Change, and Catastrophe in the Seventeenth Century* (New Haven, CT: Yale University Press, 2013), xix–xxvii, 3–24.

2 Guido Alfani and Cormac Ó Gráda, eds., *Famine in European History* (Cambridge: Cambridge University Press, 2017), 8–11; Cormac Ó Gráda, *Famine: A Short History* (Princeton: Princeton University Press, 2009), 1–7, 25–38.

3 Andrew McRae, "Husbandry Manuals and the Language of Agrarian Improvement," in *Culture and Cultivation in Early Modern England: Writing and the Land*, ed. Michael Leslie and Timothy Raylor (Leicester: Leicester University Press, 1992), 35–62.

4 Paul Slack, *The Invention of Improvement: Information and Material Progress in Seventeenth-Century England* (Oxford: Oxford University Press, 2015), 4–5, 56–60, 231–235; Paul Warde, "The Idea of Improvement, c. 1520–1700," in *Custom, Improvement, and the Landscape in Early Modern Britain*, ed. Richard W. Hoyle (Farnham, UK: Ashgate, 2011), 127–148.

progress proved to be a salient force in the efforts of agrarian reformers to make agriculture more productive. Improvers who sought greater social stability often argued that a reliable, experimentally grounded agriculture provided the foundation for all other forms of improvement.

For many reform-minded natural philosophers and agrarian reformers during this period, the theories and practices of alchemy—or more capaciously, chymistry—offered the key to accomplishing these goals.[5] By the early modern period, alchemy had moved on from its earlier incarnation as a secretive, scholarly art based on laboratory practices concerning metallic transmutation to become a commercial and entrepreneurial enterprise where individuals from many social strata applied its basic practices to support the burgeoning mining, metallurgical, and pharmaceutical industries.[6] Common laborers, farmers, and everyday alchemists fought to legitimize their commercially oriented trades against accusations of fraud from suspicious elites and incredulous merchants while creating a patchwork, vernacular scientific community outside formal institutions such as universities or princely courts.[7] Although the creation of the philosophers' stone and the transmutation of base metals into gold remained their primary desiderata, many alchemists also sought pragmatic applications of their skills and used alchemy as a tool to accomplish a number of mundane goals such as developing chemical medicines, smelting metals more effectively, improving the power of incendiary weapons, or preserving food for longer periods of time.[8] For these historical figures, alchemy

5 For more on "chymistry" as a catchall term to describe both chemical and alchemical activities prior to c. 1720—as before this date there was little terminological distinction—see, e.g., William Newman and Lawrence Principe, "Alchemy versus Chemistry: The Etymological Origins of a Historiographical Mistake," *Early Science and Medicine* 3, no. 1 (1998): 32–65.

6 Sven Dupré, ed., *Laboratories of Art: Alchemy and Art Technology from Antiquity to the Eighteenth Century* (New York: Springer, 2014); Tara Nummedal, *Alchemy and Authority in the Holy Roman Empire* (Chicago: University of Chicago Press, 2007); Bruce T. Moran, *Distilling Knowledge: Alchemy, Chemistry, and the Scientific Revolution* (Cambridge: Harvard University Press, 2005).

7 Nummedal, *Alchemy and Authority*, 73–95; Tara Nummedal, "Words and Works in the History of Alchemy," *Isis* 102, no. 2 (2011): 330–337; Tara Nummedal, "Practical Alchemy and Commercial Exchange in the Holy Roman Empire," in *Merchants and Marvels: Commerce, Science, and Art in Early Modern Europe*, ed. Pamela H. Smith and Paula Findlen (New York: Routledge, 2002), 201–222; Deborah Harkness, *The Jewel House: Elizabethan London and the Scientific Revolution* (New Haven, CT: Yale University Press, 2007), 1–56, 211–253; among others.

8 On the rise in historical works about the practical and applied arts of alchemy and of alchemy as a vernacular and entrepreneurial activity, see, e.g., Nummedal, "Words and Works," 330–337. For examples of this in practice, see, e.g., Pamela H. Smith, *The Business of Alchemy: Science and Culture in the Holy Roman Empire* (Princeton: Princeton University Press, 1994), 173–176, 206–214, 226–227.

offered a comprehensive, explanatory worldview to interpret the physical or chemical transformation of *any* natural substance. Agrarian reformers who adopted the techniques of alchemists did so to accelerate germination, develop larger and hardier crops, and rejuvenate exhausted soils, transforming them into fertile land once more. This vernacular epistemology—which made many farmers and landowners the alchemists of their own soil—underpinned their agrarian improvement projects.[9]

Just as alchemists had promised to produce cornucopian quantities of gold from base metals, those who advocated alchemically manipulating soils promised to make exhausted soil deliver cornucopian harvests. The perceived crisis of husbandry, the propagation of agricultural improvement projects to counter this, and the rise of vernacular alchemical interpretations of natural change all intersected in the seventeenth century. In this context, agrarian reformers argued that soil was a transmutable substance and that transmutation was not only possible but necessary for the improvement of agriculture.[10] Precisely how to do this became one of the most vexed questions of the time and the subject of vigorous debate. Was there a vital force in soil that made plant growth possible, and if so, could it be harnessed? How did fertility occur in the earth without human intervention and could this be artificially replicated? Could one type of soil be transmuted into another?

The focus of this chapter is on the vernacular alchemical interpretations of and agrarian interventions with a particular type of soil called marl (Figure 8.1).[11] Marl, sometimes referred to as marlstone or mudstone, is a dense, clay-like, silt- and lime-rich sedimentary soil full of carbonate compounds that can be found throughout the British Isles and was prized for centuries for its supposed long-term fertilizing properties (Figure 8.2). This type of soil develops through the erosion of older sedimentary soils that deposited material high in calcium carbonate such as the fossilized shells of prehistoric mollusks, coral, bivalves,

9 On vernacular epistemology and alchemy, see Pamela H. Smith, *The Body of the Artisan: Art and Experience in the Scientific Revolution* (Chicago: University of Chicago Press, 2004), 141–149, 165–181.

10 Dana Jalobeanu and Oana Matei, "Treating Plants as Laboratories: A Chemical History of Vegetation in 17th-Century England," *Centaurus* 62, no. 3 (2020): 542–561.

11 I use the British English definition of "marl"—as my historical actors did—roughly meaning "any clay-like soil containing calcium carbonate." In North American English, "marl" usually refers to calcareous lake sediment or bedrock made of sedimentary mudstone. These are sometimes, but not always, overlapping geological categories. See, e.g., William D. Shannon, "'An Excellent Improver of the Soil': Marl and the Landscape or Lowland Lancashire," *Agricultural History Review* (hereafter *AgHR*) 68, no. 2 (2020): 141–167, on 165.

FIGURE 8.1 A typical piece of marlstone or dried marl
SOURCE: PIECE OF MARL, WIKIMEDIA COMMONS (CC BY-SA 4.0), UPLOADED BY TASHKOSKIM, DECEMBER 13, 2015, HTTPS://EN.WIKIPEDIA.ORG/WIKI/MARL#/MEDIA/FILE:PIECE_OF_MARL.JPG

and other hard-bodied sea creatures.[12] Records from the later Middle Ages indicate that farmers went to great lengths and expense to procure marl.[13] In England, large marl pits in Kent, Lancashire, Cheshire, Shropshire, and Sussex were well known in the twelfth century and still regularly exploited in the mid-sixteenth century.[14] Unlike more accessible soils like loam, humus, or the dark topsoil found on especially fertile ground, marl had emerged as an integral

12 Sam Boggs, *Principles of Sedimentology and Stratigraphy*, 4th ed. (Upper Saddle River, NJ: Pearson Prentice Hall, 2006), 172; Harvey Blatt and Robert J. Tracy, *Petrology: Igneous, Sedimentary, and Metamorphic*, 2nd ed. (New York: W. H. Freeman, 1996), 217. On the importance of this process for the creation of calcium carbonate-rich soils, see F. Wolfgang Tegethoff, ed., *Calcium Carbonate: From the Cretaceous Period into the 21st Century* (Basel: Birkhäuser, 2001), 276. For one of the earliest attempts to bridge the historical understanding of marl use with knowledge from modern chemistry, see, e.g., G. E. Fussell and K. R. Fussell, "Marl: An Ancient Manure," *Nature* 183, no. 4656 (1959): 214–217.

13 See, e.g., Joan Thirsk and H. E. Halam, *The Agrarian History of England and Wales*, vol. 2, *1042–1350* (Cambridge: Cambridge University Press, 1988), 435–441; Shannon, "'Excellent Improver of the Soil,'" 146–147.

14 W. M. Mathew, "Marling in British Agriculture: A Case of Partial Identity," *AgHR* 41, no. 2 (1993): 97–110; G. E. Fussell, "Crop Nutrition in Tudor and Early Stuart England," *AgHR* 3,

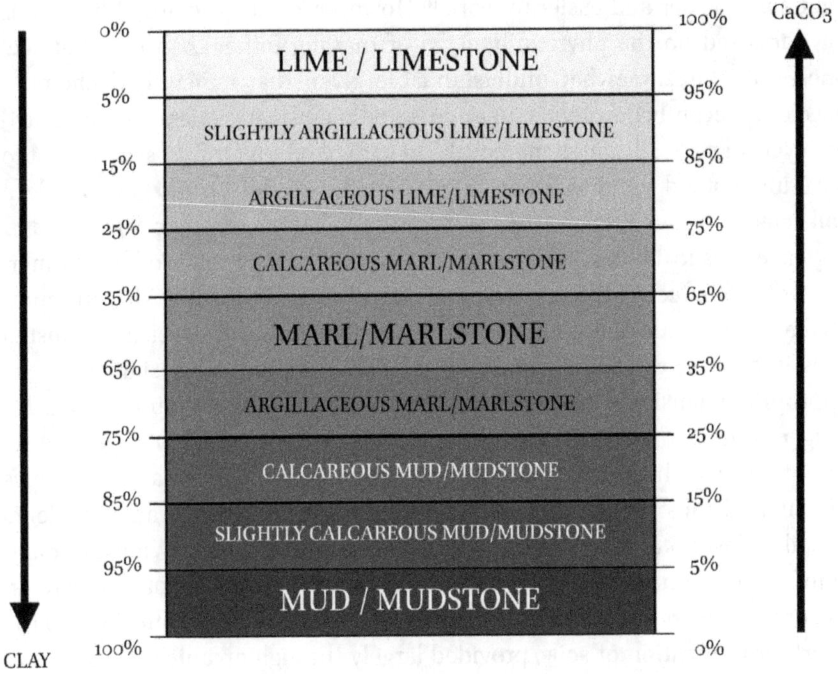

FIGURE 8.2 Transitional lithotypes from limestone to mudstone based on clay and calcium carbonate content. Marl or marlstone is a roughly equal mixture of argillaceous and calcareous soils.
SOURCE: FIGURE ADAPTED FROM "SCHEME OF TRANSITIONAL LITHOTHYPES [...]," WIKIMEDIA COMMONS (CC BY-SA 4.0), UPLOADED BY FOXBAT DEINOS, JANUARY 23, 2010, HTTPS://EN.WIKIPEDIA.ORG/WIKI/MARL#/MEDIA/FILE:MARL_VS_CLAY_&_LIME_EN.PNG

component of agricultural management through centuries of traditional practices and as a particularly mobile soil that farmers, merchants, landowners, and their laborers excavated in specific locations and transported to add to other soils in an effort to increase fertility. This practice was particularly prevalent in Britain and other places in northern Europe beginning in the Middle Ages, and was still quite common well into the nineteenth century.[15]

Historians of agriculture have largely addressed "marling"—the process whereby marl is mixed with animal manures and lighter soils directly on tilled fields—as a time-honored, customary activity employed by farmers to make

no. 2 (1955): 95–106, on 99–101; Eric Kerridge, *The Agricultural Revolution* (London: Allen and Unwin, 1967), 246–249.

15 Shannon, "'Excellent Improver of the Soil,'" 141–167; Mathew, "Marling in British Agriculture," 97–110.

light soils denser and easier to work.[16] However, most agricultural historians have focused on the physical benefits of marling rather than the chemical ones.[17] This is somewhat understandable given that agricultural chemists have only recently developed an understanding of marl's effects on the nutrient cycle of soils. Although medieval and early modern farmers seemed not to have understood the reasons for marl's benefits, so this history goes, modern soil science has explained that the addition of marl and other calcareous or argillaceous substances like lime, chalk, and clay to exhausted soil had numerous positive effects. Mixing marl with softer topsoil reduces acidity in sandy, chalky, stony, or nutrient-depleted soils and balances the pH level. In exhausted soils, it replenishes calcium, phosphorus, potassium, and magnesium. With the appropriate density, it creates a mixture of soil capable of retaining water for longer periods of time and accelerates the water uptake. Clayey soils contract in hot, dry weather and expand in cool, wet weather more than other types of soil, and thus the addition of clays and marls liberates organic nutrients in soils. Most notably, as historians and soil ecologists Verena Winiwarter and Winfried Blum have described, marls and rock powders do not intervene in the organic energy budget of soils, meaning that they do not contribute to the short-term nutrition of soils, provided largely through organic processes that depend on solar radiation and the photosynthetic energy it produces; rather, they contribute to the orogenic cycle, the process by which the weathering of rock parent materials provides mineral nutrients and energy to soils over longer periods of time.[18]

Though the scientific reasons for this have only been discovered in the modern era, this fact was not lost on late medieval and early modern English

16 Mathew, "Marling in British Agriculture," 97–110.

17 On this historical error, see Malcolm Thick, "Sir Hugh Plat and the Chemistry of Marling," *AgHR* 42, no. 2 (1994): 156–157. Notably, this has been known since at least the early work of historian of alchemy Allen Debus. See his "Palissy, Plat, and English Agricultural Chemistry in the Sixteenth and Seventeenth Centuries," *Archives internationales d'histoire des sciences* 21, no. 1 (1968): 67–88; and *Chemical Philosophy: Paracelsian Science and Medicine in the Sixteenth and Seventeenth Centuries* (New York: Science History Publications, 1977), 410–425.

18 Verena Winiwarter and Winfried E. H. Blum, "From Marl to Rock Powder: On the History of Soil Fertility Management by Rock Materials," *Journal of Plant Nutrition and Soil Science* 171 (2008): 316–324, on 316–318; Winfried E. H. Blum, "Soil as an Open, Complex System of Exogenic Biotic and Abiotic Interactions—Energy Concept," *Mitteilungen der Österreichischen Bodenkundlichen Gesellschaft* 55, no. 1 (1997): 13–15. On the release of organic nutrients, see A. D. Harley and R. J. Gilkes, "Factors Influencing the Release of Plant Nutrient Elements from Silicate Rock Powders: A Geochemical Overview," *Nutrient Cycling in Agroecosystems* 56, no. 1 (2000): 11–36.

farmers, who regularly touted marling as the best way to ensure decades of fertile soil. While manuring with animal dung required multiple applications throughout the growing season, early modern farmers reported that good marl provided benefits for three to five years and excellent marl might last for decades.[19] Even in the early modern period, many natural philosophers, agrarian reformers, and alchemists were clearly aware not only of the physical importance but also of the chemical importance of marl, and established research programs and experimental regimens to explain these astounding benefits. These investigations, the matter theory undergirding them, and the practical effects of the use of marl among agrarian reformers in seventeenth-century England provide evidence for this.

Marling has a long, if somewhat disjointed, history in Western Europe. In ancient Rome, it received its most extensive treatment in Pliny the Elder's *Natural History* (77–79 CE), where he called marl the "fat of the earth" and classified it according to color, texture, and use. Among its agricultural uses, Pliny suggested mixing it with animal dung and spreading it upon tilled earth shortly after plowing. According to him, dry or rough marls were useful in making waterlogged soils easier to cultivate, and "fatty marls," presumably meaning thick, absorbent clays, were best to spread on drought-stricken land.[20] Notably, Pliny mentioned that marl was virtually nonexistent in Italy but could be found abundantly in Britain and Gaul, where local farmers had been making use of it for a long time.[21] In his *De re rustica* (60–65 CE), the Roman agricultural writer Moderatus Columella wrote that his uncle, Marcus Columella, the owner of an estate near Cadiz, Spain, spread it on the gravelly ground in his wheatfields and vineyards when he could procure it; and when he could not, he substituted it with chalk and crushed limestone.[22] While commanding an army stationed in Transalpine Gaul during Caesar's civil war, the Roman military leader and later agricultural writer Marcus Terentius Varro described farmers in the Upper Rhine Valley using it on their fields.[23]

---

19 G. E. Fussell, *Crop Nutrition: Science and Practice before Liebig* (Lawrence, KS: Coronado Press, 1972), 68; Gervase Markham, *The Inrichment of the Weald of Kent* (London, 1625).

20 Pliny the Elder, *Natural History* 17.4.42ff.; Winiwarter and Blum, "From Marl to Rock Powder," 318.

21 Fussell, *Crop Nutrition*, 27–28. Pliny also mentioned its use in the Greek city of Megara.

22 Moderatus Columella *De re rustica* 2.16; Fussell, *Crop Nutrition*, 27.

23 Varro, *Res rusticae* 1.7; Fussell, *Crop Nutrition*, 27. Varro specifies *candida fossica creta* as the type of marl in use, a natural mixture of clay and lime, later known to the French as "marne" and sometimes used interchangeably with the English term "marl." The French river Marne is so named due to heavy deposits.

There is little evidence of a continuation of these practices in Western Europe during the early Middle Ages,[24] but marling had again become an integral component in agricultural soil management by the twelfth century at the latest. Lease agreements in Holland from the twelfth and thirteenth centuries required tenants to marl fields every twelve years, while those in Normandy specified an interval of every fifteen to eighteen years.[25] Records of demesne use from the manor of Ebony in Kent in 1256 reported that oats on land that had not been marled yielded 1,230 kilograms per hectare, while marled land yielded over 1,900 kilograms in the same space.[26] Records from Christ Church Estates in 1309 showed a similar ratio of productivity between the two.[27] The thirteenth-century Anglo-Norman agricultural writer Walter of Henley, who penned the agricultural treatise *Le Dite de Hosebondrie* in 1280, wrote of the practice of mixing heaps of manure with soil and marling one's "sheepfold ... every fortnight with clay land or with good earth, as the cleansing out of ditches, and then strewing it over" one's farmland.[28]

The medieval practice of marling developed locally, and most early records suggest that landowners excavated marl on their property or on nearby wasteland. In the late Middle Ages, a market developed for agricultural marl. One of the earliest examples comes from 1310, when a land grant permitted one landowner in Lancashire to have sole rights over marl pits on common land in Thornton.[29] Late fifteenth- and early sixteenth-century estate records from landowner Humphrey Newton show that he undertook a huge marling project outside of Manchester, where for ten weeks in 1502 numerous laborers in his employ dug large pits and transported up to twelve metric tons of

24 Absence of evidence is, of course, not necessarily evidence of absence. On the possibilities of early medieval marling, see Charles Parain, "The Evolution of Agricultural Technique," in *The Cambridge Economic History of Europe*, vol. 1, *Agrarian Life in the Middle Ages*, ed. M. M. Postan (Cambridge: Cambridge University Press, 1966), 125–179, on 132. He cites a Frankish edict from 864 enforcing the practice of regularly marling land.

25 Fussell, *Crop Nutrition*, 49.

26 Lord Ernle, *English Farming, Past and Present*, 6th ed. (London: Heinemann Educational Books, 1961), 10. See also Jules N. Pretty, "Sustainable Agriculture in the Middle Ages: The English Manor," *AgHR* 38, no. 1 (1990): 1–19, on 10.

27 Mavis Mate, "Medieval Agrarian Practices: The Determining Factors?," *AgHR* 33, no. 1 (1985): 22–31, on 23.

28 Elizabeth Lamond, trans. and ed., *Walter of Henley's "Husbandry": Together with an Anonymous "Husbandry," "Seneschaucie," and Robert Grosseteste's "Rules"* (London, 1890), 19; Ruth E. Harvey, "En sa veillesce set li prodhom," in *Anglo-Norman Anniversary Essays*, ed. Ruth E. Harvey (London: Anglo-Norman Texts Society, 1993), 159–178.

29 Land Grant, LA DDM/49/12, The National Archives, Kew, quoted in Shannon, "'Excellent Improver of the Soil,'" 147.

marl to his estate in Highfield. According to his record books, in the following growing season his yield of barley on the marled land more than doubled.[30] According to agricultural historians, this form of marling was typical in the late Middle Ages for large landowners and for farmers with smaller plots, assuming they could afford the labor and—if marl could not be found on one's own property—the transportation costs, which could be quite high.[31] Those who could not access or afford marl might make do with liming, chalking, or spreading pulverized rock like limestone or granite on their soil, which provided similar benefits on a smaller scale.

Prior to the mid-sixteenth century, few agrarian writers or naturalists had attempted a causal explanation of the importance of marling. Pliny's references to marl, for instance, were largely descriptive and taxonomic, and many of the later medieval textual examples of marling tended to be practical or pedagogical. This changed with the intervention of alchemists, who began to wrestle with the matter theory underlying soil composition and the notion that soil, like other "base materials," could be transmuted from one form to another. For many of these alchemists, marl emerged as a "perfected" form of soil, or a soil that nature had transmuted without human intervention. For example, in his *Discours admirables* (1580), the French potter, agriculturist, and amateur alchemist Bernard Palissy had called the "seminal cause" of growth and fertility the quintessence, or "fifth element," described as a "generative water mingled ... with common waters," which rain, rivers, streams, natural springs, and floods redistributed around the world.[32] In nature, common water evaporated and returned to the sky in the form of steam, which formed the

---

30 Winiwarter and Blum, "From Marl to Rock Powder," 320–321; Deborah Youngs, "Servants and Labourers on a Late Medieval Demesne: The Case of Newton, Cheshire, 1498–1520," *AgHR* 47, no. 2 (1999): 145–160.

31 Deborah Youngs, "Estate Management, Investment and the Gentleman Landlord in Later Medieval England," *Historical Research* 73 (2000): 124–141; Fussell, *Crop Nutrition*, 49–50; Shannon, "'Excellent Improver of the Soil,'" 146–147.

32 Bernard Palissy, *Discours admirables, de la nature des eaux et fontaines, tant naturelles qu'artificielles, des metaux, des sels et salines, des pierres, des terres, du feu et des maux* (Paris, 1580), 306–309. Translation adapted from Aurèle la Rocque, trans. and ed., *The Admirable Discourses of Bernard Palissy* (Urbana: University of Illinois Press, 1957). On Palissy's contributions to early modern chemistry, see, e.g., Smith, *The Body of the Artisan*, 70–90, 149–153, 224–227; Debus, "Palissy, Plat, and English Agricultural Chemistry in the Sixteenth and Seventeenth Centuries"; and various essays in Frank Lestringant, ed., *Bernard Palissy, 1510–1590: L'Écrivain, le réforme, le céramiste* (Mont-de-Marsan: Coédition Association Internationale des Amis d'Agrippa d'Aubigné, Éditions SPEC, 1992), esp. Bernard Rivet, "Aspects économiques de l'œuvre de B. Palissy," 167–180, and Frank Lestringant, "L'Eden et les tenèbres extérieures," 119–130.

clouds and started the entire process again, but "generative waters" slowly congealed and eventually hardened to form marl.[33] This particularly vital substance accounted for the fertility of marl. Notably, unlike many references to marl found in both ancient and medieval texts, which often derived their authority from eyewitness accounts of farmers, Palissy claimed his own work was necessary to explain natural causes to "doctors and all physicians, philosophers and naturalists" because the "farmers who use it do not care to know why it makes the ground fertile."[34] This is not to say that Palissy had no respect for the knowledge of farmers; merely, that he considered a more sophisticated knowledge of chymistry necessary for understanding a causal account of soil fertility.

Palissy's exhortation to seek these causes did not fall on deaf ears. A decade and a half later, in a commentary on Palissy's work, the English agricultural experimenter Hugh Plat wrote that marl was "transmuted soil," containing the congealed quintessence, that "waxed hard, and became white by vertue" of this congelation. When it was dispersed upon cultivatable land, Plat wrote, seeds planted in it "glut [themselves] with this generative, and congelative water."[35] Referring to Palissy's chemical descriptions of marl, Plat noted that marl was cold and dry—both in the tactile sense and in its fundamental Aristotelian qualities—and had an "inward heat" that could only actuate fertile ingredients in the soil when "stirred up by a counter-heate, wherein consisteth the seminal act." Pinpointing the source of this heat, or the "essential cause that moveth" plants to grow, was the basic task of the husbandman endeavoring to improve soil fertility.[36] Clearly, both Palissy and Plat were motivated to understand not just the physical properties but the chemical composition of marl, its potential to "transmute" poorer soils, and its status as a "natural, and yet divine soile … [that] giveth a generative vertue to all Seedes that are sown" upon it.[37]

To the alchemist-cum-agrarian-reformer, marl, as both natural and divine, was analogous to gold in the transmutational alchemy of metals or *chrysopoeia*.

33 Winiwarter and Blum, "From Marl to Rock Powder," 318.

34 Palissy, *Discours admirables*, 297.

35 Hugh Plat, *Diverse New Sorts of Soyle not yet Brought into any Publique use* […] (London, 1594), 23–24; Ayesha Mukherjee, "'Manured with the Starres': Recovering an Early Modern Discourse of Sustainability," *Literature Compass* 11, no. 9 (2014): 602–614, on 605. For more on Plat's views of marl, see Ayesha Mukerjee, *Penury into Plenty: Dearth and the Making of Knowledge in Early Modern England* (London: Routledge, 2014), 105, 108–111, 113, 136, 142n22. On Plat's work on agricultural chemistry more generally, see Malcolm Thick, *Sir Hugh Plat: The Search for Useful Knowledge in Early Modern London* (Totnes, UK: Prospect Books, 2010), 80–114.

36 Plat, *Diverse New Sorts of Soyle*, 21–23; Mukherjee, "'Manured with the Starres,'" 605.

37 Plat, *Diverse New Sorts of Soyle*, 28.

Gold and marl were both described as "perfect" substances forged underground by nature. Human intervention might hasten the transformation, but labor could only ever aid what was fundamentally a natural process. This mirrored the belief common among many early modern alchemists that the transmutation of metals produced in the laboratory was not "unnatural"—it was simply the human-induced acceleration of what nature generated. That is, given the proper geological conditions and enough time, base metals like lead would eventually transform themselves into gold underground.[38] Similarly, marl might be a naturally fertile soil, but other infertile soils could become "marl-like" through human intervention. For these other soils, Plat claimed, it was essential to provide a "vegetable and fructifying Salt of Nature," which existed in marl.[39]

Plat adopted the idea of a vegetable salt from the early sixteenth-century physician and alchemist Paracelsus, via Palissy, whom he quoted in the opening passage of his *Diverse New Sorts of Soyle* (1594) as an authority on "all sorts and kinds of marl, or soyle whatsoever, either known or used already for the manuring or bettering of all hungry and barren grounds ... [which] draw their fructifying vertue from that vegetative salt."[40] Though some of his descriptions are clearly metaphorical, Plat seems to have regarded this vegetative salt as a physical substance, and even though he wrote that the "seminal virtue" that circulated throughout the cosmos eternally was immaterial, he argued, like Palissy, that it "congealed" into saline material that could be used as fertilizers, manures, and seed steeps, just as generative waters ultimately congealed into marls. This vegetative salt, he wrote, "may become pregnant from the heavens, and draw abundantly that celestiall and generatiue vertue into the Matrix of the earth," which would no doubt prove that the agricultural improver was, as he put it, "the true and philosophicall Husbandman."[41] From these writings, it is clear that Plat considered the vegetable salt found in manures, plant compost, and topsoil to be equivalent to the vitalizing power of naturally fertile

38 On the common early modern alchemical belief that mineral and metallic transmutation was a natural, geological process, see, e.g., William Newman and Lawrence Principe, *Alchemy Tried in the Fire: Starkey, Boyle, and the Fate of Helmontian Chymistry* (Chicago: University of Chicago Press, 2005), 249; Lawrence Principe, *The Secrets of Alchemy* (Chicago: University of Chicago Press, 2013), 35–37; Moran, *Distilling Knowledge*, 70–98.

39 Plat, *Diverse New Sorts of Soyle*, 3. See also Antonio Clericuzio, "Plant and Soil Chemistry in Early Modern England: Worsley, Boyle, and Coxe," *Early Science and Medicine* 23, nos. 5–6 (2018): 550–583, on 555.

40 Plat, *Diverse New Sorts of Soyle*, 1.

41 Hugh Plat, *Floraes Paradise* (London, 1608), 9. For more on this concept of fertilizers, see Mukherjee, "Manured with the Starres,'" 604–605.

marl.[42] Between the generative waters contained in marl and the vegetable salt contained in manure and other fertile soils, Plat suggested that English farming might be drastically improved by an application of this knowledge of chymistry.

How did this knowledge affect farming practices and what did laborers do with this knowledge? For everyday farmers, understanding soil fertility was a tactile, empirical experience. Those who penned agricultural improvement manuals often described in detail procedures used by common farmers, even as they credited them anonymously and revealed little of the historical development of these ideas. Like Pliny and other Roman agricultural writers, these authors reported, for example, how rural farmers dubbed fertile soils "fat" or "fatted" and experimented with various methods to test for "fatness" on their farms. A mixture of custom, folklore, oral history, and centuries of inherited advice, this traditional ecological knowledge contributed to intricate methods of soil analysis.[43] According to several manuals, farmers sifted handfuls of dirt between their fingers to test for moisture, clay or silt content, color, temperature, and density. Clammy soils, like marl, suggested especially fertile soil if found in less obviously arable land such as highlands, heathlands, and forests. Farmers regularly "transplanted" soils from one location to another and examined whether the new soil "split" apart from the old. They sought plots of land where elm, sloe, bullace, and crab apples grew, as these were considered signs of good farming ground. These techniques had developed independently of contemporaneous alchemical analyses, but all of them were tangible experiments lay farmers performed to gain elementary knowledge about the qualities of soil, which later developed new meanings within an alchemical context.[44] Alchemists, who began to intervene in these debates in the mid-sixteenth

42 Plat, *Diverse New Sorts of Soyle*, 22–25; Clericuzio, "Plant and Soil Chemistry in Early Modern England," 556–557.

43 On "traditional ecological knowledge" (TEK), see Verena Winiwarter, "The Art of Making the Earth Fruitful: Medieval and Early Modern Improvements to Soil Fertility," in *Ecologies and Economies of Medieval and Early Modern Europe*, ed. Scott G. Bruce (Leiden: Brill, 2010), 93–116, on 97–98. On the origins of the concepts, see International Council for Science, *Science, Traditional Knowledge, and Sustainable Development*, ICSU Series on Science for Sustainable Development 4 (Paris: ICSU, 2002).

44 See, e.g., Thomas Hyll, *A Most High and Pleasant Treatyse for Teaching How to Garden* […] (London, 1563), 15; Barnabe Googe, *Foure Books of Husbandry* […] (London, 1577), 18; Gervase Markham, *The English Husbandman* (London, 1613); Walter Blith, "Epistle Dedicatory," in *The English Improver Improved, or the Survey of Husbandry Surveyed* (London, 1652), 4–5. For a discussion of these and other forms of traditional chemical soil analysis, see Simon Schaffer, "The Earth's Fertility as a Social Fact in Early Modern Britain," in *Nature and Society in Historical Context*, ed. Mikuláš Teich, Roy Porter, and Bo

century, often drew upon all of these sources as grist for their experimental mills.

Given the perceived crisis in husbandry in the second half of the sixteenth century, chymists and agrarian reformers alike began calling for an overhaul of husbandry practices to prevent famine and create abundance. A "revival" of marling was among the most common suggestions. As early as 1523, writers of husbandry manuals were lamenting that soil improvement through marling had fallen out of fashion, despite evidence of the practice continuing at the turn of the sixteenth century. In his *Boke of Surveyeng and Improvmentes*, John Fitzherbert of Derbyshire complained that much arable land and many marl pits went unused, though he suggested that the reason for this was that prudent tenants did not want to improve the economic output of their farmland because then their rents would be raised.[45] Other writers on the possible improvements to husbandry came to similar conclusions.[46] In the early seventeenth century, Gervase Markham, whom agricultural historian G. E. Fussell called an "enthusiast" for marl, also wrote about the "revival" of an interest in marling. He noted that "trees 200, or 300, years old now doe grow upon" marl pits, which suggested they had been abandoned for at least that long, and he quoted Pliny, Walter of Henley, and agricultural records from the reign of Edward II (r. 1307–1327) in support both of its economic benefits and the chymical theory substantiating this.[47]

Perhaps those most engaged in this sort of applied agricultural chemistry were those affiliated with the Hartlib Circle, a loose network of natural philosophers, social reformers, and agrarian experimenters, many of whom hoped to transform husbandry through an application of cutting-edge knowledge about plant and soil chymistry. Samuel Hartlib's unpublished memoranda, letters, journals, treatises, and other manuscripts—totaling over 25,000 folios—provide some of the best evidence of these alchemical theories of marl in practice during the mid-seventeenth century. In journals (*ephemerides*) from 1648 and 1653, Hartlib argued that farmers across England should be directed to

Gustafsson (Cambridge: Cambridge University Press, 1997), 124–147, on 127–128; Fussell, *Crop Nutrition*, 57–59; Fussell, "Crop Nutrition in Tudor and Early Stuart England," 103.

45 John Fitzherbert, *Boke of Surveyeng and Improvmentes* (London, 1523), 82. On Fitzherbert's discussion of marl, see Fussell, "Crop Nutrition in Tudor and Early Stuart England," 99; Fussell, *Crop Nutrition*, 66; Shannon, "'Excellent Improver of the Soil,'" 148.

46 John Norden, *The Surveiors Dialogue* (London, 1618), 222–223; Blith, *English Improver Improved*, 139.

47 Markham, *Inrichment of the Weald in Kent*, 3–4 (virtually the entirety of this short, twenty-page pamphlet is about the benefits of marl). See also Gervase Markham, *A Farewell to Husbandry* (London, 1638), 47–55, 65, 74, 156–157.

search their property for "marl and mines and mineral, esp[ecially] on heathes and bad groundes" as a way to avoid wasting valuable field space.[48] He wrote that marl was a "rich compost w*hi*ch might be had everywhere in Engl*and* if the Inhabitants were or could bee made sensible of it w*hi*ch was the great neglect of Marle to bee had in all Counties of Engl*and*."[49]

His associates heartily agreed with him. One example comes from the agrarian reformer Richard Weston, who had purchased large tracts of barren heathland in Sussex in the 1630s with the expectation of improving the land through experimental engagement. In an unpublished pamphlet later collected by Hartlib, Weston described his travels in Flanders, where he had observed husbandry practices quite different from those in England. While he mentioned that animal manures remained the best option for cheaply fertilizing large parcels of land, because they "had the virtue of fatten[ing], sweeten[ing] and reclaim[ing] all Barren grounds," he also noted that "unslakt Lime and Marl are of as great an Efficacie, beeing proportionably tempered with Earth and Ashes, and of longer continuance to enrich Land."[50] Weston suggested that marling, liming, and dunging land, combined with the common practice of "devonshiring" (denshiring)—in which grasses and heath were scorched and the ashes tilled into the ground—proved to be the surest way to improve soil.[51] Marling, Weston noted, provided the same nutrients as denshiring, but the benefits lasted much longer.

Similarly, in an anonymous manuscript in Hartlib's possession containing notes on the French alchemist Blaise de Vigenère's 1608 work *Traicté du feu et du sel* (Treatise on fire and salt), the author wrote that marl "in the dry places" of the Ardennes Forest fertilizes with the same capacity as the ashes of trees burned at seven or eight years of age, but that these ashes would not be as fruitful as either local marl, chalk, or salts, which are "the cause of production, because they warm and fatten the land."[52] Much as Palissy and Plat had equated the vegetable salt in manures with the vital properties of marl, experimental husbandmen often wrote of marls and "salts" in the same breath. "Salt," in these cases, did not necessarily translate as common salt; rather, it was understood to be any saline substance that could be crystallized upon evaporation, calcination, or distillation. Plat had listed a number of salts in

48 Samuel Hartlib, Ephemerides 1648 [June/July–December], 31/22/28A, Hartlib Papers (hereafter HP), University of Sheffield Library.

49 Samuel Hartlib, Ephemerides 1653 [May–September 2], HP 28/2/63A–B. The italicized letters in quotations taken from the Hartlib Papers are additions made by digital archivists, who have made these heavily abbreviated manuscripts more legible.

50 Richard Weston, "A Discourse Concerning Husbandry," HP PAM/34/10 and 13.

51 Weston, HP PAM 34/10 and 13. Denshiring is also referred to as "burn-beating."

52 Anonymous, "Copy Extract of 'On Fire and Salt,'" undated (after 1608), HP 52/176A.

his *Jewell House*, including copperas, alum, niter, sal ammoniac, and various vitriols.[53] Other alchemists influential among Hartlibian agrarian reformers, such as Joseph Duchesne, had described marl in similar terms as containing "fertilising salts" that "aninmateth, fortifieth, and giveth power" to the earth.[54] Agrarian reformers and experimental chymists like Benjamin Worsley had written to Hartlib detailing experiments conducted with "Minerall Salts" to aid in fertility, including "stones or lime of marle, chalke, fullers earth, [and] Vitrioll."[55] Another chymically knowledgeable experimental agriculturalist, Robert Child, argued in a letter to Hartlib that physical salts like table salt and saltpeter were quite different from the vegetable salt of alchemists, but nevertheless connected the fertility provided by saltpeter with that in marl. "The cause of fruitfulnes," he wrote to Hartlib in 1652, "is not only the *vita media* in dung, for when it is totally Corrupted & the *vita media* gone, it is very fruitfull," whereas "chalke, marle, nitre, which are exceeding[ly] fruitfull," do not have this property otherwise provided by dung and are even better fertilizers.[56] Like earlier observers, Child noted that the most important factor in marl's fecundity was how long lasting it was. In a letter likely to Hartlib, apple orchard owner and horticulture expert John Beale called marl "richer then [*sic*] dung for some uses, and more durable."[57] Elsewhere, citing Paracelsus, he described the "peculiar vigour … of fructification" in ground where beans had grown, writing that

> ashes of thiese being mingled w*i*th fat mold, & true marle … must needes give the Water, w*h*ich is impregnated fro*m* heaven, & then draynd through it, a nitrous vigour, & fit it as well to make plants sprout up to heaven, as to bee o*u*r pulvis pyrius [gunpowder].[58]

---

53 Hugh Plat, *The Jewell House of Art and Nature* (London, 1594), 10. See also Anna Marie Roos, *The Salt of the Earth: Natural Philosophy, Chemistry, and Medicine in England, 1650–1750* (Leiden: Brill, 2007); Clericuzio, "Plant and Soil Chemistry in Early Modern England," 556. These referred, respectively, to iron(II) sulfate ($FeSO_4$); hydrated double sulfate salt of aluminum, usually potassium alum ($KAl(SO_4)_2 \cdot 12H_2O$); potassium nitrate (KNO_3); ammonium chloride (NH_4Cl); and various metallic sulfates.

54 Joseph Duchesne, *The Practice of Chymicall and Hermeticall Physicke, for the Preservation of Health*, trans. Thomas Tymme (London: Thomas Creede, 1605), sig. O3r. On Duchesne and mineral salts in marl, see Roos, *Salt of the Earth*, 21, 226.

55 Benjamin Worsley to Samuel Hartlib, May 16, 1654, HP 66/15/1B.

56 Robert Child to Samuel Hartlib, February 2, 1652, HP 15/5/19A.

57 John Beale to Samuel Hartlib [?], undated, HP 25/6/1A.

58 John Beale to Samuel Hartlib [?], September 23, 1658, HP 52/13A–B. Here, Beale referred to saltpeter, the primary ingredient in gunpowder, which conferred potassium and nitrogen to plants when dissolved in water or absorbed by soil.

Seventeenth-century agrarian reformers who praised marl did so largely for its chymical properties.

Many early modern alchemists and agrarian reformers were convinced that the same natural processes were at work in both the formation of metals and minerals and in the growth of plant life. For those reformers who adopted alchemical theory, the potential for alchemical transmutation went far beyond the possibility of discovering the philosophers' stone and changing base metals into gold. Indeed, these theories suggested that transmutation entailed the conversion of *all* natural material from one thing into another with the proper chymical engagement. Although the ambitious goal of transforming lead into gold promised more consequential and more lucrative results than transforming soils, it also remained practically infeasible, a fantastical chimera for alchemists who otherwise accepted the notion that transmutation was possible. Figures like Palissy, Plat, and the agrarian reformers associated with the Hartlib Circle concentrated their efforts on more prosaic and utilitarian, though no less important, aims. Understanding how one material transformed into another was among the most crucial goals for these reformers, and the alchemical concept of transmutation provided a heuristic tool for evaluating these processes. For the agrarian reformer, alchemy promised solutions to the issue of soil exhaustion and the augmentation of soil fertility through new chemical knowledge derived from experiments conducted on the fields and farms of England. As a naturally fertile soil that provided long-lived fertility without human intervention, marl offered an example of a naturally "perfected" soil, one that alchemically minded agrarian reformers could emulate. The evidence presented in this chapter, drawn from agricultural records, husbandry manuals, alchemical treatises, and experimental reports, suggests that they did just that.

CHAPTER 9

Mastering the Soil and Measuring Flow in Early Modern Istanbul

Deniz Karakaş

1 Introduction

If, in early May 1709 (1121 AH), we had traveled beyond the city walls through Istanbul's western hinterland towards the Belgrad Forest (about twenty kilometers north of the city land walls), we would have come across Oṣmān Āġā, the superintendent of the imperial water mains (*ḫāṣṣa ṣu nāẓırı*) and several royal water technicians (*ṣuyolcular*, sing. *ṣuyolcu*). They were surveying and measuring the newly completed waterworks that had been commissioned by the administrator (*mütevellī*) of the charitable endowment (waqf, in Turkish: *vakıf*) of Muṣṭafā Āġā (deceased), simultaneously estimating and calculating the amount of water that the new waterworks diverted to the main waterways (*ṣuyolu*) of Istanbul.[1] The royal water technicians with their rolled-up baggy trousers descended by rope to a horizontal tunnel now cleaned of all dirt, mud, and silt, and then climbed up another shaft after carefully measuring its segments. They recorded the tunneling work (*laġm*) as about two hundred and twenty-two *ḳulac* (~416 m) with twenty-one shafts at intervals, ending with a termination point (*laġm aġżı*) where 3850 water supply pipes (*künk*) connected the *laġm* to the Çaşnigir aqueduct (*kemer*) outlet.[2] Or if we had gone

1 The title *ḫāṣṣa ṣu nāẓırı* was an official designation in the Ottoman Empire, comparable to "water master" (*maestro d'acqua*) in Italy. Until the early seventeenth century, this office was a steppingstone to the position of imperial chief architect (*miʿmār āġā*). For an overview of the superintendent's responsibilities, see Gülru Necipoğlu, *The Age of Sinan: Architectural Culture in the Ottoman Empire* (London: Reaktion Books, 2005), 140–42, 171–72, appendix 4.5, 565.

2 I follow the archival system used in the cataloging of the *Şerʿiyye Sicilleri* (court registers) of the Presidency of the Republic of Turkey, Directorate of State Archives, Ottoman Archives (BOA), Istanbul. Hereinafter, references to the İstanbul *Şerʿiyye Sicilleri, Ḫāṣlar ḳażāsı* registers are abbreviated as İSTM.ŞSC.15.d, followed by the register number, the page number (with a and b indicating the recto [right] and verso [left] side of the page), and the individual case number on that page. İSTM.ŞSC.15.d 126 76b/1, date: rebīʿü'l-āḫir 1121 AH. Regarding the units of measurement used in building trades and in the building of hydraulic structures, *ẕirāʿ*, *ʿarşın/ʿarşun, ḳulac, şāṭrāncī, mesāḥa, ḳarış, usbūʿ*, and *barmaḳ/parmaḳ* were the common

FIGURE 9.1 Open hilly area where the Güzelce Aqueduct spans the Alibeyköy Stream near Cebeci village
SOURCE: PHOTO BY KÂZIM ÇEÇEN, COURTESY OF İTÜ MUSTAFA İNAN LIBRARY, PROF. DR. KÂZIM ÇEÇEN SLAYT ARŞİVİ, ISTANBUL

further back in time to early summer in 1607 (1016 AH), and this time walked from the Cebeciköy aqueduct (known also as Güzelce) towards the south end of Cebeci village and on to the Sırıklıçayır meadow, we would have seen the repairman (*meremmetcī*) Trandafil, son of Yorgi, riding on horseback. Hired by the administrator of the waqf of Mesīḥ Meḥmed Pāşā (d. 1592), he rode across the open mostly treeless grassy plains broken only by scattered vegetable gardens and vineyards to clean an underground conduit about five hundred and eighty *ḳulac* (~1084 m) long (Figure 9.1).[3]

basic units. *Ḳulac* is the unit measure of length extending between the tips of two stretched arms. Gül Kale, "Intersections between the Architect's Cubit, the Science of Surveying, and Social Practices in Caʿfer Efendi's Seventeenth-Century Book on Ottoman Architecture," *Muqarnas* 36 (2019): 131–77. The *ẕirāʿ* (and *arşun*) measure varied over time and place. See Alpay Özdural, "Sinan's *Arşın*: A Survey of Ottoman Architectural Metrology," in *Muqarnas*, ed. Gülru Necipoglu (Leiden: Brill, 1998), 15: 101–115, Neslihan Sönmez, *Osmanlı Dönemi Yapı ve Malzeme Terimleri Sözlüğü* (Istanbul: Yem Yayınları, 1997), 25; and Ünal Taşkın, "Osmanlı Devleti'nde Kullanılan Ölçü ve Tartı Birimleri" (MA thesis, Fırat Üniversitesi, Turkey, 2020), 141–153. It is hard to know the precise length of *ḳulac* used at that time. I take the round figure of 75 cm for the *ẕirāʿ* as the basis of measurements and 1 *ḳulac* equivalent to 2.5 *ẕirāʿ* (*ʿarşun*).

3 BOA, TS.MA.e 834–9, date: 20 rebīʿü'l-āḫir 1016 AH and BOA, TS.MA.e 180–25, date: 1 receb 1007 AH.

Drawn from the records of the judgeship of the Ḫāṣlar *ḳażāsı* in Istanbul's countryside, imaginary excursions offer a glimpse into the diversity of spaces that constituted Istanbul's hydraulic infrastructure.[4] They also indicate the range of different investigative practices used, including quantitative scientific methods such as inspection (*keşf*), survey or mensuration (*mesāḥa*), practical geometry (*hendese*), examination (*naẓar*), quantitative estimation (*taḫmīn-i sahih*), calculation (*ḥisāb*), report (*taḳdīr*), explanation (*beyān*), verification (*taḥḳīḳ*), division (*taḳsīm*), and certification (*burhān, sened, ḥuccet*), that both precede and coexist alongside the trials and inquiries.[5] Moreover, these examples show the fluidity between what is frequently defined as lay knowledge and practice vs. that of the expert or professional.

By focusing on the earthworks involved in maintaining the water systems in the countryside of early modern Istanbul, this chapter argues that the emphasis on expertise often found in history of science and technology research tends to obscure "common knowing practices."[6] Judicial proceedings and courtrooms have been crucial for the study of early modern natural and medical knowledge. Recent scholarship has noted the commonality in observing, describing, explaining, and arguing in the courtrooms, where natural and medical knowledge was routinely presented and contested in trials.[7] Similarly, this

4 Ḫāṣlar ḳażāsı (district) was a judicial and administrative unit, whose center was the town of Eyüp, comprising the villages along the western shores of the Bosporus and Istanbul's northern and western hinterland outside the land walls up to Silivri and Çatalca. A vast portion of the district was originally the sultan's private imperial property (*ḫāṣṣa*) and, over time, much of this was incorporated into revenue-producing holdings of sultanic charitable endowments.

5 Information on legal cases is provided in the records (*şer'īyye sicilleri*) of the judgeship of the Ḫāṣlar district and those of the *Dīvān-ı Humāyūn* (Imperial Council), the highest-ranking judicial and administrative officials in Istanbul, which contained the registers of *mühimme* (roughly, important affairs) and *şikāyet* (books of complaint).

6 On the role of early modern experts in government consultation, legal proceedings, and in the resolution of scientific controversies, see Eric H. Ash, ed., "Expertise: Practical Knowledge and the Early Modern State," *Osiris* 25 (2010). On knowing and expertise in early modern empires, see J. A. Mendelsohn and A. Kinzelbach, "Common Knowledge: Bodies, Evidence, and Expertise in Early Modern Germany," *Isis* 108, no. 2 (2017): 259–279, on 278.

7 The scientific value of juridical proceedings and courtrooms for the study of early modern medicolegal practice and theory has been the topic of extensive research for a while now. See, among others, Silvia De Renzi, "Witnesses of the Body: Medico-legal Cases in Seventeenth-Century Rome," *Studies in History and Philosophy of Science* 33, no. 2 (2002): 219–242; Cathy McClive, "Blood and Expertise: The Trials of the Female Medical Expert in the Ancien-Régime Courtroom," special issue, *Bulletin of the History of Medicine* 82, no. 1 (2008): 86–108; Michael Stolberg, "Learning from the Common Folks: Academic Physicians and Medical Lay Culture in the Sixteenth Century," *Social History of Medicine* 27, no. 4 (2014): 649–667.

chapter shows that, from the 1570s through to the 1750s, the knowledge and practical skills of lower-ranking common laborers in the building trades were recognized as comparable to those of elite groups of experts, and the participation and authority of these experienced earth workers became an accepted part of managing the hydraulic projects in the countryside of Istanbul and the demands of the administrative system. Laborers appeared as expert witnesses in the Ottoman courts and were asked to testify, inspect, report, and certify in legal disputes and proceedings, and the records of these offer a window into their knowledge and know-how.[8]

In this chapter, I focus on the many kinds of knowledge-making deployed by low-ranking laboring groups who mostly derived their livelihood from creating and maintaining Istanbul's water conduits. Through the investigation of how knowledge was produced in water sciences and hydraulic engineering, this chapter aims to uncover a diverse host of practices reconfiguring soil, and to address a set of shared skills related to the observation and manipulation of the subsoil in situ. Exploring how knowledge about soil led to the concept of a ground (*yir/yer, zemīn*) composed of layers (*ṭabaḳa*), I will thus build on other chapters in this volume by connecting this narrative to the creation and dissemination of knowledge about soil in spaces outside of agriculture.[9]

Two sets of secondary literature inform this study. As Justin Niermeier-Dohoney and Aleksandar Shopov's introduction to this volume outlines, over the last three decades much of the scholarship in the history of science has moved beyond a dichotomy between *episteme* (knowledge) and *techne* (craft/art), that is, mind versus hand, to exploring the role of artisanal know-how and embodied, tacit knowledge in a variety of unconventional epistemic contexts, from mining operations to kitchens. This reframing allows us to trace how modern science and technology emerged through interactions between practical and learned knowledge during the early modern period. Today, a growing body of scholarship embraces sites not previously thought of as conducive to scientific inquiry, as well as a larger cast of actors.[10] In more recent studies,

---

8 On *keşf* and trial documentation, see Ömer Korkmaz and Nasi Aslan, "Osmanlı Ceza Muhakemesinde Olay Yeri Ön İnceleme-Keşif ve Tahrir Raporları," *İslam Hukuku Araştırmaları Dergisi* 33 (2019): 239–259. On the term "earth workers," see p. 212.

9 See Niermeier-Dohoney and Shopov, this volume.

10 See Sonja Brentjes and Robert Morrison, "The Sciences in Islamic Societies (750–1800)," in *The New Cambridge History of Islam*, vol. 4, *Islamic Cultures and Societies to the End of the Eighteenth Century* (Cambridge: Cambridge University Press, 2010), 564–639. For a more extended discussion on recent trends in the history of Islamicate science, see Nahyan Fancy et al., "Current Debates and Emerging Trends in the History of Science in Premodern Islamicate Societies," *History of Science* 61, no. 2 (2023): 123–178.

Lissa Roberts, Seth Rockman, and Alexandra Hui have called to extend the impulse brought into new materialisms through labor history to a more inclusive and globally expansive history of science, expanding the analytical lens of a science-knowledge continuum beyond the world of artists, scholars, and elites to include "workers once known."[11] Ottoman earth workers are examples of such actors, whose webs of experience, experimentation, and observation have gone largely unseen.[12] Their labor also resonates with recent studies by Johannes Mattes and Patrick Anthony, the latter of whom has argued that "the embodied experience of vertical travel was itself constitutive of a geographical imaginary that spanned mineshafts and mountain summits."[13] The work and experience of those who explored nature's horizontal and vertical axes, encompassed a rich spectrum of knowledge. Following up on the broader literature on the spatial dimensions of science, the recent surge of interest in early scientific ventures and expeditions in extreme environments shows how epistemic, technical, and bodily challenges, such as difficulty of access and limited

11 Alexandra Hui, Lissa Roberts, and Seth Rockman, "Introduction: Launching a Labor History of Science," in "Let's Get to Work: Bringing Labor History and the History of Science Together," Focus section, *Isis* 114, no. 4 (2023): 817–826; on the expression "workers once known," see Gabriela Soto Laveaga, "Worker Once Known: Thinking with Disposable, Discarded, Mislabeled, and Precariously Employed Laborers in History of Science," *Isis* 114, no. 4 (2023), 834–840. See also Roberts, Rockman, and Hui, "Historiographies of Science and Labor: From Past Perspectives to Future Possibilities," *History of Science* 61, no. 4 (2023): 448–474; Rockman, Roberts, and Hui, "Joining Forces: Labor History and the History of Science," *Labor: Studies in Working-Class History* 21, no. 1 (2024): 1–9. For a survey of the labor history of science in natural history, see Patrick Anthony, "Introduction to 'Working at the Margins: Labor and the Politics of Participation in Natural History, 1700–1830,'" *Berichte zur Wissenschaftsgeschichte* 44, no. 2 (2021): 115–136. In connection to this line of inquiry in the premodern Islamicate world, see Shireen Hamza, "Vernacular Languages and Invisible Labor in Ṭibb," *Osiris* 37 (2022): 115–138; Duygu Yıldırım, "Ottoman Plants, Nature Studies, and the Attentiveness of Translational Labor," *History of Science* 61, no. 4 (2023): 497–521.

12 Roberts, Rockman, and Hui, "Historiographies of Science," 464; Laveaga, "Worker Once Known."

13 Patrick Anthony, "Mines, Mountains, and the Making of a Vertical Consciousness in Germany ca. 1800," *Centaurus* 62, no. 4 (2020): 612–630, on 614. Johannes Mattes, "Mapping the Invisible: Knowledge, Credibility and Visions of Earth in Early Modern Cave Maps," *British Journal for the History of Science* 55, no. 1 (2022): 53–80. For a recent discussion on the role of verticality in past scientific work, see Wilko Graf von Hardenberg and Martin Mahony, "Introduction—Up, Down, Round and Round: Verticalities in the History of Science," in "Verticality in the History of Science," special issue, *Centaurus* 62, no. 4 (2020): 595–611.

vision, influenced how knowledge was produced.[14] Keeping in mind this "vertical turn," let us now interrogate further the experience of verticality—moving through the earth, seeing and knowing from within—through the lenses of the Ottoman earth workers.

In the old walled city, where Istanbul's population of about 300,000 people was concentrated from the sixteenth century through the seventeenth century, water was piped in from the Ḫāṣlar district outside the city walls.[15] The astonishingly complex topography bordering the city proper to the north and west in Thrace—rolling hills, large marshy tracts along the rivers, lakes, seacoast, valleys, and grass-covered plains—was rich in water resources (Figure 9.2). The capital city continued to draw on these water supplies for several centuries.[16]

---

14 See, e.g., Lachlan Fleetwood, "'No Former Travellers Having Attained Such a Height on the Earth's Surface': Instruments, Inscriptions, and Bodies in the Himalaya, 1800–1830," *History of Science* 56, no. 1 (2017): 3–34; Thomas Morel, *Underground Mathematics: Craft Culture and Knowledge Production in Early Modern Europe* (Cambridge: Cambridge University Press, 2022); Rebekka von Mallinckrodt, "Exploring Underwater Worlds: Diving in the Late Seventeenth-/Early Eighteenth-Century British Empire," in *Empire of the Senses*, ed. Daniela Hacke and Paul Musselwhite (Leiden: Brill, 2017), 300–332.

15 Mehmet Öz, "The Population of Istanbul from the Conquest to the end of the 18th Century," in *From Antiquity to the 21st Century: History of Istanbul / Büyük İstanbul Tarihi Ansiklopedisi*, ed. Coşkun Yılmaz (Istanbul: Türkiye Diyanet Vakfı İslâm Araştırmaları Merkezi [hereafter İSAM], 2015), https://istanbultarihi.ist/461-the-population-of-istanbul-from-the-conquest-to-the-end-of-the-18th-century. On villagers from the Hāslar district, see Stefanos Yerasimos, "15. Yüzyılın Sonunda Haslar Kazası," in *18. yüzyıl Kadı Sicilleri Işığında Eyüp'te Sosyal Yaşam*, ed. Tülay Artan (Istanbul: Tarih Vakfı Yurt Yayınları, 1998), 86. On their roles as repairers of the Kırkçeşme waterways, see Ömer Lütfi Barkan, "Fatih Camii ve İmareti Tesislerinin 1489–1490 Yıllarına Ait Muhasebe Bilançoları," *İstanbul Üniversitesi İktisat Fakültesi Mecmuası* 23/1–2 (1962–3): 297–341 and BOA, TT.d 370, date: 973 AH. In the mid-sixteenth century, twenty-seven households from the two villages of Orta Belgrad and Kömürlü Belgrad in the Hāslar district, were responsible for the upkeep (e.g. *inşā ʾāt, taṭhīr, tamʿīr ve termīm*) of the city's water supply network. See İsmet Binark, ed., *3 numaralı mühimme defteri, 966–968/1558–1560* (Ankara: Başbakanlık Devlet Arşivleri Genel Müdürlüğü, Osmanlı Arşivi Daire Başkanlığı, 1993), 682, decree 1538. Likewise, in the 1560s, the villagers of Akpınar were granted exemption from extraordinary levies in recognition of their role in maintaining the Kağıdhane waterways. See Aleksandar Shopov, "Between the Pen and the Fields: Books on Farming, Changing Land Regimes, and Urban Agriculture in the Ottoman Eastern Mediterranean ca. 1500–1700," (PhD diss., Harvard University, 2016), 129–130. By the early eighteenth century, the number of villages in the Hāslar supplying labor to maintain the city's water supply network had risen to twelve, and they were known as the imperial waterway villages (*mīrī ṣu yolu ḳaryeleri*). BOA, MAD.d 9892/216/1, date: 25 muḥarrem 1116 AH.

16 On Istanbul's dependence on outside water sources during the Byzantine period, see Cyril Mango and Gilbert Dagron, "The Water Supply of Constantinople," in *Constantinople and Its Hinterland: Papers from the Twenty-Seventh Spring Symposium of Byzantine Studies, Oxford, April 1993* (Aldershot: Variorum, 1995), 9–18; James Crow, Jonathan Bardill, and Richard Bayliss, *The Water Supply of Byzantine Constantinople* (London: Society for the

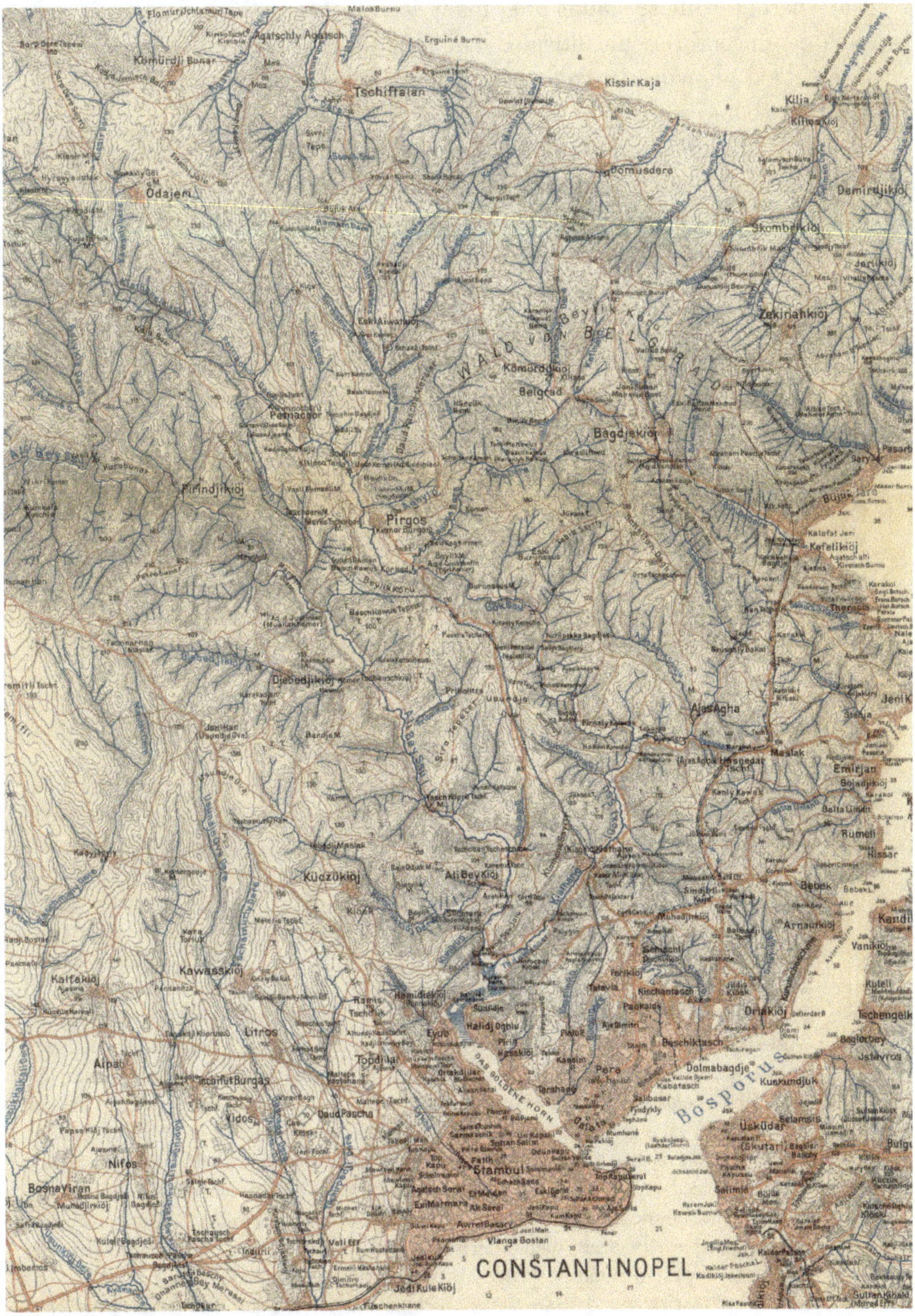

FIGURE 9.2 Detail from "Karte der Umgegend von Constantinopel" (Map showing Istanbul and surroundings), 1897, by C. Frh. v. d. Goltz (Pascha)

SOURCE: SALT RESEARCH APLGOKP 001, ISTANBUL, HTTPS://ARCHIVES.SALTRESEARCH.ORG/HANDLE/123456789/102289

The sharing of these waters took many forms. During the seventeenth and eighteenth centuries, uncultivated land that did not belong to anyone, or had been ruled unowned in court proceedings, constituted a fertile ground for hydraulic enterprises.[17]

Knowledge of soils and their uses echoed in almost every step of the construction and maintenance of the hydraulic infrastructure: from finding water and building the complex aqueduct system (vast underground channels, dams, ditches, and trenches) to the making and laying of pipes, the ongoing dredging, cleaning for removal of silts and sediments, and the control of erosion and siltation. And those who came to labor directly on the construction sites of Istanbul's hydraulic infrastructure tended to know the most about soil.

Contemporary writings offer additional insight into practices of the time. For example, Muḥammed bin Muṣṭafā's 1590 (998 AH) Ottoman Turkish translation of *Kitāb al-filāḥa*—the seminal twelfth-century agricultural text by the Andalusian scholar Ibn al-ʿAwwām—detailed methods for finding groundwater based on the knowledge of the soil.[18] Another example was the field inspection (*keşf*) for the construction of an artificial lake (*göl*) near Mostar in the province (*sancaḳ*) of Bosnia one decade earlier in 1576 (984 AH).[19] A century later, doing much the same thing, commoner trench- (*ḫendeḳ ḳāzıcı*)

Promotion of Roman Studies, 2008), 9–19; Knut Olof Dalman, Paul Wittek, and Martin Schede, *Der Valens-Aquädukt in Konstantinopel* (Bamberg: J. M. Reindl, 1933).

17 Deniz Karakaş, "Water Resources Management and Development in Ottoman Istanbul: The 1693 Water Survey and Its Aftermath," in *Istanbul and Water*, ed. Paul Magdalino and Nina Ergin [Macaraig] (Leuven: Peeters, 2015), 177–205. A fundamental reference work on Istanbul's water supply is Kâzım Çeçen, *İstanbul'un Osmanlı Dönemi Suyolları*, ed. Celâl Kolay (Istanbul: İstanbul Büyükşehir Belediyesi, İstanbul Su ve Kanalizasyon İdaresi [hereafter ISKI], 1999). See also, Filiz Karakuş et al., "The Historical Water Systems of Istanbul and Their Preservation Problems: The Case of the Kırkçeşme Water System," *Gazi University Journal of Science* 31, no. 2 (2018): 368–379.

18 See Ayfer Aytaç, "Kitābü'l–Felāḫat Tercümesi: Giris–Metin–Dizin" (PhD diss., Çanakkale Onsekiz Mart Üniversitesi, Turkey, 2015). Alongside the original, I consulted the French translation and Aytaç's transcription and index for the English translation. For example, taste, smell, color, and texture were important indicators (*ʿalāmet*) and he writes: "meṣelā ṣol ṭopraḳ ki yağlı gibi olup iṣṭilāḥ-ı felāḥatda arż-ı desime dėrler ve levni siyāh olup veyāḫūd ġubār renginde olup cüzv-ī ṣu ṭoḳınsa ele yapışur gibi olsa yėre yaḳın degüldür" (If, when mixed with a little water the surface had a black-colored soil or dust (*toz*), [and] if this loamy soil sticks to the hands or if it is the color of greasy soil, as it is known by the finest experts in agriculture, then water was not close to the surface). Beyazıt State Library, Istanbul, Veliyüddin Efendi, MS 2534, 31b; Aytaç, "Kitābü'l–Felāḫat Tercümesi," 118.

19 BOA, A.DVNSMHM.ZYL.d 3, decree 121, date: 20 muḥarrem 984. The Governor of the Province (*beg*), the local judge (*qadi*), and superintendent at the imperial mines (*maʿden nāẓırı*) made their first report on all matters of concern—the distances and heights for safely moving the waters. The survey was conducted by local miners (*maʿdenciler*), tunnel

and tunnel-digger (*lağımcı*) recruits from well-known mining centers and the capital city, Istanbul, became central to military campaigns. The Ottoman State also sought out certain individuals, not as a source of manual labor, but for their skills and knowledge in overcoming the technical challenges of excavating.[20] For example, in his account of the long siege and subsequent conquest of the Candia [Tur.: Kandiye] fortress (1648–1669), Ḥāsan Efendi, the former chief clerk of the Janissary corps, noted that "a massive rock (*yekpāre ḳayā*) resembling a mountain" was a formidable obstacle.[21] However, Ḥāsan Efendi continued, a recently converted blacksmith (*demirci*) helped manufacture the twisted augers (*burgu*) and other drills fit for drilling through stone (*taş*) and rock (*ḳayā*) beneath the ground. Those new tools allowed the sappers to drill a wide tunnel down into the rock reaching a depth of ten *ḳulac*, where they deposited a great quantity of barrels of powder.[22]

Within the specific context of each worksite and the nitty-gritty details, the various tasks performed by well diggers, trench diggers, miners, tunnel diggers, sappers, surveyors (*messāḥ*), and blacksmiths reveal both surface and subsoil posed challenges and the extent to which hydraulic projects required a range of practical skills and experience with soils. *Lağımcı* was one of the most common designations for laborers working on Istanbul's water system in the seventeenth and eighteenth centuries.[23] The presence of such workers may well go back to the sixteenth century, when they started to take contracts or were recruited by the imperial chief architect to build hydraulic works—digging (*ḥafr*) wells (*kuyū*) and tunnels to divert groundwater, streams, and

---

diggers (*lağımcılar*), and surveyors (*ölçüçeke ve yararlar*). This "expert" local laboring group noted the kinds of soils and offered a lithological description of the terrain.

20 On the digging of trenches (sing. *siper*, also *ḫendeḳ*), tunnels (*lağm*, also pl. *meterīs*) and assessment of the soils (*türāb,* also *zemīn*) during the siege of Candia (present-day Heraklion) in Crete, see Hasan Efendi, *Tevârîh-i Cezîre-i Girid*, ed. Hasan Ali Cengiz (İstanbul: Paradigma Akademi Yayınları, 2024).

21 Efendi, *Tevârîh-i Cezîre-i Girid*, 221–222.

22 Efendi, 199.

23 On *lağımcı*, see Mehmet Akkuş, ed., *Eyyûbî, Menâḳıb-ı Sultan Süleyman* (*Risâle-i Pâdişâh-Nâme*) (Ankara: Kültür Bakanlığı, 1991), 196–197; and Evliyā Çelebi [b. Derviş Muhammed Zıllî], *Evliya Çelebi Seyahatnâmesi: Topkapı Sarayı Kütüphanesi Bağdat 304 Yazmasının Transkripsiyonu, Dizini*, ed. Robert Dankoff, S. Ali Kahraman, and Yücel Dağlı (Istanbul: Yapı Kredi Yayınları, 2006), 1:321, 323. On their military service, recruitment, and organization, see Filiz Yıldırm "Osmanlı Devleti'nin Yeraltı Savaşçıları: Lağımcılar," *Fırat Üniversitesi Sosyal Bilimler Dergisi*, 29, no. 2 (2019): 385–408; Abdulkadir Özcan, "Lağımcı Ocağı," *Türkiye Diyanet Vakfı İslâm Ansiklopedisi* [hereafter TDV] (Ankara: TDV Yayınları, 2003), 27:49–50. My reading of this material along with the archival sources suggests the changing working lives of *lağımcı* and what such practitioners should know in theory and practice.

rivers to covered channels underground to reach Istanbul proper and in elsewhere.[24] From 1594 to 1667, the record of the occasional conscription of *lağımcı* to join the ranks of the permanent imperial regiments yields the names of no less than one hundred Armenians (*Ermenī*) working in this trade in the capital (Figure 9.3).[25] There are records of several directives ordering *lağımcı* to commence with repairs and cleaning after the usual spring flood damage, such as mud-filled pipes and debris-clogged channels.

In a similar way, the technical workings and upkeep of the system were entrusted to a corps of water technicians under the supervision of the superintendent of the imperial water mains, who had access to an additional workforce when necessary. By 1614, 359 peasants, mostly Greek Orthodox Christians from villages in the Ḫāṣlar district, had been exempted from paying poll taxes and extraordinary taxes ("ʿavārīż-ı dīvāniye ve teḳālīf-i ʿörfiye") in exchange for being included on the list of those who could be called on to work as *ṣuyolcular* (water technicians) or *yamakān* (auxiliary artisans).[26] By the early eighteenth century, this number had risen to 450, not including those receiving a regular

24 For details, see Rıfat Günalan, Mehmet Canatar, and Mehmet Akman, *Istanbul Kadı Sicilleri Üsküdar Mahkemesi 51 Numaralı Sicil H.987–988/M. 1579–1580* (Istanbul: İSAM, 2010), 86, decree 55; Rıfat Günalan, Mehmed Akan, and Mehmed Canatar, *Üsküdar Mahkemesi: 84 Numaralı Sicil (h. 999–1000)* (Istanbul: İSAM, 2010), 519, decree 1018 and Rasim Erol, Mehmed Akman, and Fikret Sarıcaoğlu, *İstanbul Mahkemesi 12 Numaralı Sicil (H.1072–1074/ M.1663–1664)* (Istanbul: İSAM, 2010), 241, decree 227. Hereinafter, references to the *Mühimme* registers follow the classification system of the BOA, using the abbreviation A.DVNSMHM.d for *Mühimme Defteri*, followed by the register number and the decree number. BOA, A.DVNSMHM.d 67, decrees 243 and 273, date: 15 cemāẕiyü'l-evvel 999 AH. See also Caroline Finkel and Aykut Barka, "The Sakarya River-Lake Sapanca-İzmit Bay Canal Project: A Reappraisal of the Historical Record in the Light of the Morphological Evidence," *Istanbuler Mitteilungen* 47 (1997): 429–442.

25 Luigi Marsili (sometimes Marsigli) (1658–1730), a polymath, passionate bibliophile and diplomat, observed closely the skills of the Ottoman sappers at the siege of Vienna (1683) and during his temporary enslavement in an Ottoman Pasha's household at the Habsburg–Ottoman frontier. In his description of the making of trenches, he drew attention to the fact that the smaller the tunnel is, the easier it is to build. But whether the tunnel is small or wide, shallow or deep, it must be lined. This the sapper does by placing wooden planks either vertically or horizontally (*sertek*) against the earth at the sides and roof of the tunnel. Luigi F. Marsili, *L'État militaire de l'Empire ottoman, ses progrès et sa décadence* (La Haye, Amsterdam: P. Gosse et al., 1732), 37–40.

26 BOA, KK 7422, date: rebīʿü'l-āḫir 1022 AH. Conflicts among the villagers sometimes arose and ended up in court when one refused to pay taxes, claiming exemption due to service. Such instances offer some insights into the lives of the villagers and suggest that not all inhabitants of the villages were involved in the repair and care work of the city's water supply infrastructure. See e.g., Rıfat Günalan, Mehmed Akman, and Mustafa Oğuz, *Bâb Mahkemesi 3 Numaralı Sicil (H. 1077 / M. 1666–1667)* (Istanbul: İSAM, 2011), 95, decree 10.

FIGURE 9.3 Illustration of a *lağımcı* (*operari miniere*)
SOURCE: DETAIL FROM LUIGI FERDINANDO MARSILI, *L'ÉTAT MILITAIRE DE L'EMPIRE OTTOMAN, SES PROGRÈS ET SA DÉCADENCE* (LA HAYE, AMSTERDAM: P. GOSSE ET AL., 1732), PLATE XIV. BIBLIOTHÈQUE NATIONALE DE FRANCE

stipend working for waqfs in other parts of the city.[27] Additionally, palace pages (*ġılmān-ı ḫāṣṣa*, also *ġılmān-ı 'acemīyan*) were assigned to serve the superintendent of the imperial water mains and the imperial chief architect outside of the Topkapı Palace (*çıkma*).[28] The 450 Ḫāṣlar peasants thus made up only a portion of the *ṣuyolcular*.

Various documents of the Ottoman central administration related to the design, construction, and repair of Istanbul's aqueduct-fed underground water supply network repeatedly refer to the worker categories *lağımcı* and *ṣūyolcu* or *rāḥ-ı ābī*. However, these categories were fluid and are found along with, *erbāb-ı* or *ehl-i vuḳūf* (professional practitioners), *ser-rāh-ı āb* or *serbölük* (chief

27 BOA, MAD.d 8484/32, date: 22 şevvāl 1112 AH; BOA, MAD.d 9892/216/1, date: 25 muḥarrem 1116 AH. In my earlier publications, I mistakenly stated the total number as 450.

28 BOA, A.DVNSMHM.d 4, decrees 922 and 979, date: 967 AH.

of water technicians for the branch area), *üstād* or *ustā* (master), *ırġād* (agricultural worker, hired laborer), *yamakān*, and *reʿāyā* (peasant)—suggesting occasional and shifting occupations, which makes it difficult to say anything very definite about these workers' identities or affiliations with guild structures or organizations. Put simply, during the period under consideration, the workers who engaged in hydraulic activities in the capital were either contractors (in building trades or in agriculture, etc.), had enough skills to shift between professions, or engaged in multiple activities simultaneously. Furthermore, some organized themselves into guilds (*eṣnāf*), each with their own guild head—as in the case of the tunnel diggers' guild (*lağımcı ṭāʾifesi*) in Üsküdar by the early eighteenth century.[29] All played a crucial role in the smooth functioning of Istanbul's water supply and distribution. Together with the *künkcū* (the artisans who made the *künk*), these laboring groups performed the actual construction and maintenance of hydraulic works and came in regular contact with soil between and throughout the subterrain and the surface.

Looked at this way, these low-ranking workers quite uniquely fell into the category Lydia Barnett has named "earth workers," a term I use in this chapter to highlight the informal and seemingly mundane but nevertheless important role they played in shaping knowledge about the soils and physical landscape of Istanbul. Barnett calls for us to revisit the practice of "acknowledging" in eighteenth-century natural history in order to think through how various worksites doubled as scientific sites of fieldwork and how European naturalists kept their low-ranking discoverers—including the shell diggers and fossil finders on whom they heavily relied—well-hidden and marginalized.[30] This was no isolated occurrence. Low-ranking workers engaged in the construction and repair of Istanbul's water supply network were rarely recognized by the Ottoman scholars and bureaucrats. The renowned chief physician from Chios, Sakızlı İsa Efendi (d. 1649) devoted a few lines to the process through which "the buckets lift the water entering the waterwheel and drop it abundantly into a marble slab laid beneath [*altına mermer döşeyüp*] with such mighty force that … water returns [*rucūʿ*] to its original qualities [*aṣil*] and becomes [*ṣāfi*] better [*aʿlā*] and lighter [*ḫiffet*]."[31] Yet he chose to employ the anonymizing

29 In 1727, a group of tunnel diggers in Üsküdar chose a new headman (*lağımcı-başı*) by the name of Vasil. BOA, C. BLD.141-7034, date: ṣafer 1140 AH.

30 Lydia Barnett, "Showing and Hiding: The Flickering Visibility of Earth Workers in the Archives of Earth Science," *History of Science* 58, no. 3 (2019): 245–274.

31 Sakızlı İsa Efendi served as the chief imperial physician (*ḥekīmbaşı* or *reʾīs ül-aṭibbā*) at the court of İbrāhīm IV (r. 1640–1648), having been deposed and reappointed to the position four times. Sakızlı İsa Efendi, *Nizâmü'l-Edviye*, ed. M. Bedizel Aydın and Sibel

pronoun *derler* and offered no recognition of the laborers actually doing the work of digging out the silt below the waterwheel to keep the water flowing with such great force.

Soil formed the meeting ground—both theoretical and practical—for a range of officials, scholars, elite patrons, artisans, and various high- and low-ranking members of the Ottoman building trade. The voluminous, complicated legal records (*şerʿīyye sicilleri, şikāyet defterleri*), surveys (*tevzīʿ ve taḳsīm defterleri*), and administrative and financial registers (*mühimme, keşf-u taʿ mirāt, vāridāt-u maṣraf*, and *muḥāsebe defterleri*) are sources of material on the manual or tacit knowledge, know-how, and experience of these low-ranking groups of laborers that has often been neglected in historical accounts. In other words, apart from recording the construction of specific buildings, these archival documents are components of a broader, methodically organized body of knowledge, which encompasses information concerning groups of workers, including those involved in hydraulic construction.

While such workers exercised agency of their own, they were subjects of the soil and the continuous interaction between the climate and subterranean hydrolithic formations (aquifers), as well as the hydraulic structures and building foundations that crossed all of the different subterranean layers. The earth itself became a conduit for the flow of water, and, whether rising or falling, the surrounding terrain was susceptible to being inundated, destabilized, and transformed by water. It should come as no surprise that the soil classification of the Ottoman administration often went hand in hand with concerns about its porosity and issues raised by subterranean construction—in modern parlance, soil mechanics. What sorts of problems were the earth workers called upon to solve? What challenges did they face in diverse field conditions? How did they work on practical problems? And how did soil management inform techniques of hydraulic and structural engineering? Evidence in archival records answers some of these questions. The following section focuses on solutions found to the problem of Istanbul's muddy streets in the late sixteenth century. The continuous flow provided by city fountains spilled into the streets raising concerns about the accumulation of mud, as well as about the waste and uneven distribution of water.

Murad (Ankara: Türkiye Bilimler Akademisi, 2019), 868. Arranged in alphabetic order, *Niẓāmü'l-Edviye* (*The Order of Remedies*) was a medical text comprised mostly of drug recipes. This is from his entry on water (*māʾ*).

2 Trouble with Mud

In the early modern period, Istanbul's relationship with mud was far from simple. The dictionaries of the time list four words that were all in some way related to mud: *bālçıḳ*, *çāmur*, *ṭīn*, and *gil*. Himmet Taşkömür's contribution to this volume highlights a sixteenth-century Persian-to-Turkish lexicon entry for *gil-gūy*, which describes a type of black mud (*ḳarabālçıḳ*) formed by the constant trampling of sheep. *Çāmur* is not to be found in the seventeenth-century trilingual dictionary *Risāle-i miʿmāriyye* (A Book on architecture) written by Caʿfer Efendi (d. after 1633), but the entry for *bālçıḳ* (*ṭīn* in Arabic and *gil* in Persian) is also among the entries related to building materials. If *bālçıḳ* is mixed with straw to plaster walls, Caʿfer writes, it is called *samanlı sıvā bālçıġı*; if it is placed inside walls, it is called *helik*. He also places the word among his entries on tools; *bālçıḳ* is listed together with a pickaxe (*ḳazma*) used for dredging.[32] *Bālçıḳ* is also listed in *Lehçetü'l Luġat*, written between 1723 and 1732 by Esʿad Meḥmed Efendi (1685–1753), who also expanded the definitions by listing various types. For instance, his entry for *ḳarabālçıḳ* denotes a black mud which remains after water recedes and dries, or one found at the bottom of a pool (*żavīṭa*). Other types include *revġa*, a dark and watery (*ṣulu*) form of *bālçıḳ*; *ṣalṣāl*, a mixture of *bālçıḳ* with sand (*ḳum*) used in pottery; and *ḥulb*, a sticky, black, and viscous (*özlü ḳara*) form of *bālçıḳ*. For *çāmur*, Esʿad did not provide a separate definition but simply noted that it is the same as *bālçıḳ*.[33]

Indeed, in a wide variety of archival records, *çāmur* and *bālçıḳ* were used interchangeably until the late eighteenth century. Still, whether separated or placed side by side, their placement shows the intersection of labor and

32 Caʿfer Efendi, *Risâle-i miʿmâriyye* (in romanized Ottoman Turkish; with a facsimile of MS. Yeni Yazma 339, ca. 1614, Topkapı Palace Museum Library), ed. I. Aydın Yüksel (Istanbul: İstanbul Fetih Cemiyeti, 2005), fol. 74b, 101; fol. 75b, 103; and fol. 79b, 113. For a comparison, see Celal Esad Arseven, *Istılahât-ı Miʿmâriyye*, ed. Şeyda Kalay (Istanbul: Kaktüs Yayınları, 2017), 43, 69. According to the popular Persian-to-Turkish lexicon *Luġat-i Niʿmetu'llāh* written by Niʿmetullāh Aḥmed al-Rūmī (d. 1561), a Nakşibendī sufi writer, *gil* is *bālçıḳ* in Turkish, and the entry for *kah-gil* defines this as *bālçıḳ* mixed with straw for plastering walls. Niʿmetu'llâh Ahmet, *Lügat-i Niʿmetu'llâh*, ed. Adnan İnce (Ankara: Türk Dil Kurumu Yayınları, 2015), 410, 444. See also Franciszek Mesgnien Meninski's entry "[چامور] *ćiamur* [طين] *tȳn* [كل] *gîl* [بالچق] *balćik*" in *Thesaurus linguarum orientalium Turcicae-Arabicae-Persicae* [...] (Vienna, 1680), 1562–1563, https://archive.org/details/bub_gb_7MBIAAAAcAAJ/page/n815/mode/2up.

33 Şeyhülislām Esʿad Efendi, *Lehçetü'l Luġat* (*alt. sp. Lügat-i Niʿmetu'llâh*) (Istanbul: Matbaʿa-i ʿāmire, 1853–1854), 249–250, 350. On his brief biography, see H. Aksoy, "Esad Mehmed Efendi (Ebû İshâk-zâde)," *Türk Dili ve Edebiyatı Ansiklopedisi* 3 (Istanbul: Dergâh Yayınları, 1979) 86–87.

knowledge production. In the records, mentions of *çāmur* and *bālçıḳ* are to be found in three types of situations. The first are situations related to reports of excess water in fountains or illegal water diversion that required an immediate response. The second involved the reporting of, or response to, the building and repair of tunnels (*laġm*), covered channels (pl. *ḳanavāt*), ditches (*ḫendeḳ*) and laying pipes. The third arose when *çāmur* and *bālçıḳ* were used to manufacture the water supply pipes (*künk*). Reports from the late sixteenth through to the early eighteenth century testify to techniques and operational steps as well as to the material forces that shaped earth workers' practices of observation, manipulation, and experimentation with physical nature, including topographic and geological features and types of soil such as silt (*lıġ*), sand, stone, gravel, and of course, various kinds of mud. By the late sixteenth century, the addition of seventeen water spouts brought the total amount of water flowing in Istanbul's aqueduct-fed hydraulic infrastructure to eighty-one *lüle*—providing enough water to fill more than 1.5 Olympic-sized swimming pools.[34] The increase was due to the completion of the Kırkçeşme and Kağıdhane waterways (*ṣu yollarī*), with new sources from the Belgrad Forest and village of Cebeciköy (about 2.5 km northeast of present-day central Sultangazi) added under the royal prerogative of Sultan Süleymān I (r. 1520–1566).[35] In the fall of 1564 (972 AH), the new water supply significantly changed Istanbul's physical

34 According to the earliest distribution register (*tevzīʿ defteri*) of the capital's water supply network, dating from 1568–1569 (976 AH). This document is published in Kâzım Çeçen, *Mimar Sinan ve Kırkçeşme Tesisleri* (Istanbul: İstanbul Büyükşehir Belediyesi, 1988), 165–169. For a comparison, see also Kâzım Çeçen, *İstanbul'un Osmanlı Dönemi Suyolları*, 107. A *lüle* is both a measuring unit of waterflow and a waterspout or pipe. One *lüle* equals the amount of water that flows through a standard pipe in twenty-four hours (approximately 52 m^3 per day). This capacity and velocity requires a pipe of thirty *dirhem* (about 96.3 gr), that is, with an internal diameter of roughly 74 mm, through which a lead ball of thirty dirhem could pass. For further details, see A. Àndréossy, *Constantinople et le Bosphore de Thrace pendant les années 1812, 1813, et 1814* (Paris, 1828), 390; Kâzım Çeçen, "Osmanlı Suyollarında Künkler, Debi Ölçme Tertibatı ve Su Terazileri," in *Osmanlı İmparatorluğu'nun Doruğu 16 Yüzyil Teknolojisi*, ed. Kâzım Çeçen (Istanbul: ISKI, 1999), 59; Metin Sözen and Uğur Tanyeli, "Lüle," in *Sanat Tarihi Kavramları ve Terimleri Sözlüğü* (Istanbul: Remzi Kitabevi, 2010), 193.

35 During the Ottoman period in Istanbul, the term waterway (*ṣu yollarī*) was used to describe elements of the water supply system. The word *yol* (pl. *yollar*) meant both "road" and "way" (*ṭarīḳ*), and it also referred to the act of laying stone on soil, covered channels (*ḳanavāt*), or, more commonly, an underground conduit (*kārīz*). See Cengiz Orhonlu, *Osmanlı, İmparatorluğunda Şehircilik ve Ulaşım Üzerine Araştırmalar* (Izmir: Ege Üniversitesi Edebiyat Fakültesi yayınları 31, 1984), 28. James W. Redhouse, *A Turkish and English Lexicon*, 4th ed. (Istanbul: Çağrı Yaynları, 2011), 2218; and Şemseddin Sami, *Kamus-ı Türki* (Istanbul: İkdam Matbaʿası, 1899), 1566–1567.

landscape, as it reached all districts within the city walls. It was still essentially based on water sources at elevations between seventy and one hundred and fifty meters, but a succession of conduits raised by siphons (*terāzū*) meant that water coursed to the lowest levels: ranging from the low-lying parts of the city (lower levels of the first hill) that ran from west to east to the land walls at the gate of Eğrikapı, and from areas along the bases of the five hills overlooking the Golden Horn, to the valley of the Lycus river, the Avrat Pazarı public market, and the Yedikule fortress in the southwestern extremity.[36] Initially, at least one hundred and forty fountains were fed by the Kırkçeşme and Kağıdhane waterways; before the beginning of the seventeenth century, it would be more than three hundred.[37]

However, this physical expansion exposed a major flaw in the water resource utilization. The continuous flow from fountains turned unpaved streets into sheets of viscous mud. An illustration in the travel album by European traveler, collector, and mayor of Gdansk Bartholomäus Schachman (1559–1614), who visited Istanbul in circa 1588–1589, vividly captured this scene.[38] Folio 27v (Figure 9.4) shows waterspouts set in a reddish stone wall, and in front of them a number of men are seen in the act of ablution before prayer; and to the right side, uneven patches of water-saturated mud.

A muddy landscape represented disorder. It was a cause for concern and a threat to the healthful ideals of Ottoman urbanization. In the well-known narrative of ʿĀşıkpāşāzāde's *Chronicles of the House of Osman*, written in the 1480s, the completion of infrastructure projects like the bridge over the Ergene in 1443 (situated today in Turkish Thrace near the Greek border) held a dual role. The whole terrain was in need of improvement, because, as ʿĀşıkpāşāzāde explains, the prevailing muddiness (*çāmur*), swampy (*çökek*) forest, and unruly highwaymen (*ḥarāmī adamlar*) were obstacles to the economic and moral imperatives of cultivated land. Thus, thanks to the piety of Sultan Murād II

36 Kate Ward, James Crow, and Martin Crapper, "Water-Supply Infrastructure of Byzantine Constantinople," *Journal of Roman Archaeology* 30 (2017): 179–182.

37 "Maḥrūse-i Istanbul'da üç yüzden mutecāviz çeşmelerin taʿ mīr ve termīm lāzim gelen iḫrācāt cumle-i evḳāf mezbureden olduğundan." BOA, A.DVNSMHM.d 73, decree 1305, date: 13 ramażan 1003 AH. On the list of fountains supplied from the Kırkçeşme and Kağıdhane waterways in 1677, see BOA, TS.MA.e 560-87, date: ẕi'l-ḥicce 1087 AH.

38 This album is now part of the collection of the Lusail Museum in Doha-Qatar. See Tadeusz Majda, Magdalena Mielnik, and Piotr Łuba, *Bartholomäus Schachman (1559–1614): sztuka podróży/Bartholomäus Schachman (1559–1614): The Art of Travel*, exh. cat. (Gdańsk: Muzeum Narodowe, 2012); and Olga Nefedova, Anna Frackowska, and Hyejung Yum, *Bartholomäus Schachman, 1559–1614: The Art of Travel* (Milan: Skira, 2015). Bartholomäus's Ottoman travels coincided with the reign of Murād III.

FIGURE 9.4 Turks Performing Ablution before Praying. From the Travel Album of Bartholomäus Schachman (27v), 1590, watercolour, pencil on paper, 19.8 × 13.1 cm, inscription in iron gall ink. OM. 749
SOURCE: ILLUSTRATION INCLUDED IN NEFEDOVA, FRACKOWSKA, YUM, *THE ART OF TRAVEL*, COURTESY OF PROF. OLGA NEFEDOVA, © QATAR MUSEUMS / LUSAIL MUSEUM, DOHA—QATAR

(r. 1421–1444; 1446–1451), the land was cleansed (*pāk etdürdi*) and transformed into a flourishing town (*şehir etdi*) at both ends of the bridge.[39]

In the early summer of 1568 (967 AH), to clean up the problem of muddy streets in the capital, the grand vizierate decreed that *burma lüle* (twisted pipes) were to be installed in the fountains.[40] We know little about the design of such pipes in the sixteenth century, but the decree grants them starring roles. The *burma lüle* shut off the water flow once the water level in the reservoir reached a certain level, preventing overflow and reducing the accumulation of mud on the streets.[41]

39 Here, I have used the most recent edition of ʿĀşıkpāşāzāde's chronicles. Necdet Öztürk, ed., *Âşıkpaşazâde Tarihi (1285–1502)* (Istanbul: Bilge Kültür Sanat, 2013), 102.

40 BOA, A.DVNSMHM.d 7, decree 1620, date: Ġurre-i muḥarrem 976 AH; published in Ahmet Refik, *On Altıncı Asırda Istanbul* (Istanbul: Devlet Basımevi, 1935), 17–18. The imperial order also decreed that the surplus water could serve as a supply to those willing to build a new fountain or a faucet (*muşluḳ*).

41 Indeed, the use of the term *burma* to denote the new spout (*lüle*) is not to be found in the seventeenth-century dictionaries of Meninski and Caʿfer Efendi. The definition in the

The search for a durable solution to the mud problem also led to resistance from many who feared it would negatively impact their well-being. Prominent residents such as the neighborhoods' imam repeatedly wrote to the Imperial Divan, noting: "this water has been given to us to flow to our produce gardens (*bostān*) and orchards (*bāğçe*) ... and we do not consent to the installation of a *burma lüle* even if waters flow in vain."[42] In 1583 (991 AH), Emine Ḫātūn, about whom the available record says nothing, but who must have been a woman of some means, argued her case in a petition and provided similar reasoning for her private use of fountain waters: it was beneficial to take the excess water of the old fountain into her own garden in Yeni Bāğçe near the Edirne gate. The neighboring community had no use for it, and because of her undertaking, "there was [now] no waste of water and no mud on the roads."[43] Clearly, different city inhabitants had vastly different understandings of what constituted a proper use of the city's conduit waters.

Until the first half of the seventeenth century, soil that turned to mud in the capital city was not just a background to city life but a dynamic character in changing expectations about the proper use and allocation of Istanbul's water supplies. The problem of letting water run was by no means new. However, this new measure may have been aimed at taking the regulation of water flow away from established groups at a time when Sultan Murād III (r. 1574–1595) was attempting to reassert monarchical power and marginalize the vizier

Redhouse dictionary for the more modern form refers to a screwed, twisted, convoluted spiral or something formed by screwing twisting or winding. Redhouse, *A Turkish and English Lexicon*, 396. Later construction appraisal registers indicate that the valve bodies were generally made of iron or brass which must have provided resistance to corrosion and friction.

42 "Aḳan şu bāğçelerümüze ve bostānlarumuza aḳmaḳ ecliyçün bu şu bize virilmişdür ... aḳarsa aḳsun, riżā'muz yoḳdur." Hacı Osman Yıldırım et al., *7 Numaralı Mühimme Defteri: (975–976/1567–1569) 2, Özet, Transkripsiyon, Indeks* (Ankara: Başbakanlık Devlet Arşivleri Genel Müdürlüğü, Osmanlı Arşivi Daire Başkanlığı, 1998), 224, decree 1620. See also BOA, A.DVNSMHM.d 30, decree 362, date: 25 ṣafer 985 AH. Something similar arose with water distributors (*saḳḳā*) transporting drinking water on horseback and those drawing water for private use. According to the imperial decree dated to 1572 (980 AH), the number of water distributors increased, and they began to prevent fountain-goers from accessing water. Additionally, some garden owners opened *burma lüle* at night to irrigate their gardens, while neighborhoods (*maḥalle ḫalḳī*) and bathhouse keepers colluded to divert water illegally. Mehmet Bostancı, "19 Numaralı Mühimme Defteri (Tahlil-Metin)" (MA thesis, Istanbul University, Turkey, 2002), 239–240, decree 324.

43 "yabana aḳan ṣudan yollar çāmur olmayub zā'il olmaması ecliyçün." BOA, A.DVNSMHM.d 52, decree 109, date: 14 ramażan 991 AH. The response of the imperial court came in October 1583 (991 AH). Emine Ḫātūn received a satisfactory answer.

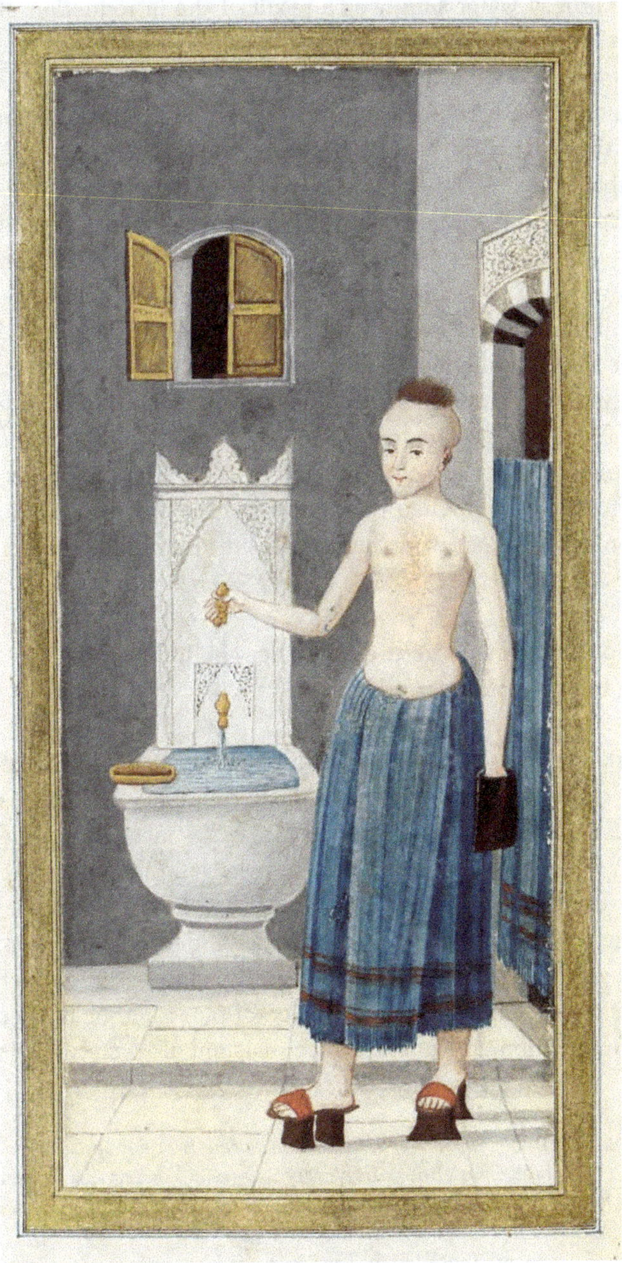

FIGURE 9.5 Illustration of a *burma lüle* in Fāẓıl Enderūnī, *Hūbannāme ve Zenānnāme* [1793]
SOURCE: ISTANBUL UNIVERSITY, RARE BOOKS LIBRARY, NEKTY 05502, FOL. 50R. IMAGE COURTESY OF İ. Ü. NADİR ESERLER KÜTÜPHANESİ, ISTANBUL

households.[44] The improved water supply as a result of the introduction of the *burma lüle* offered a mundane but effective means of establishing the palace as the sole holder of the authority to control and regulate the city's waters. Whether or not this was the real motivation behind the installation of the *burma lüle*, in the roughly thirty years between 1568 (976 AH) and 1595 (1003 AH), the available evidence shows that streets free of mud proved valuable as a means of legitimizing who could establish, manage, maintain, and regulate the capital's public water supply. Furthermore, the implementation of the twisted pipes was tied to technical adjustments, experience, and the application of hydraulic knowledge. Over the next centuries, even if people did occasionally break or leave them open to allow water to run into the streets, the continuous flow system, once common, was long forgotten. Many *lüle* were converted into *burma*, where excess water (*girü tepen ṣu*) would accumulate, and these were no longer worthy of special notice, becoming mundane fixtures of daily life (Figure 9.5). Then, their initial novelty gave way to aesthetic reinterpretation whereby *burma lüle* were transformed into decorative objects.

3 Buried Tensions: the Soil-Structure Challenges in Building and Maintaining the Water Mains and Pipes

Around 1600, concerns over muddiness in Istanbul's streets—caused by the continuous outflow from neighborhood fountains—prompted authorities to take action. The new regulations and technical measures taken to resolve the issue increasingly drew upon earth workers' knowledge in managing the behavior of water. Besides the new designs in fountains, as I discuss elsewhere, the second half of the seventeenth century saw an increase in private feeder lines (*ḳātma*) being used to pipe conduit waters directly to houses and inner courtyards—as well as the rise in the illegal diversion of pipes, adding further complexity to the construction and maintenance of Istanbul's waterways.[45]

---

44 Baki Tezcan, *The Second Ottoman Empire: Political and Social Transformation in the Early Modern World* (Cambridge: Cambridge University Press, 2010), 80, 95–104; Özgen Felek, "Re-creating Image and Identity: Dreams and Visions as a Means of Murad III's Self-Fashioning" (PhD diss., University of Michigan, 2010).

45 In the seventeenth and eighteenth centuries, the legal framework governing water allocation and use was grounded in ownership rights. Once an individual constructed a *ḳātma* and connected it to the waterways managed by the imperial waqf supplying various neighborhoods within intra muros Istanbul from the Ḫāṣlar district, they were legally entitled to the additional water as freehold property (*mülk*). On this topic, see Deniz Karakaş, "Water Resources Management and Development in Ottoman Istanbul," Gülfettin Çelik,

These developments meant that ad hoc methods would no longer suffice, and they stimulated the search for specific construction methods and hydraulic technologies to observe, measure, and stabilize the invisible underground water that ran beneath open fields and inhabited areas of the capital.

Despite the regulations in place, many water channels were covered, built over, and forgotten over time, and the filth, debris, and muck continued to return.[46] Tracing the construction of the previously mentioned new feeder line toward the end of the sixteenth century provides a telling example of the impact that changes in the use and physical layout of land, in a relatively short period, had on subsurface design and accessibility. In 1584 (992 AH), after obtaining a permit from the Sultan, the third vizier, Mesīḥ Meḥmed Pāşā purchased many plots in the Yeni Bāġçe quarter. Shortly before his death, he built a funerary mosque complex on this land and established a waqf.[47] Fourteen years later, rather than cleaning and repairing the pipes that no longer supplied an adequate flow of water, the administrator of the waqf contracted three Christian men, Koni, Dimo, and Koze, to build a new conduit to channel water from the head springs of the village of Cebeci in the Ḫāṣlar district, some eighteen kilometers north of the city.[48] Although the records are silent about the occupations and identities of these three earth workers, they were probably agriculturalists with irrigation knowledge. The administrator spent around 518,000 piasters (*aḳçe*), a large sum, to ensure that sufficient water was supplied to the mosque and its dependent facilities. However, the former vineyards outside the Edirnekapı gate (*ḫāric-i bāb-ı Edirne*) were now a large cemetery, making it difficult to dig shafts and access the subterrain pipes, thus

---

Vakıf Su Tahlilleri 2: Su Hukuku Ve Teşkilatı (İstanbul: İSKİ, 2000) and Ebu-l Ula Mardin, *İstanbul'un Vakıf Katma Suları* (Istanbul: Kenan Basımevi, 1939).

46 The regulations to protect the drinking water by integrating sewage into the urban layout came at an early point in Greater Istanbul. Compare imperial orders in Mehmet Bostancı, "19 Numaralı Mühimme Defteri: Tahlil ve Metin," 179–180, decree 376; BOA, A.DVNSMHM.d 14 decree 742, date: 29 cemāẕiyü'l-evvel 978 AH and decree 1545, date: 18 ẕi'l-ḥicce 978 AH; İsmet Binark ed., *12 Numaralı Mühimme Defteri (978–979/1570–1572): Özet-Transkripsiyon ve İndex* (Ankara: Başbakanlık Devlet Arşivleri Genel Müdürlüğü, Osmanlı Arşivi Daire Başkanlığı, 1996), 364, decree 536 and repair registers BOA, D.BŞM.BNE 387-16248, date: 08-17 ṣafer 1235 AH.

47 Said Öztürk, *İstanbul'un Tarihi Su Yolları*, 2 vols. (Istanbul: İSKİ, 2006), 1: 184, decree 134. See also BOA, A.DVNSMHM.d 58, decree 36, date: 8 rebīʿü'l-āḫir 993 AH; BOA, TS.MA.e 80-3, date: 10 rebīʿü'l-āḫir 993 AH. On the imperial order permitting Mesīḥ Meḥmed Pāşā to supply his new fountain built in the vicinity of the mosque, see BOA, TSMA 317-266, date: 10 rebīʿü'l-āḫir 993 AH.

48 BOA, TS.MA.e 834-7, date: cemāẕiyü'l-āḫir 1006 AH; BOA, TS.MA.e 180-25, date: 1 receb 1007 AH.

the administrator was forced to look for other options. Put differently, examining underground conduits and pipes was difficult without access.

In 1579 (987 AH), a case came before the Üsküdar court: three Armenian workers named Ḫudāverdi, Yaḳob, and Vasil were hired for the completion of a tunnel through existing rock cavities. But after the start of the excavation, just below the village of Bulgurlu in the vicinity of Üsküdar they discovered that the subsoil was not rock (*ḳayā*), necessitating a different construction method. They reported the need to shore up the tunnel in stone and brick masonry (*kārgīr*). However, they did not have enough experience of building this sort of tunnel, so they carried out the work in collaboration with two Armenian *lağımcı* named Simaven and Āyvād. From their testimonies, we learn that the new hires dug out five shafts at regular intervals. Such shafts (*bāca*) can typically aid in ventilation and provide entrance to a gallery or tunnel.[49] This time the shafts also exhibited a novel solution for the problem of the tunnel's inclination. For the flow of water, the tunnels were to have a continuous fall and be carefully graded. The three diggers found the proper level for the tunnel by excavating each shaft at a different depth (*ʿumḳ*), starting with 7.5 *ḳulac* (~14 m) and ending with five *ḳulac* (~9.5 m).[50]

These two hydraulic projects and their related inquiries suggest that answering the question of what to do with what was underneath called for blending technical knowledge and knowledge of the natural world. Attending to the particularities of subsoil was crucial, and these examples demonstrate that the intricate relationships between rock, soil, mud, and water left little room for inexperienced or unskilled workers. The hydraulic projects in Istanbul in this period increasingly required an ability to address problems arising from the dynamic and unstable interactions between soil and hydraulic structures. The composition of soil and subsoil varied greatly across greater Istanbul. Some of the earth workers appear to have taken advantage of the natural heterogeneities in the subsurface, such as hidden cavities or crevices in the

49 Arseven, *Istılahât-ı Miʿmâriyye*, 39–41. Redhouse gives two meanings to *bāca*: (1) Any small opening in a roof that admits air or light; a skylight; a trap door. (2) A chimney flue or other shaft of the same character as a sewer ventilator, a mineshaft. Redhouse, *Turkish and English Lexicon*, 316. It is also called *nefeslik*, which can be roughly translated as "vent." Sami, *Kamus-ı Türki*, 259. For a few extant examples in extra and intra muros Istanbul, see Selim Karahasanoğlu, *Küçükköy-Taşlıtarla, Gaziosmanpaşa: Bir İlçenin Öyküsü* (Istanbul: Gaziosmanpaşa Belediyesi Yayınları, 2013), 430–431 and Cahide Nur Özdemir, "Unesco Dünya Miras Alanları Bağlamında Süleymaniye Bölgesi Yer Altı Envanterinin Değerlendirilmesi" (PhD diss., Fatih Sultan Mehmet Vakıf Üniversitesi, 2020), 265–266.

50 Rıfat Günalan et al., *İstanbul Kadı Sicilleri Üsküdar Mahkemesi 51 Numaralı Sicil (H. 987–988/ M. 1579–1580)*, 86, decree 55.

bedrock that were not visible from the surface. This knowledge likely came to be acquired from years of practical experience and keen geological observations, allowing them to guess where rock strata might pass under their feet.[51]

A few examples illustrate this variety. A routine problem encountered in greater Istanbul was the collapse (*munhedim*) of one of the deep conduits. Some lacked strength and stability or had been damaged or shifted by the weight of the earth above them. Sometimes the source of the trouble was surface inlets (*ıskara*). These filtering structures were to prevent debris, sediment, or large particles from entering the tunnels and pipes. But when they were broken or displaced, earth, silt, and stone backfill entered the system (*memlū*), clogging the conduits and displacing the pipes, which caused the points where the pipes joined to open or break. One such example occurred in the winter of 1714 (1126 AH) near the village of İvaz (also called Avas) on one of the branch lines of the Ḥalkalı waterways (*mīrī*, also referred to as *ḫāṣṣa* or *beylik*).[52] Some portions were clogged with sand and mud (*ḳum ve çāmur ile memlū*), and some substantial underground structures were broken. The work on this was carried out in cold weather and required the damaged sections of the conduit (*ta'mīr-i laġmīyye*) to be reconstructed, which included digging a steep descent for a tunnel no less than twelve *ẕirāʿ* wide and forty *mesāḥa* (~30 m) long through solid rock (*ḳayā*), making even the measurement of inclines through the rugged terrain a challenge.[53] This conduit was connected to a fifty *mesāḥa* (~37.5 m) tunnel cleaned (*taṭhīr-i laġım*) and reinforced by inserting, at intervals, columns (*sutūn istāde*), and vertical timber bracing (*taḫta*). Finally, a four-hundred *ẕirāʿ* (about 300 m) subterranean stretch of pipes (*künk*) had

51 For example, during the cleansing and repair of a feeder line, which included a notably long 1,100 *ḳulac* tunnel cut through rock in the vicinity of the village of Cebeciköy. BOA, İSTM.ŞSC.15.d 126 76b/1, date: 1121 AH; BOA, İSTM.ŞSC.15.d 126 111a/2, date: 1121 AH; BOA, İSTM.ŞSC.15.d 127 27b/2, date: 1122 AH. Compare BOA, İSTM.ŞSC.15.d 114 42a/3, date: 22 ramażan 1113 AH; BOA, İSTM.ŞSC.15.d 118 8b-9a/3-1, date: 22 ẕi'l-ḳaʿde 1113 AH. For locations, see Cazibe Arıç, "Haliç—Küçükçekmece Gölü Bölgesinin Jeolojisi" (PhD diss., Istanbul Technical University, Turkey, 1955) and İbrahim Gedik et al., *1/100,000 ölçekli Türkiye jeoloji haritaları İstanbul—F 21 ve Bursa G 21 paftaları* (Ankara: Maden Tetkik ve Arama Genel Müdürlüğü, 2014).

52 BOA, İSTM.ŞSC.15.d 132 47b/3, date: 12 muḥarrem 1126 AH. For details on the springs feeding the channels and system configuration, see Kâzım Çeçen, *İstanbul'un Vakıf Sularından Halkalı Suları* (Istanbul: İSKİ, 1991); James Crow, "Waters for a Capital: Hydraulic Infrastructure and Use in Byzantine Constantinople," in *The Cambridge Companion to Constantinople*, ed. Sarah Bassett (Cambridge: Cambridge University Press, 2022), 67–86.

53 The *mesāḥa* measure varied over time and place. A document from 1705–1706 records 6421.5 *mesāḥa* as equivalent to 2568 *ḳulac*, which suggests that the *mesāḥa* was equivalent to the *ẕirāʿ* in Istanbul. So, here, I have used the round figure of 75.0 cm for 1 *mesāḥa*. BOA, MAD.d 1655, fols.20r–21v, date: 14 muḥarrem 1117 AH–ṣafer 1118 AH.

been displaced, resulting in a large number of broken pipes that needed to be removed, replaced, and re-laid. In addition, an eighty *mesāḥa* (~60 m) section of the pipeline that ran through the brook (*dere*) south of neighboring Muderris village had rested on a bed of stones that had been dislodged (*deruñlarında ṭāş yetmeğin*) over time. Thus, the bottom needed to be covered anew with a layer of heavy stone (*ṭāş*) to prevent the pipes shifting. The repairs were overseen by the superintendent of the imperial water mains (*ḫāṣṣa ṣu nāẓırı*) and the imperial architect both reporting to the Palace administration. It seems unlikely that they made the descent themselves. Their report refers to two Muslim *ṣuyolcular*, named el-Hāc Ḥāsan and el-Hāc Ibrāhīm, and some unnamed professional practitioners (*erbāb-ı vuḳūf*), masters (*üstādlar*), and porters (*ḥammāl*), who were tasked with preparing the ground (*ṣu laġīmlārī derūñlarī*) and doing the repairs.[54]

This case illustrates the differentiated workforces required to deal with the diverse field conditions in greater Istanbul. Although these documents do not explicitly detail the specific skills, knowledge, and experience of the earth workers involved, the required experience and know-how is implicit, allowing the modern observer to understand how they threaded the waterways through mountains, streams and rivers, and soil layers (Figure 9.6). Among the cases described above, one key skill that emerged across various periods of hydraulic projects was rock excavation. This skill seems to have been consistently sought by patrons over time, most likely because rocky substrata helped reduce costs and minimized the need for reinforcement and maintenance.[55]

Another laborious task addressed by the earth workers was digging and refilling ditches (*ḫendeḳ kuşādī*, also called *ḫendeḳ ḥafrī*). It was a crucial step in the installation of the *künk*, especially in the distribution lines in intra muros Istanbul. In the eighteenth century, this work began to be approached in a more systematic manner. Though few physical traces have been found, archival evidence reveals that, while the length of the ditches varied greatly, their widths and depths (*ʿumḳ*) were generally consistently between two and three *arşun* (~1.5–2.25 m). For example, in 1732 (1144 AH), repair of a now-lost pipe ditch (*ḫendeḳ*) supplying a fountain endowed in 1669 by Seḥrāb Ḥātun revealed a trench seventy *arşun* (~52.5 m) long, two *arşun* wide, and

54 BOA, İSTM.ŞSC.15.d 132 47b/3, date: 12 muḥarrem 1126 AH.

55 On this, see the claims made by three *ṣuyolcular* at the court who were employed by a certain ʿAlī Beg, son of Ömer, for a hydraulic project to supply water to his residence in Eyüp. According to their statements, during the construction of the tunnel, they faced significant challenges due to the absence of rock through which to excavate. As a result, the tunnel required additional structural support (*binā' muḥtāc olmāġla*) leading to an increase in costs. BOA, İSTM.ŞSC.15.d 114 42a/2, date: ramażan 1113 AH.

FIGURE 9.6 A reconstruction of the steps involved in the early modern tunnel construction process. From left: determining the safety area around a shaft with a pickaxe, deepening a shaft, digging out soil to expand the tunnel, filling a bag with soil, dragging a sack of soil and straw to level the tunnel floor
SOURCE: DRAWING BY DENIZ KARAKAŞ, AFTER THE SKETCHES OF CHARLOTTE KENDE AND MOSTAFA SHAFIEE KADKANI IN ALI ASGHAR SEMSAR YAZDI, MAJID LABBAF KHANEIKI, *QANAT KNOWLEDGE, CONSTRUCTION AND MAINTENANCE* (DORDRECHT: SPRINGER, 2016)

three *arşun* deep.[56] Similarly, repairs almost a century later, extending from the distribution center (*maḳsem*) feeding Maḥmūd Çelebi Lodge (neighboring the Edirne gate) toward a branch pipe connected to the fountain of Lālī Muṣṭafā Efendi,[57] included a ditch three *arşun* deep, and another two *arşun*

56 The non-extant fountain was in the Atik Nişancı Pāşā neighborhood (present-day Kumkapı to the south of Sultan Ahmet Mosque along the northern shore of Marmara Sea). The registers often made reference to the length and width (*ṭūlān ve ʿarẓen*) of the ditch. BOA, EV.HMH.d 3365, date: 1144 AH. For the depth in this case, I consulted later repair records. Ahmet Tabakoğlu et al., *İstanbul Su Külliyâtı XXVII: Vakıf Su Defterleri: Su Keşif Defteri 1 (1842–1862)* (Istanbul: İSKİ Yayınları 44, 2003).

57 BOA, D.BŞM.BNE 387-16248, date: 8–17 ṣafer 1235 AH,1. The fountain, located within the present-day Edirnekapı Cemetery, was commissioned in 1736 (1148 AH) by Lālī Muṣṭafā Efendi (d. 1741–1742), a calligrapher (*ḫaṭṭāṭ*) and the steward (*kāhya*) of Princess Hadice Sulṭān, in memory of his deceased wife, Fatma. On its layout and its decoration, see İbrahim Hilmi Tanışık, *Istanbul Çeşmeleri* (Istanbul: Maarif Matbaʿası, 1943), 1:148. On its present condition, see also Su Vakfı, "Lâli Hacı Mustafa Efendi Çeşmesi (H. 1148–M. 1735)," last accessed April 23, 2025, https://www.suvakfi.org.tr/cesme/adalar/lali-haci-mustafa-efendi-cesmesi-h-1148-m-1735.

deep. Ditches also varied in form and purpose. The former, then known as a split ditch (*yarma hendek*), measured one hundred and ten *mesāḥa* (~82.5 m). The shallower trench, referred to as the earth-dug ditch (*türāb-i ḥafr hendek*), which was also one hundred and ten *mesāḥa* long, was employed in less demanding conditions.[58]

After the 1570s, these sets of regulations became familiar, and earth workers knew how to do the kind of hydraulic work required by pipe ditches. For example, the practice of maintaining the slope of pipes buried in ditches for hydraulic efficiency was of constant interest and reassessment. However, the hilly terrain, excessive rainfall, and geological conditions of intra muros Istanbul presented additional challenges. For example, the topography divided by the land walls and the flat valley of the Lykus stream around the Topkapı gate was bracketed by the steep slopes of the seventh hill (68.32 m) and the gentler slopes of Koca Muṣṭafā Pāşā Hill (39.9 m), only one kilometer distant. These conditions, exacerbated by the underlying sedimentary stratum of permeable limestone and marl made the area prone to flooding, and trenches vulnerable to seepage and collapse.[59] Istanbul's high water table further complicated matters, particularly in areas where groundwater levels neared the surface.[60] Digging sewage latrines (*ḥalā*ʾ, also recorded as *kenīf*) and channels for other liquid waste at any depth inevitably led to disputes and conflicts among

58 BOA, D.BŞM.BNE 387-16248, date: 8–17 ṣafer 1235 AH.

59 BOA, A.DVNSMHM.d 26, decree 125, date: 1 rebīʿü'l-evvel 982 AH; BOA, A.DVNSMHM.d 26, decree 128, date: 3 rebīʿü'l-evvel 982 AH. The drains (*menfez̠*) on the land walls failed after heavy rains and caused surge to overtop the defense. Residents brought complaints to the state council, but five years later, the area—the Belġrādlu neighborhood—flooded again. BOA, A.DVNSMHM.d 36, decree 55, date: 7 z̠i'l-ḳaʿde 986 AH; BOA, A.DVNSMHM.d 36, decree 416, date: 1 ṣafer 987 AH.

60 In 1623, the depth of the well dug for a woman named Şemsimāh in the inner courtyard of her rented house, situated uphill from the Golden Horn, was seven *ḳulac* (about 13.125 m). Mehmed Akman et al., *Istanbul Kadı Sicilleri Evkaf-ı Hümâyûn Müfettişliği 1 Numaralı Sicil* (*H. 1016–1035/M. 1608–1626*) (Istanbul: İSAM, 2019), 372, decree 321. In the same area, further up hill, another well rebuilt in the grounds of shaykh Yavsī Tekke (present-day Çarşamba neighborhood, Fatih) had a depth of sixteen *ḳulac* (about 30 m). M. Akman and Rıfat Günalan, *Istanbul Kadı Sicilleri Rumeli Sadâreti Mahkemesi 40 Numaralı Sicil* (*H. 1033–1034/M. 1623–1624*) (Istanbul: İSAM, 2019), 161, decree 146. These cases suggest that the water table on higher elevations throughout the city is no less than thirty meters deep. This observation agrees with the account of Ahmed Efendi, the secretary of the comptroller (*Binā Kātibi*) of the Nuruosmaniye Mosque, recorded during the mosque's foundation construction in the 1760s. Ali Öngül, "Târih-i Câmi-i Nuruosmânî," *Vakıflar Dergisi* 24 (1994): 131. The variation in water table depth likely posed significant challenges and probably prompted subtle modifications and refinements in ditch construction methods.

neighbors, especially when both wells and pipe ditches became polluted.[61] In 1646 (1056 AH), for instance, when residents who rented rooms at the Kurşuncu lodgings (*evleri*) went to fetch water from the fountain in the yard it was dry. The fountain depended on surplus water (*burma lüle*); however, as they elaborated in their petitions, a broken sewage line (*kārīz*) under Atmeydanı, the main public square of the city, was causing a nauseating backup of waste and clogging the main pipe that supplied the water to the fountain.[62] It appears that, in the 1640s, the constant loss of water from the fountain was as worrying as a leaking sewage line. In the decades to come, these difficulties likely spurred modifications and improvements in ditch construction practices. The depth and width came to matter more. A cursory glance at the list of materials and the terminology used in ditch building quickly suggests that new construction methods and techniques prioritized stability, support, and waterproofing both above, below, and alongside the pipe.[63]

All the cases mentioned above and countless other records of the construction and maintenance of the subterranean pipeline infrastructure make visible how local factors—soil type, weather, topography, geology, water sources, flood zones, flora and fauna above and below ground—required earth laborers to undertake a variety of specialized tasks, which over time increasingly merged with soil management techniques as they responded to and affected the geophysical environment of the capital city.

4 *Künk*, the Intersection of Earthen Labor and Knowledge of Clay

In this final part, let us shift the focus to the other role of earth and soil: the making of *künk*, the durable water supply pipes. It was often clay, rather than stone, wood, or lead, that would be used to craft the *künk* that became the mainstay of the water infrastructure until the late nineteenth century, sustaining the capital city over centuries. As early as the mid-sixteenth century, *künk*

61 For examples from different periods, see Mustafa Karaca, "68 Numaralı Mühimme Defteri (Tasnif-Transkripsiyon)," (MA thesis, Gazi Üniversitesi, Turkey, 2000), 81–82, decree 86, date: 999 AH; BOA, A.DVNSŞKT.d.111, 230, date: cemāẕiyü'l-āḫir 1139 AH. On later repairs due to wastewater seepage to the Kātib Meḥmed Efendi fountain near the Topkapı gate, see BOA, C.BE 62-3063, date: 29 receb 1154 AH.

62 Mertol Tulum et al., *Mühimme Defteri 90* (*Redaksiyon ve Sadeleştirme*) (Istanbul: Türk Dünyası Araştırmaları Vakfı Yayınları, 1993), 53, decree 163, date: 1056 AH.

63 For an overview, see Meḥmed Ḥulūsī, "Fennī tevzīʿ-i miyāhdan Ḳırkçeşme ve Ḥalkalı ṣuları," *ʿUmūr-ı Nāfiʿa ve Zirāʿat Mecmūʿası* 114 (1896): 363–368.

already had standardized dimensions.[64] Yet analysis of pictorial and documentary sources gives some indication that, over the next century, they became more varied and came to be distinguished by their dimension and weight.[65] By the 1680s, despite differing diameters, the pipes maintained a uniform length of 0.4 *arşun* (about 30 cm), with an average capacity of thirty-six liters of water per minute. What caused this relatively short, standardized length is not evident. It seems probable that using longer pipes would have been too costly, because fewer of them could be placed in the kiln simultaneously, and a bigger kiln would have required a longer firing process and more wood. The *künk* were cylindrical in shape with a socket (*zebāne*) at one end, so that they joined easily. They were made in a variety of diameters. The most commonly used *künk* had an in internal diameter of six (*altı*) (~18.9 cm) or seven (*yedi*) *parmaḳ* (~22.1 cm).[66] Small *künk*, such as the *altı yolluḳ* (also *pulluḳ*) measured 3.5 *parmaḳ* (~11 cm), and *eski birerlik* (also *bezirlik*) *künk* had a diameter of four *parmaḳ* (~13 cm), see Figure 9.7. These small *künk* were probably used as branch pipes, intended to measure or limit the amount of water removed from the main supply (*ana yol*).[67] The *künk* with the largest diameters were the eighteen-*parmaḳ* (~56.7 cm) *künk* and *Süleymaniye künk* (also referred to as *kebir* or *battal*) with an internal diameter of eleven *parmaḳ* (~34.7 cm).[68]

---

64 On the three *künk* types (1.5 *fülüs*, 3 *fülüs*, 6 *fülüs*) used in the building of the Süleymaniye Mosque complex, see Serpil Çelik, "Mevcut Belgeler Işığında Süleymaniye Külliyesinin Yapım Süreci" (PhD diss., Istanbul Technical University, Turkey, 2001), 121.

65 For example, the 1640 price lists (*narḫ defterleri*) refer to *künk-i altı parmaḳ*, weighing 4.5 *vuḳıyye*; *künk-i eski bezirlik*, measuring 3.5 *parmaḳ* and weighing 3 *vuḳıyye* 300 dirhem; *künk-i üç parmaḳ*, weighing 3.5 *vuḳıyye*; and *künk-i altı parmaḳ-ı yol*, weighing 2 *vuḳıyye* 30 dirhem. See Mübahat S. Kütükoğlu, *Osmanlılarda Narh Müessesesi Ve 1640 Tarihli Narh Defteri* (Istanbul: Enderun Kitabevi, 1983), 321.

66 On the dimensions of each pipe, see BOA, İSTM.ŞSC.15.d 556 9a/2, date: 4 muḥarrem 1246 AH. A *parmak* (also *barmaḳ*) is a unit of length which, just like the *ẕirā'* (also *arşun*) and *mesāḥa*, varied over time and place. For the second half of the sixteenth century, it is estimated to have measured around 3.05 cm. According to Ca'fer Efendi, the builder's cubit (*bennā' ẕirā'*) was equal to twenty-four fingers (*barmaḳ*). Ca'fer Efendi, *Risâle-i Mi'mâriyye*, fol. 61a, 79. See also Cengiz Kallek "Parmak," *TDV Ansiklopedisi*, vol. 34 (Istanbul: TDV Yayınları, 2007): 172–173. I calculate the *parmaḳ* as 31.5 mm.

67 See, e.g., BOA, TS.MA.e 1252-53, date: 29 rebī'ü'l-āḫir 1119 AH; BOA, C. BLD 109-5414, date: ramażan 1191 AH. For a comparison, see Feyza Aykutlu, "Şehzade ve Süleymaniye Külliyeleri'nde Su Mimarisi" (MA thesis, Fatih Sultan Mehmet Vakıf Üniversitesi, Turkey, 2014), 70–71.

68 During the construction of the imperial mosque of Ahmed I, other types mentioned in the documents are: *künk-i meyāne-i kebīr* and *saġīr* and *temel-i künk* or *ḥadā'iḳ-ı birūn-ı sarāy künk* referring to the places being used. See Aliye Öten, *Arşiv Belgelerine göre Sultan Ahmed Külliyesi ve İnşâsı* (Ankara: Atatürk Kültür Merkezi Yayınları, 558), 182–183, 498.

FIGURE 9.7 Examples of remains of Ottoman-period earthenware pipes (*künk*) discovered on the Valens Aqueduct in the Fatih District, Istanbul, before 1997
SOURCE: PHOTO BY KÂZIM ÇEÇEN, COURTESY OF İTÜ MUSTAFA İNAN LIBRARY, PROF. DR. KÂZIM ÇEÇEN SLAYT ARŞİVİ, ISTANBUL

As for the clay used in the production of *künk*, the type of soil used was a crucial factor, as it directly affected their strength, durability, and ability to resist pressure. Although codified as little as that of other crafts at the time, the practical knowledge of selecting and preparing the clay belonged to the *künkcū*, the specialized potters who made the *künk*. How did they find and identify the clay to use?

During the seventeenth and eighteenth centuries, two striking pieces of evidence suggest that *künkçü* over time, by trial and error, seemed to have come up with a method that maximized the local loam's inherent properties. The first, recorded between 1792–1798 during repair works, starts in early 1792 when the water supply to the Anapa fortress, located on the northeastern shore of the Black Sea (about forty-five to fifty kilometers southeast of the mouth of the Kuban River) proved to be insufficient.[69] The fortress inhabitants petitioned the Porte, asking for relief. Soon after, a proposal to bring water from a more distant spring about thirteen kilometers over the plain was

69 For the strategic importance and rebuilding history of the Anapa fortress, see Cengiz Fedakâr, *Kafkasya'da Imparatorluklar Savaşı: Kırım'a Giden Yolda Anapa Kalesi, 1781–1801* (Istanbul: Türkiye İş Bankası Kültür Yayınları, 2014), 187–222.

provisionally approved, but progress was slow.[70] There was one big technical roadblock that impeded the work: how to transport the urgently needed *künk* to this remote spot.[71]

The records of the Imperial Divan show that, instead of shipping large numbers of such pipes from Istanbul, which would have required substantial investment, Seyyid Muṣṭafā Pāşā, the guard (*muḥāfiẓ*) of the fortress, wrote to the Porte about the possibility of manufacturing facilities to produce (*iʿmāl*) *künk* at the fortress, and asked for two or three skilled masters who knew how to make *künk* and fire a kiln (*fırın yakmasını bilür*).[72] But when the *ṣuyolcular* and *künkcū* arrived from the capital to investigate the matter (*lede'l-istinṭāḳ*), they were quick to note the poor quality of soil: "the environs of Anapa was seashore [*leb-i deryā*] and sandy beach [*ḳūmluk*], and even in the surrounding farmlands [*tarla*], there was no soil suitable [*elvirir türāb olmayub*] for crafting *künk*, and that in any case [*beher-ḥāl*], they needed fatty soil [*yaġlu türāb*] like that found in the bed of the Kağıthane river [Istanbul]," they told the commission.[73] This brief report is a rare firsthand account from *ṣuyolcular* and *künkcū*, whose voices are difficult to come by. A specification of the qualities and attributes (*keyfiyyet*) of soil required for the production of *künk* to transport water into the fortress sheds a telling light on the broader applicability of their insights and know-how.

The fragmentary records reveal that clay reserves, used in a variety of items ranging from the manufacture of decorated pottery and earthenware pots (*sifāl*) to the casting of cannons, were readily available in the immediate surroundings of Istanbul and fetched a good price. On September 30, 1637 (1047 AH), clay (*bālçıḳ*) from the town of Eyüp located a few kilometers from the city's land walls along the southern shore of the Golden Horn (known by the same name today), was the subject of unpaid debts. Almost a year earlier, two Armenian roof tile makers (*kiremitci*) named Avanos and Asvazador entered into a contract with two boatmen (*kayıkcıyān*) promising to pay them nine *aḳçe* in cash for every boatload (*pereme*) delivered.[74] The same clay was used for purposes other than tile roofing. Evliyā Çelebi's remarks help us make sense of the rather limited data furnished by the registers. In making water jugs called *kūze*, potters used *ṭīn -i Ensārī*, a soft dough-like clay found in the vicinity

70 For details, see Cengiz Fedakâr, "Anapa Kalesi: Karadeniz'in Kuzeyinde Son Osmanlı istihkâmi (1781–1801)" (PhD diss., Mimar Sinan Üniversitesi, Turkey, 2010), 53–55, 199, 215–216.

71 BOA, C.BLD 93-4604, date: 29 ṣafer 1208 AH.

72 BOA, C.NF 7-334, date: 20 ṣafer 1208 AH.

73 BOA, C.NF 7-334, date: 20 ṣafer 1208 AH.

74 Salih Kahraman et al., *Eyüb Mahkemesi (Havâss-ı Refîa)—37 Numeralı Sicil (H. 1047/ M.1637–1638)* (Istanbul: İSAM, 2011), 93, decree 56.

of Eyüp.[75] Evliyā Çelebi went on to note that the prized water jugs and vessels (*kāseler*) were made of a clay smelling like musk (*ṭīn-i mumessek*) dug out of the land around Sarıyer (a coastal neighborhood along the western shore of the Bosphorus) and of clay taken from the depths of the Golden Horn's seabed between Eyüp and Hasköy.[76] In September 1656 (1066 AH), Muṣṭafā Āġā, the superintendent of the Imperial Arsenal (*Tophāne-i ʿĀmire*) arranged the purchase of soil (*türāb*) from Ayastefanos (present-day Yeşilköy neighborhood) and soil and mud (*çāmur*) from the Kağıthane river to make clay models for cannon barrels.[77] In June 1664 (1074 AH), a large order of mud balls (*çāmur-ı top*) from Eyüp were loaded onto a galleass (*mavna*) and transported to the Imperial Arsenal in Tophane across the Golden Horn.[78] In 1719 (1131 AH), during the fountain construction in the new library (*kitāb-ḫāne -i hümāyūn*) at Topkapı Palace, the plasterers (*sıvacıyān*) used mud from Eyüp (*Eyüp çamuru*) for some of their work.[79] Those directly involved in quarrying and utilizing clay seemed to possess an understanding of the local environment, which enabled them, among other things, to identify clay substrata by examining the color and hardness of stones lying on the ground above.

Regarding the source of clay, the Ottoman administrative apparatus of the seventeenth and eighteenth centuries has left us very little direct evidence. However, when combined with contemporary written sources, the available evidence provides insights into the properties and taxonomy of the clay, but also about its qualities, including specific types such as fatty soil (*yağlu türāb*) and *ḳarabālçıḳ*, a dark, viscous, and sticky clay. A brief and incisive geological reading further prompts us to trace the interchangeable ways of imagining and examining nature through the lens of earth workers' experiences. The territory outside the city walls rests on a complex sequence of sedimentary deposits. The area northeast is occupied by sediments forming the rich alluvial plain of

75 "*Ṭīn -i Ensārī* derler bir maʿcūn- misāl ṭīn -i mulāyimdir." Çelebi, *Evliya Çelebi Seyahatnâmesi*, 1:29.

76 Çelebi, 1:29. Notably, there has been a growing scholarship on pottery in Eyüp Istanbul. See Örcün Barışta, "1996–1998 Yılları Arasında İstanbul EyüpSultan'da Yapılan Kazı ve Yüzey Araştırmaları Bulguları ve Ardından Düşündürdükleri," in *x. Eyüpsultan Sempozyumu* (Istanbul: Eyüpsultan Belediyesi Kültür Yayınları, 2006), 32–46 and Filiz Yenişehirlioğlu, "Eyüp Çömlekçiler Mahallesi Araştırmaları," *Tarihi, Kültürü ve Sanatıyla III. Eyüpsultan Sempozyumu* (İstanbul: Eyüpsultan Belediyesi Kültür Yayınları, 2000), 42–51.

77 BOA, İE.AS 5-372, date: 25 cemāẕiyü'l-evvel 1066 AH. The bill paid by the cannon foundry to the Galata boatmen amounted to 400 piasters (*aḳçe*). My sincere gratitude is owed to Ali Çakır (at Vakıflar Genel Müdürlüğü, Ankara, Turkey) for his help in reading documents written in *siyāḳat* script and deciphering difficult handwriting.

78 BOA, İE.AS 5-429, date: ẕi'l-ḳaʿde 1074 AH.

79 BOA, TS. MA.d 2003-1 f. 17, date: 27 rebīʿü'l-evvel 1131–9 ẕi'l-ḳaʿde 1131 AH.

FIGURE 9.8 The sandstone-shale sequence that dominates the Thracian Formation, as observed in the Alibeyköy area
SOURCE: IMAGE COURTESY OF THE DEPREM RİSK YÖNETİMİ VE KENTSEL İYİLEŞTİRME DAİRE BAŞKANLIĞI DEPREM VE ZEMİN İNCELEME MÜDÜRLÜĞÜ, ISTANBUL

the Alibeyköy and Kağıthane rivers. The substratum is primarily clay dating to the late Quaternary period.[80] Around the village of Cebeci, the terrain near old stream beds and valleys features strata of alluvium, gravel, sand, silt, marl deposits, and calcareous clay—typically ranging from eight to twenty meters in thickness—on a basement of fossiliferous limestone and tertiary rocks (Figure 9.8).[81] The area around Ayastefanos is a combination of alluvial and lacustrine deposits, which consist primarily of sand and silt with some clay layers (Quaternary).[82]

80 E. Togrol, "Golden Horn: A Historical Survey of Geotechnical Investigations," *Proceedings of the International Conference on Soil Mechanics and Geotechnical Engineering* 4, no. 15 (2002): 2445–2472.

81 Nejdet Özgül et al., *İstanbul İl Alanının Jeolojisi* (Istanbul: İstanbul Büyükşehir Belediyesi, 2011), https://depremzemin.ibb.istanbul/uploads/prefix-ist-5000-jeoloji-rapor-66a3951b76d99.pdf. For more recent maps showing the main units in the territory outside the city walls, see Gedik et al., *1/100,000 ölçekli Türkiye Jeoloji haritası.*

82 See Istanbul Büyük Şehir Belediyesi: Deprem Risk Yönetimi ve Kentsel İyileştirme Daire Başkanlığı Deprem ve Zemin İnceleme Müdürlügü, *Microbölgeleme Rapor ve Haritalaırının Yapılması Avrupa Yakası (Güney)*, map of Istanbul 1/25,000, 2007, http://

This means that obtaining the clay for the production of *künk* in the capital's semirural and rural environs required ingenuity and inventiveness on the part of the artisans. Yet not all the soils that were quarried and selected for use could be used in the raw forms in which they were found. Proper clay selection was only the first step, and just as farmers amended soil, the earth workers needed to prepare the clay through treatment and mixture. It is hard to find much evidence on the proportions and preparation methods of clay used for the making of *künk*. Records document the formula of a special glue called *lökün* used to bind the pipes which was valued enough to be preserved, but the formula for the pipe's clay seems to have been kept a secret.[83]

How, then, did the *künk* makers acquire their *ḳarabālçıḳ*? Indirect evidence indicates that, in 1655, clay (*bālçıḳ*) taken out of pits made up part of the tax levy in the aforementioned village of Ayastefanos. With the arrival of the plague, the indebtedness of eighteen peasants shows that these levies could temporarily not be met; however, the state made up for any shortfall caused by the plague with remarkable speed.[84] A glimpse of the activities of the peasants digging the famed medicinal clay known as Lemnian earth, or *ṭīn-i maḫtūm* (sealed earth) in Lemnos (an island in the northern Aegean Sea), gives us an insight into how the peasants might have done the job in extra muros Istanbul.[85] In his notes on Lemnian earth, Jacopo Soranzo (1518–1599), a Venetian statesmen and ambassador to Constantinople wrote that first the waters of the spring were diverted, then a pit was dug in the abandoned natural channel to remove the clay found there.[86] Writing more than a century later, John Covel (1638–1722), chaplain to the English Levant Company in Istanbul, visited Lemnos in 1677

www.ibb.gov.tr/tr-TR/SubSites/DepremSite/Documents/EK2_Jeoloji_Haritasi_1440x910_25000_300DPI.pdf.

83 For a document concerning the purchase of the ingredients to repair the waterways of the old palace, see BOA, MAD.d 4752-1135, fols. 24v, 28v–29r, date: rebīʿü'l-āḫir 1130–7, rebīʿü'l-āḫir 1131 AH.

84 Baki Çakır and Coşkun Yılmaz, *İstanbul Kadı Sicilleri Eyüb Mahkemesi (Havâss-ı Refîa) 61 Numaralı Sicil (H. 1065–1066/M. 1655)* (Istanbul: ISAM, 2011), 280, decree 322. On the plagues that hit Istanbul particularly badly during the seventeenth century, see Nükhet Varlık, "Plague Epidemics in Istanbul," in Yılmaz, *History of Istanbul*, https://istanbultarihi.ist/558-the-history-of-water-in-ottoman-istanbul.

85 On the rediscovery of Lemnian earth by the Ottomans and how the term *ṭīn-i maḫtūm* (sealed earth) and calques (such as *terra sphragis*) came to be the official terms under the Ottomans, see Henryk Jaronowski, "'An Earth by Any Other Name': Pre-Ottoman Sources and Names for Lemnian Earth," *Helenika* 58, no. 1 (2008): 48–70.

86 During the years 1566–1568, Jacopo Soranzo had been in Istanbul for peace negotiations in the office of a *bailo*. He served as ambassador to Constantinople in 1575 during the reign of Murād III. F. W. Hasluck, "Terra Lemnia," *Annual of the British School at Athens* 16 (1909): 220–231, on 225.

and gave further information about the pit. He described it as being dug near the eastern side of a small clear spring before sunrise. The villagers extracted soft, loamy earth—some of it like butter—and removed twenty to thirty *kintal* (quintal, ~100 kg) before refilling the pit, he explained.[87]

In addition to taking out clay from pits or obtaining it from surface outcrops, another method was also used in the capital city. Evliyā Çelebi recounts that many divers (*ġavvāṣ*) retrieved a clay he describes as *ṭīn-i siyāh* (used synonymously with *ḳarabālçıḳ*) from the bottom of the Golden Horn.[88] During the seventeenth century, then, *künk* makers developed and maintained relationships with varied suppliers. The clay taken from the riverbanks, sea floor, or pits, was purchased from either peasants, divers, or middlemen.

The actual processing stages of earthenware (clay, temper, firing procedures) were handled by *künkcū* assigned to the potters (*çanaḳ çömlekci*) guild. When other potters or tile makers attempted to take advantage of this prerogative, conflicts quickly arose. For example, in early summer of 1720 (1132 AH), the senior masters of the potters guild (*çömlekci ṭā'ifesi*) and their warden filed a complaint claiming that, since ancient times (*ḳadīmden muġāyir*; standard guild parlance to defend their rights and privileges), only four shops had been used to manufacture water supply pipes (*dört bāb dükkān ṣu künkleri işleyegelüb*), and they sued Serābyun (سرابين). The son-in-law of the Armenian *künk* master Zaġran (ذغرن), Serābyun had sought to convert his workshop from one producing dishes and vessels to one manufacturing earthenware pipes (*künk işlemek*), but he did so without requesting permission. Of the forty-two pottery workshops (*kārḫāne*), four that lined both sides of Çömlekciler Street in the Eyüp district were given a monopoly on the making and sales of *künk*.[89] It appears that the main potters guild exerted pressure on the dependent *künk*

87 James Theodore Bent, *Early Voyages and Travels in the Levant* (1893; repr., Farnham, UK: Routledge, 2010), 283. For more details on Lemnian earth and the rituals accompanying its extraction, both in antiquity and in the early modern period, see Effie Photos-Jones and Allan J. Hall, "Lemnian Earth, Alum and Astringency: A Field-Based Approach," in *Medicine and Healing in the Ancient Mediterranean*, ed. Demetrios Michaelides (Oxford: Oxbow Books, 2014), 183–189.

88 Çelebi, *Evliya Çelebi Seyahatnâmesi*, 1:29. Eremya Çelebi (1637–1695) made a similar observation and further added that peasants in Hasköy on the north shore of the Golden Horn spent 3 months during the summer diving there to collect mud from the sea floor for manufacturing roof tiles. Eremya Çelebi, *İstanbul Tarihi: XVII. Asırda İstanbul*, trans. and ed. Hrand D. Andreasyan and Kevork Pamukciyan (Istanbul: Eren Yayıncılık ve Kitapçılık, 1988), 32.

89 BOA, C. İKTS 33–1624, date: 8 şaʿbān 1132 AH. One of the key advantages likely held by the manufacturers in the town of Eyüp was proximity to the Golden Horn, thus, clay, which would have reduced transportation costs and ensured a steady supply.

artisans when new market spaces opened up. Fragmentary evidence suggests that, until the first quarter of the eighteenth century, the *künk* masters were a relatively small and socially homogenous group of non-Muslim Armenians. However, there is also evidence of change in the first half of the eighteenth century. For example, one *künk* maker, named Süleyman, was a Muslim.[90] Clearly, by the 1740s, the potters guild showed more flexibility, allowing new members to acquire the right (*gedik*) to open a shop and practice their craft through purchase rather than inheriting their parents' *gedik*.[91]

Nevertheless, this seemingly static picture of *künk* workshops obscures a more dynamic situation than one might suppose. The divide between pottery manufacturers and *künk* makers was not absolute. In the seventeenth and eighteenth centuries, there were mainly two types of construction: the large-scale projects and private feeder lines. For example, during the construction of the Sultan Ahmed Mosque, the construction accounts referred to the purchase of no fewer than 2,015 *künk-i kebīr* and *künk-i sağīr* pipes.[92] In the summer of 1705 (1117 AH), the appraisal prepared by the aforementioned superintendent of the imperial water mains Oṣmān Āġā, estimated 11,575 *künk* for the construction of new waterways to bring the waters from the right bank of the stream near Piyāle Pāşā Mosque in Kasımpaşa (present-day Kurtuluş and Tatavla) to the fountains commissioned by Gülnuş Sultan (d. 1715), wife to Sultan Meḥmed IV (r. 1648–87) in Galata.[93] Later, in 1801 (1218 AH), when Beyhān Sultan (1766–1824), the daughter of Muṣṭafā III (r. 1757–1774), commissioned a freestanding fountain in Arnavutköy, its water was brought from the Belgrad Forest to the shore of the Bosphorus using 35,000 to 40,000 *eski birerlik künkler*.[94] In such projects, the central administration seems to have continued to speed things up by bypassing the *künk* makers' monopoly, temporarily granting other potters the right to produce pipes.[95] Yet the *künk* makers

90 Abdulkadir Altın and Salih Kahriman, *Eyüb Mahkemesi (Havâss-ı Refîa)—182 Numaralı Sicil (H. 1154–1161/M. 1741–1748)* (Istanbul: İSAM, 2019), 305, decree 468, n.d.

91 A document from the first years of the nineteenth century lists five workshops whose proprietors were all Muslims. BOA, İSTM.ŞSC.15.d 325 47a/2, n.d.

92 Aliye Öten, *Arşiv Belgelerine göre Sultan Ahmed Külliyesi ve İnşâsı*, 182.

93 BOA, TSMA E. 0101 0002 037, date: 8 ṣafer 1117 AH. For more details on the construction process of Gülnuş Sultan's waterways to supply her new mosque and five fountains in Galata, which lasted for a decade, see Muzaffer Özgüles's excellent analysis in *The Women Who Built the Ottoman World* (London and New York: I. B. Tauris, 2017), 140–156.

94 BOA, İSTM.ŞSC.15.d 325 13b/2, date: 10 cemāziyü'l-āḫir 1218 AH.

95 For example, in the 1800s, during the repair of conduits supplying the Ottoman Imperial Arsenal (*Tophāne-i ʿĀmire*), the central administration granted permission to a jug maker (*destīcī*) named Ömer Ustā and a roof tile maker (*kiremitci*) from Hasköy, Artin Ustā, to manufacture *kebīr künk*. See BOA, İSTM.ŞSC.15.d 325/47b, decree 92, n.d. See also Nevzat

defended their preexisting monopolistic rights over their crafts and trades through collective legal actions.[96] The number of *künk* workshops was not constant, however. While other potters were prohibited from manufacturing earthenware pipes, the growing demand must have created new opportunities for others. Although further research is needed for conclusive evidence, illicit production and distribution may have crossed guild boundaries.

During the period 1675–1702 (1086–1114 AH), even the building of a single feeder line could necessitate the production of no fewer than two hundred and as many as two thousand one hundred *künk*. In the spring of 1631, for a project undertaken to restore fully the feeder line belonging to Meryem Ḫātūn (daughter of the deceased Meḥmed Pāşā), buried beneath the grassy meadows near Aḳyār aqueduct, north of the village of Muderris (north of Istanbul's land walls), the construction crew purchased a total number of seven hundred *künk*.[97] In 1714, the aforementioned maintenance work near the village of İvaz required a new nine hundred *Süleymaniye künk* and two hundred *altı pulluḳ künk*.[98] In early summer 1726 (1139 AH), the repairs in the walled city required 11,475 *künk* to be re-laid in the ditches.[99] In light of the foregoing, the *künk* makers of the period adapted their recipes to local clay to maintain a uniform shape, size, and weight, and continued to meet the required standard in large quantities.

5 Conclusion

This chapter has explored some of the ways in which Ottoman Istanbul's earth workers handled soil. The dynamic materiality of the sedimentary traces played into the building and maintenance processes of the water supply infrastructure in the capital city, which fashioned the landscape of silt, mud, sand, rock, and clay, physically as well as institutionally. Soils were seen as both quotidian substances and substances of inquiry, both obstacles to and agents of

Sağlam, "Havâss-ı Refi'a Mahkemesi 369 nolu Kadı Siciline Göre 1815–1820 Tarihlerinde Eyüp'de Sosyal ve İktisadi Hayat" (MA thesis, Marmara University, Turkey, 1994), 11–12, decree 11, date: 15 cemāẕiyü'l-evvel 1231 AH.

96 Nevzat Sağlam, "Havâss-ı Refi'a Mahkemesi." In this later legal case, the *künk* makers filed a petition against the guild of pot makers (*saksıcı eṣnāfı*), who refused to respect guild limitations on production.

97 Kahraman, *Eyüb Mahkemesi (Havâss-ı Refîa)—37 Numeralı Sicil*, 344, decree 435, date: şevvāl 1047 AH.

98 BOA, İSTM.ŞSC.15.d 132 47a/2, date: 12 muḥarrem 1126 AH.

99 BOA, C.BLD 72-3579, date: 17 şevvāl 1139 AH.

transformation. Their entanglement with early modern sciences of water and technical knowledge offers a diverse understanding of craftsmen's and artisans' practices in "soil science," hydraulic engineering, and architecture. Most of the innovations involving soil unfolded within closed communities and were preserved and passed down by nontextual channels, making such sweeping changes difficult for historians to track. By gathering and comparing archival traces about worksites and "earthy" labor, we are afforded a vivid portrayal of these lower-ranking laborers who ventured into the field, as well as their labor, the challenges they faced, their local knowledge in Istanbul, and their contributions to the science of soil. By the eighteenth century, these workers brought to their enterprises more sophisticated traditions of hydraulic and structural engineering—a case of "episteme from below," to use a phrase coined by Pietro Daniel Omodeo.[100] Early modern Istanbul relied on local knowledge to find drinking water, to divert it into channels and buried pipes for irrigation, mills, domestic uses, and city supplies, and to make repairs and manage the water network. The material realities of the earth workers' practices explored here—practices of reshaping nature, observing, tasting, smelling, surveying, digging, measuring, and (re)building to supply the city with drinking water—offer an articulation of how such workers labored with soil, and of their forms of knowledge, too often relegated to the margins of social, scientific, and historical narratives.

Acknowledgments

This chapter has been in the making for the past four years, and it would be impossible to imagine its completion without the support, patience, and insightful comments of the editors, Aleksandar Shopov and Justin Niermeier-Dohoney. I am deeply grateful to them for giving me the chance to participate in the unique workshop "Towards a Global History of Soil: Sciences, Practices, Materialities, and Mobilities, 1100–1700" at the Max Planck Institute for the History of Science (MPIWG), Berlin (July 22–23, 2021). Conversations at the workshop and the reading group meetings that followed opened up new and exciting avenues for my research as I had not previously thought of the relationship between water and soil. The unwavering guidance I received in shaping my initial presentation was invaluable and has resulted in this work.

100 Pietro Daniel Omodeo, "Hydrogeological Knowledge from Below: Water Expertise as a Republican Common in Early-Modern Venice," *History of Science and Humanity* 45, no. 4 (2022): 538–560, on 539.

I would also like to thank Stephanie Porras for her continuous encouragement and for facilitating my access to the Tulane Library. I owe special thanks to the Publications team of Department Artifacts, Action, Knowledge (AAK)—Gina Partridge-Grzimek, Melanie Luise Glienke, Spencer Forbes, and Rebecca Schmitt—as well as Benjamin Carter, for their careful reading and invaluable editorial support. My foremost thanks go to the MPIWG and Managing Director Dagmar Schäfer for hosting me as a short-term fellow in Department AAK and creating an intellectual home where I had the opportunity to meet with numerous scholars from around the world. I owe heartfelt thanks especially to Melissa Charenko, Sarah Schneewind, and Serafina Cuomo who shared their time, work, creative suggestions, and valuable feedback, and who inspired me to read widely and adopt a more holistic approach. I also thank to Paulina Banas who heard different versions of this research and made intellectual journeys enjoyable, and Şebnem Gümüşçü, Feyda Sayan-Cengiz, and Şebnem Yardımcı, my online writing companions, who made writing and rewriting more rewarding. Any errors and omissions are, of course, my own.

CHAPTER 10

Three Seventeenth-Century Ottoman Books on Flowers (Şükūfe-nāme), Flower Breeding (*terbiye-i ezhār*), and the New Science of Soil in Istanbul

Aleksandar Shopov and Himmet Taşkömür

This chapter examines early modern Ottoman knowledge about soil as articulated in books on flowers called Şükūfe-nāme, which were written by flower breeders in mid-seventeenth-century Istanbul for the community of flower breeders in the city. During this period, Ottoman scholars, bureaucrats, and artisans were fashioning themselves as experts on flower breeding, an increasingly lucrative enterprise that resulted in the creation of many new, signature varieties of flowers such as tulips, narcissi, hyacinths, and roses. The Şükūfe-nāme books classify flower varieties according to color, size, form, and techniques of cultivation. Some of the extant manuscripts contain intricate watercolor illustrations of new varieties. Previous scholarship has considered the flower industry in Istanbul as part of discussions about new cultural and economic values attached to flowers, particularly tulips, in Ottoman society.[1] However, there has been little exploration of the relationship between the Şükūfe-nāme books and agricultural practices and knowledge.[2] These books attest to a living discourse of the period on the technologies and practices

1 See, e.g., Turhan Baytop, "The Tulip in Istanbul during the Ottoman Period," in *The Tulip: The Symbol of Two Nations*, ed. Michiel Roding and Hans Theunissen (Utrecht: Houstma Stichting, 1993), 5–56; Yıldız Demiriz, "Tulips in Ottoman Turkish Culture and Art," in Roding and Theunissen, *Tulip*, 57–75; Ariel Salzmann, "The Age of Tulips: Confluence and Conflict in Early Modern Consumer Culture (1550–1730)," in *Consumption Studies and the History of the Ottoman Empire (1550–1922)*, ed. Donald Quataert (Albany: State University of New York Press, 2000), 83–106.

2 An exception can be found in the early twentieth-century writings of Cevat Rüştü, the late Ottoman and early Republican agronomist. See Cevat Rüştü, *Türk çiçek ve ziraat kültürü üzerine: Cevat Rüştü'den bir güldeste*, ed. N. Hikmet Polat (Istanbul: Kitabevi, 2001). A large volume compiling illustrations and transcriptions from a number of Şükūfe-nāme works has recently been published: Seyit Ali Kahraman, *Şükûfenâme: Osmanlı Dönemi Çiçek Kitapları* (Istanbul: İstanbul Büyükşehir Belediyesi Kültür A.Ş., 2015). For a discussion of knowledge about seeds in Şükūfe-nāme literature, see Aleksandar Shopov, "Flower Breeding in Early Modern Istanbul: A Science of Seeds," *Isis* 113, no. 3 (2022): 588–596.

DOI:10.1163/9789004748484_012

involved in flower breeding (*terbiye-i ezhār*), which, by the mid-seventeenth century, had been recognized as a science (*ʿilm*). Of particular significance are descriptions of techniques related to soil, which also became recognized as a distinct science in the same period. In addition to various writings on agriculture and archival documents, the following pages examine three texts, two of which are Şükūfe-nāme: one authored by ʿAbdullāh Çelebi (d. 1669), who hailed from a family of Ottoman bureaucrats; the other by ʿAli Çelebi (d. 1678), a lumber merchant (in this period, "Çelebi" was a title used for an urban gentleman and learned person). The third text is a biographical dictionary of Istanbul flower breeders by ʿUbeydī, a seventeenth-century preacher. Until now these authors have not been aknowledged as scholars because, as self described flower breeders, their writings are seen as falling outside what is perceived as established Ottoman scholarship. Nevertheless, these works attest to the advent of a community of practitioners who saw technical know-how as specific to the experimental practices and agricultural spaces of individuals. These works attest to the advent of a community of practitioners who saw technical know-how as specific to the experimental practices and agricultural spaces of individuals. Şükūfe-nāme literature was thus part of an emerging discourse that emphasized the know-how of practitioners as the founders of various schools of thought about soil and that elevated knowledge about soil into a science. Disregarding earlier Islamic agricultural authorities, Şükūfe-nāme books expound new ways of improving soil. We see these innovations as originating in the concrete spaces available to specific practitioners. The creative drive related to the improvement of soil for flower cultivation could be seen as a response both to the demand for desirable soil in Istanbul—where agriculture was nonetheless widely practiced—and to a burgeoning urban culture that placed value on novelty, sociability, visibility, and spectacle. It is within this context that an entirely new science of flower breeding developed.

Soil is given a very prominent place in texts on flower breeding. Indeed, as we will discuss in the case of the *Şükūfe-nāme* of ʿAbdullāh Çelebi, some of the seventeenth-century flower breeders in Istanbul and its surroundings are actually characterized as founders of schools on how to treat the soil in which flowers were to be planted. In 1641 (1051 AH), a letter of appointment (*berat*) was given to Sarı ʿAbdullāh Efendi, a well-known scholar and mystic, which refers to him as

> the indisputable authority in the asserting of and distinguishing among various classes of flowers [*tabakat-i ezhār*], one who is superior to his peers in solving matters pertaining to the science of flowers [*ʿilm-i ezhārʾda fāikʾul-akrān*] … In the sciences of gardening, he is an arbiter, he

> is the leader par excellence, and appraiser among the artisans and flower breeders who are busy cultivating the gardens and flowers [*terbiye-i riyaz ve ezhār*] in the royal domains [of the Ottoman Empire].[3]

Sarı ʿAbdullāh Efendi is also described as someone with deep knowledge in the science of music who could solve difficult matters of music writing and was known as the Aristotle of his time for his ability to solve astrological puzzles. The letter of appointment clearly describes Sarı ʿAbdullāh Efendi as an authority in matters of flower breeding, and the wording recognizes flower cultivation as a science (*ʿilm-i ezhār*, a science of flowers). He is given a leadership role (*mümeyyiz ve başbuğ*) as the head of all of the flower breeders in the entire Ottoman domain. The letter of appointment further clarifies that Sarı ʿAbdullāh Efendi was to be the sole arbiter in matters related to flowers and trees and given the authority to identify (*teşhis*), to distinguish one from another (*temyiz*), and to examine their qualities and features (*tedkik*). This terminology indicates the existence of techniques for the observation and classification of plants. ʿUbeydī's *Netāyicü'l-Ezhār* (1698–1699) records the letter as a document which shows that, in 1641, flower breeders were recognized as a distinct guild (*taife*) whose appointed leader possessed scientific qualifications. The creation of this position in 1641 was preceded by the involvement of high-ranking Ottoman officials in matters related to flower breeding, and by the growing communities of flower breeders from different social classes. ʿAli Çelebi, in the introduction of his *Şükūfe-nāme*, writes that, during the term of the grand vizier Hadım Gürcü Mehmed Paşa (in office, 1622–1623), renowned and ordinary gardeners from the city were summoned by him twice a week to discuss flowers. During his term, he also organized banquets and distributed generous gifts to the flower breeders.[4]

Flower breeding is generally treated as evidence of changing cultural attitudes in the opening decades of the eighteenth century, a period the early twentieth-century Ottoman historian Ahmed Refik dubbed the Tulip Age (*Lale Devri*).[5] While he rightly highlighted the cultural value of tulips in this period, Ahmed Refik, and those who followed his conceptualization of this era as the beginning of Westernization/modernization, overlooked that the

3 Kahraman, *Şükûfenâme*, 88–89. This letter of appointment is included in the biographical entry on Sarı ʿAbdullāh Efendi in ʿUbeydī's *Netāyicü'l-Ezhār*, a biographical dictionary of Istanbul's flower breeders written in 1698–1699. Unless otherwise noted, all translations by the authors.

4 Kahraman, *Şükûfenâme*, 29. According to ʿAli Çelebi's treatise, even such a person as Sultan Ahmed I (r. 1603–1617) sowed seeds to grow flowers and cultivated them.

5 Ahmed Refik, *Lale Devri (1130–1143)* (Istanbul: Kitabhane-i Askeri, 1915), 43–46.

Şükūfe-nāme literature had appeared earlier in the seventeenth century, when flower breeding was already recognized as a separate science (*ʿilm*).[6] Another school of thought reevaluates flowers in terms of their economic significance as early modern commodities.[7] Both the cultural and economic approaches have largely overlooked the texts themselves and their construction of knowledge about flowers, including knowledge about soil and other material objects related to flower breeding. These omissions have contributed to the common assumption that the Ottoman study of plants and soil commenced in the nineteenth century with the establishment of the modern schools of agriculture following a Western European curriculum. However, our close comparison of two books on flower breeding demonstrates that methods of soil improvement were already being discussed and implemented in Istanbul and its surroundings in the seventeenth century. Groups of practitioners with a common interest in hands-on examining of soil began to share the results of their experiments and disseminated this knowledge within the city's growing flower-breeding community.

1 Classifying and Experimenting with Soil in ʿAbdullāh Çelebi's *Şükūfe-nāme*

An Ottoman treatise in verse on the art of tending fruit trees, vegetables, and flowers, dated 1638, dedicates its first chapter to the "technique of examining the soil" (*ṣanʿat-ı tecribe-i zemi̱n*). The author, whose name is given as Kemānī, described a method for evaluating the quality of soil: Dig up one *ẕirāʿ* of soil and mix it with rainwater. Then pass the mixture through a piece of delicate cloth into a bottle "until it becomes mud" (*balçık ola*) and leave outside "overnight in the frost and dew" (*bir gece ṭursun ayazda*) in order for the water to separate again from the soil. In the morning, examine the "sky-colored and clear" (*göm gök ola*) water for a foul odor, as this indicates that the fruit grown from it will be spoiled and lacking in taste.[8] Although this method can be

6 One of the first to point out the anachronistic characterization of this period as the Tulip Age was Cemal Kafadar, who noted that the image of the tulip was very common to the period before 1718. See Cemal Kafadar, "The Myth of the Golden Age: Ottoman Historical Consciousness in the Post-Süleymânic Era," in *Süleymân the Second and his Time*, ed. Halil Inalcik and Cemal Kafadar (Istanbul: Isis Press, 1993), 37–48, on 40.

7 Salzmann, 83–106.

8 Kemānī, *Ġars-nāme*, MS Hacı Mahmud Efendi no. 5612, fol. 2b, Süleymaniye Kütüphanesi, Istanbul. This source is discussed in Aleksandar Shopov, "Between the Pen and the Fields: Books on Farming, Changing Land Regimes, and Urban Agriculture in the Ottoman Eastern Mediterranean ca. 1500–1700" (PhD diss., Harvard University, 2016), 388–400.

found in earlier agronomic treatises in Arabic and Ottoman Turkish, in these it is described as a *ka'ide*—"rule" or "custom." Kemānī's chapter title, in contrast, emphasizes "technique" (*ṣan'at*). While the word *tecribe* in the title could have various meanings, including "experience," the detailed description given for this technique for testing soil suggests that it meant something like "testing" or "examining." This further distinguishes the text from an earlier sixteenth-century Ottoman Turkish agricultural text in which the section on soil is simply titled "chapters on soil."[9]

Nothing concrete is known about Kemānī, whose name means "master archer" and who does not feature in the historiography of Ottoman science.[10] Nevertheless, the Kemānī case exemplifies a notable trend in seventeenth-century Ottoman society in which certain members of a new group of urbanites who had not pursued scholarly careers gained recognition as learned individuals who cultivated expertise and technical knowledge. The seventeenth-century Ottoman bureaucrat and traveler Evliyā Çelebi (d. c. 1682) documented two such figures, Hazarfen Hüseyin (d. 1691) and Lāgarī Hasan Çelebi, Istanbulites who conducted public demonstrations in which they attempted to fly over the Bosporus or launch themselves into the air with a rocket.[11] Practical know-how also began to appear in the written discourse that was intended to be shared with a broader, urban audience.

Kemānī's decision to write in verse renders the text accessible to a broader audience through the mediums of recitation and memorization. His treatise also breaks from tradition by not citing any older agronomic authorities, focusing instead on the author's own practical knowledge.[12] As early as the sixteenth century, the dissemination of vernacular know-how outside of institutions of higher learning emerged in certain spaces in Ottoman cities, especially in Istanbul's coffee houses, taverns, promenades, and public places. Istanbul's large produce garden complexes, or *bostāns*, were prime examples of such spaces.[13]

9 Zafer Önler, ed., *Revnak-ı Bustan* (Ankara: Turk Dil Kurumu, 2000).

10 For a brief entry on the available manuscript copies of this work, see Ekmeleddin Ihsanoğlu, ed., *Osmanlı Tabii ve Tatbiki Bilimler Literatürü Tarihi* (Istanbul: IRCICA, 2006), 1:83.

11 Evliyā Çelebi, *Evliya Çelebi Seyahatnâmesi: Topkapı Sarayı Bağdat 304 Yazmasının Transkripsiyonu-Dizini*, ed. R. Dankoff, S. A. Kahraman, and Y. Dağlı (Istanbul: Yapı Kredi Yayınları, 2006), 1:353–354.

12 See Shopov, "Between the Pen and the Fields," 388.

13 Aleksandar Shopov, "When Istanbul Was a City of *Bostāns*: Urban Agriculture and Agriculturists," in *A Companion to Early Modern Istanbul*, ed. Shirine Hamadeh and Çiğdem Kafescioğlu (Leiden: Brill, 2021), 279–307; Aleksandar Shopov, "Grafting in Sixteenth-Century Mamluk and Ottoman Agriculture and Literature," in *Living with Nature and*

Who were Kemānī's readers? Who among his contemporaries might have found his verses on the "technique of examining the soil" important and relevant? One of the surviving copies of Kemānī's treatise, held today in Istanbul in the library of the renowned nineteenth-century scholar Bağdatlı Hacı Mahmud Efendi, is bound together with a copy of a treatise on flower breeding authored by the Ottoman polymath Hacı Ahmed Çelebi.[14] The binding together of these works makes sense, as a significant portion of Kemānī's treatise is devoted to the cultivation of flowers. During Kemānī's lifetime, Istanbul and its neighboring towns witnessed the appearance of the first treatises on flower breeding, which served as precursors to a distinct genre. The earliest dated work in the eponymous genre is the 1667 *Şükūfe-nāme* by ʿAli Çelebi. Another was authored in the same period by ʿAbdullāh Çelebi, a contemporary of ʿAli Çelebi who is mentioned in the latter's work as the creator of a new variety of narcissus.[15] It is in ʿAbdullāh Çelebi's *Şükūfe-nāme* that we find a reflection on the science of soil. Let us look at how ʿAbdullāh Çelebi conceptualized the knowledge of breeding flowers. In his introduction, he establishes his authority on flowers by emphasizing his "long-term" (*müddet-i medide*) involvement in flower breeding (*terbiye-i ezhār*) and designing flower gardens (*tarh-i lale-zar*). He quotes Qur'anic verse and notes that God taught Adam the names of all things. This he considers as giving rise to known sciences (*ulum-u zevahir*), but he also explains that there are sciences yet to be established (*ulum-u bevatin*). Here, *bevatin* does not refer to mystical knowledge; it is rather a reference to "innate" or *potential* knowledge waiting to be uncovered.[16] To justify the novel undertaking of writing a treatise on flowers, the introduction also tells the story of a narcissus whose seeds were brought from Algiers by a person named Ahmed Çelebi (who must have lived at the beginning of the seventeenth century) and planted in Üskudar, a town located across the Bosporus from Istanbul.[17] ʿAbdullāh Çelebi cites this moment as marking the inception of flower breeding in Istanbul. He claims that the Istanbul Sufi master Hüdai (d. 1628) had instructed Ahmed Çelebi to save the seeds of this flower, which nobody in Istanbul had previously known could produce its own seeds.[18] Variations of this story appear in other treatises, including the 1667 *Şükūfe-nāme* by ʿAli

---

Things: Contributions to a New Social History of the Middle Islamic Periods, ed. Bethany J. Walker and Abdelkader Al Ghouz (Bonn: University of Bonn Press, 2020), 381–406.

14 Hacı Ahmed Çelebi, *Filahatname Şukufe-i Beçli Hacı der tertibi zerrin-kase*, MS Hacı Mahmud Efendi 5612, fols. 12b–17b, Süleymaniye Kütüphanesi, Istanbul.

15 Shopov, "Flower Breeding," 593.

16 Kahraman, *Şükûfenâme*, 117.

17 Kahraman, 121.

18 Kahraman, 117.

Çelebi.[19] The story associates the beginning of flower breeding in Istanbul with a Halveti Sufi network.[20] It also characterizes Istanbul as a place where plants from other regions and climes arrive, become objects of study, are propagated, and defy older Avicennian ideas that plants change their qualities when moved from one region or clime to another.[21] Following his introduction, ʿAbdullāh Çelebi offers chapters devoted to various aspects of breeding flowers: how to preserve, transport, and store seeds and bulbs; methods of irrigation; and the twenty-three characteristics a flower should have in order to be valuable and marketable.

The first chapter, however, is a chapter on soil. This opens as follows:

> There is no limit to the types of soil, but yellow fleshy [soft] soil, black fleshy [soft] soil, yellow compact soil, and clay soil are good. Among these, yellow compact soil is very good for yellow flowers, but it should not be taken from places not suitable for cultivation and left fallow. This is everyone's opinion.[22]

In emphasizing the limitlessness of soil types, ʿAbdullāh Çelebi departs from the most authoritative Ottoman Turkish work on farming, the sixteenth-century *Revnaḳ-ı Būstān* (Splendor of the garden), which had identified only two types of soil with their subtypes, one pure and the other mixed with sand.[23] He also uses the phrase *cümlenin kavli* (everyone's opinion), which is a concept belonging to Islamic legal theory that means an indisputable position reached through deliberation and consensus.[24] One example of how such consensus was achieved among seventeenth-century Istanbul flower breeders is illustrated by a story told in this same chapter. ʿAbdullāh Çelebi writes that he was summoned by an unnamed friend who had a garden (*bağçe*) in Yenibağçe, a

19 Kahraman, 29. See Shopov, "Flower Breeding," 593.

20 Shopov, "Flower Breeding," 594.

21 Aleksandar Shopov, "In the Balsam Orchard with Ṣāliḥ Çelebi Celālzāde (d. 1565): First-Person Narrative and Knowledge in Ottoman Egypt," in *Crafting History: Essays on the Ottoman World and Beyond in Honor of Cemal Kafadar*, ed. Ilham Khuri-Makdisi, Rachel Goshgarian, and Ali Yaycıoğlu (Boston: Academic Studies Press, 2023), 255–276.

22 "Hākın envāına had yoktur, lakin sarı et toprak ve sarı kesme toprak ve kıl toprak ve rakik kum toprak iyidir. Bunlardan sarı kesme toprak sarı çiçek içün gayet ile iyidir, amma pek kıraç yerden olmaya. Cümlenin kavli bunun üzerinedir." Kahraman, *Şükûfenâme*, 120.

23 On the sixteenth-century agricultural and economic concerns of Ottoman society as reflected in the *Revnaḳ-ı Būstān*, see Aleksandar Shopov, "The Vernacularization of Sixteenth-Century Ottoman Agricultural Science in its Economic Context," in Walker and Al Ghouz, *Living with Nature and Things*, 639–681.

24 Kahraman, *Şükûfenâme*, 29.

district in the walled city of Istanbul located along the Bayram Paşa stream and famous for its *bostāns* since the early sixteenth century.[25] This person had a specific issue in his garden: he was unable to grow yellow flowers. ʿAbdullāh Çelebi "examined" (*nazar eyledim*) the soil and concluded that the soil in his friend's garden was a flour-like "refined sandy soil," which caused each bulb to crack into pieces. He gave his opinion that "yellow-cut firm soil" was the most suitable for his friend's garden, and in this way was able to share his expertise. For ʿAbdullāh Çelebi, knowledge about soil was a prerequisite for cultivating flowers.

ʿAbdullāh Çelebi gives several examples of how mid-seventeenth-century Istanbul flower breeders constituted a community of practitioners. They operated within a frame of principles and methods that had to be validated through experimentation. Soil was one of their primary concerns. Mola Çelebizade, a gentleman scholar from Istanbul's suburban area of Fındıklı who bred the peach-yellow (*şeftalu sarı*) and golden-yellow narcissus flower varieties,[26] improved the soil by mixing it with grape pulp (*cibre*): "He takes a bucket of grape pulp and puts it in a place and pours four buckets of soil and makes a heap [*tepe*], and then after two years he plants flowers and rotates the soil between the planting row."[27] Another technique is attributed to the flower breeder Ulvanzade, who, depending on the size of the planting row, mixes three or four buckets of aged manure, which he had already used in the planting beds and which has been sifted, with garden soil, plants the bulb, and covers it with old soil taken from a flower planting row.[28] Another example is Hasan Kapudanzade, son of a ship captain, who turns out to be ʿAbdullāh Çelebi's father-in-law. He also used a grape pulp fertilizer, but black snails the size of olive stones appeared and ate the flowers. They also ate the leaves. This experiment, according to ʿAbdullāh Çelebi, was clearly a failure.[29]

A more successful experiment in soil improvement mentioned by ʿAbdullāh Çelebi was one conducted by Cüce Çelebi (d. 1651), who is recorded as being the commander of a fortress overlooking the harbor of Istanbul and who bred a variety called Cüce's plum-shaped tulip (*Cüce ablağı*), a depiction of which can be found in ʿAli Çelebi's *Şükūfe-nāme* (Figure 10.1).[30]

25 On the Yenibağçe produce garden complex in Istanbul, see Shopov, "When Istanbul Was a City," 283–284.

26 Kahraman, *Şükûfenâme*, 38, 42.

27 Kahraman, 120.

28 Kahraman, 120.

29 Kahraman, 120.

30 Kahraman, 46, 48.

FIGURE 10.1 *Cüce ablağı* variety of tulip in ʿAli Çelebi's *Şükūfe-nāme*
SOURCE: NURUOSMANIYE KÜTÜPHANESI, ISTANBUL, MS 4077, FOL. 27B–28A

Cüce Çelebi's technique is to dig a hole two to three yards deep in a spot in his garden. He takes the fresh soil from this hole and plants the flowers in it. The soil that is already in the flower bed after being used is mixed with manure (*gübre*) at a ratio of four to one. He fills the same pit with this mixture, and after three years he repeats the cycle.[31]

ʿAbdullāh Çelebi specifies the ratio of manure to soil for those who wish to repeat the experiment. At the end of the chapter on soil, he further specifies the location of Cüce Çelebi's soil improvement—an enclosed room in the garden—where this experiment took place, and where the author was invited personally to observe the four wells dug on each side of the room for this purpose.

31 Kahraman, 121.

Methods for improving soils and manure had been incorporated in earlier Arabic and Persian writings on agriculture.[32] The novelty of ʿAbdullāh Çelebi positioning himself as a follower of the soil-improvement methods of Cüce Çelebi is his association of soil-improving techniques with contemporary individuals. This practice is not evident among earlier agricultural writings in Arabic, Persian, or Turkish. ʿAbdullāh Çelebi employs the term *mezheb*, which can be translated to "school of thought" but is also an Islamic legal term used for the Islamic schools of law, which have distinctive methods of legal reasoning and substantive law established by medieval eponyms.[33] Each school is identified with a founder scholar who laid out the methodology and principles in interpreting legal matters and coming up with legal solutions to new cases by deliberating on sources of Islamic law. ʿAbdullāh Çelebi's analogy of the methods for improving soil with *mezheb* and the science of jurisprudence elevated the study of soil to an independent discipline of study or *ʿilm* with distinct founders and practitioners who were his contemporaries. In the sixteenth- and seventeenth-century Ottoman genre of science classification, including the foundational works of Aḥmed ibn Muṣṭafa Taşköprülüzāde (d.1561) and Kātib Çelebi (d. 1657), the study of soil did not constitute an *ʿilm*.[34]

This new epistemology of soil was related to new ways of evaluating the qualities of manure. In earlier works on farming, manure's quality is defined according to the animal from which it comes. For example, the author of the aforementioned sixteenth-century Ottoman agricultural treatise *Revnaḳ-ı Būstān* had claimed that the best manure came from sheep and goats, followed by the manure produced by horses, cows, donkeys, and mules.[35] In contrast, according to ʿAbdullāh Çelebi, the best manure for the purpose of flower breeding is cow dung, as he considered sheep manure to be of lesser quality due to the presence of saltpeter. For ʿAbdullāh Çelebi, the chemical composition of the manure is as important as the source. In this sense, the author may have been relying on practical know-how about saltpeter production, which was widespread in the Ottoman Empire due to its use in the making of gunpowder.

---

32 For an overview on the methods of improving soils with manure in agronomic writings in Arabic between the ninth and seventeenth centuries, see Daniel Varisco, "Zibl and Zirāʾa: Coming to Terms with Manure in Arab Agriculture," in *Manure Matters: Historical, Archaeological and Ethnographical Perspectives*, ed. Richard Jones (London: Routledge, 2012), 129–143.

33 Kahraman, *Şükûfenâme*, 120: "Bu hakir Çelebi Cüce mezhebindeyim."

34 Aḥmad ibn Muṣṭafā Ṭāshkubrī'zādah, *Al-Shaqāʾiq al-Nuʿmānīyah fī ʿUlamāʾ al-Dawlat al-ʿUthmānīya* (Istanbul: Jāmiʿat Istānbūl, Kulliyat al-Ādāb, Markaz al-Dirāsāt al-Sharqiyah, 1985).

35 Önler, *Revnak-ı Bustan*, 25.

Furthermore, ʿAbdullāh Çelebi makes a distinction between the quality of cow dung from the city and from the countryside, stating that the dung of cows fed with bran and millet is of lower quality than that of grass-fed cows that grazed in the countryside; this latter type, he adds, is difficult to obtain and is not harmful to flowers.

Interest in the relationship between flowers and soil quality can be discerned among Ottoman scholars at the end of the sixteenth century. The Ottoman scholar Nevʿī Efendi (d. 1599), in his highly influential work on sciences, *Netāyicü'l fünūn*, gives three cases that illustrate this. In the first, on the quality of soil, he states:

> One sign of [good soil] is if the flowers are red and grasses numerous. Another sign is that the flowers are red and there are no varieties of grass. And [as for] useless soil, its sign is that the flowers are white and the land is salty and the grasses are weak and short.[36]

This method of determining the productivity of soil according to wild flower growth is not found in the earliest work on farming written in Ottoman Turkish, the *Revnaḳ-ı Būstān*, in which the color of soil rather than color of plants is an indicator of fertility. Between the *Revnaḳ-ı Būstān* written in the mid-sixteenth century, and the first of the Şükūfe-nāme writings, flowers began to be used as a metric of good or bad soil. Was Nevʿī Efendi aware of this use of flowers in the rich literature of earlier Arabic and Persian agricultural manuals? No such reference exists in the Ottoman Turkish translation made in 1590 of the influential twelfth-century Andalusian scholar Ibn al-ʿAwwām's work on farming, *Kitāb al-filāḥa*.[37] Ibn al-ʿAwwām extensively discusses matters of soil in his work by summarizing earlier literature, which makes its way into the Ottoman Turkish translation. In fact, the classification of soils was of utmost importance to the functioning of the expanding Ottoman state bureaucracy in the mid-sixteenth century.[38]

36 "Bir alameti oldur ki şükūfesi surḫ olur ve giyahi muhtelif olunur. Bir alameti oldur ki şükūfesi sefid olur ve şure olur ve giyahi zaif ve hurde olur ve muhtelif olur. Bu makule zemine hayr olmaz." Nevʿī Efendi, *Netāyicü'l fünūn*, MS Turk 50, fol. 91a, Harvard University, Houghton Library, online collection, accessed June 4, 2023, https://iiif.lib.harvard.edu/manifests/view/drs:10697031$1i/.

37 İbn Avvâm, *Terceme-i Kitâbü'l-Filâha* (*Ziraat kısmı*), trans. Muhammed b. Mustafa b. Lutfullah, ed. Mükerrem Bedizel Zülfikar-Aydın (Istanbul: Kitabevi, 2011), 112–139.

38 Ebu's-suʿud Efendi (d. 1574), the highest-ranking Ottoman jurisconsult (*şeyhülislām*), discussed the quality of soil as a marker for the amount of tithe paid by the farmers. See

During this period, a growing number of Istanbulites became directly involved in agricultural production by taking over land belonging to the customary land holders.[39] Halil İnalcık has pointed out that government circles at that time were alerted to an increase in the number of cases involving the illegal sale of state land.[40] The use of flowers to gauge the quality of soil in Nevʿī Efendi's writings also needs to be seen in the context of the profuse planting of flowers in Istanbul and its environs in the sixteenth century. Numerous shipments of flowers and their bulbs from Edirne and northern Syria were sent to Istanbul during this period as part of operations financed by the Imperial Council and provincial treasuries.[41] Their planting in the numerous newly established royal gardens required renewed consideration of materials (soil, water, seeds) as well as climatic conditions.[42] According to ʿUbeydī's *Netāyicü'l-Ezhār* (Biographical dictionary of flower breeders; written in 1698–1699), prominent scholars from this period, such as the Şeyhülislām Ebu's-su'ud Efendi, were personally involved in flower breeding.[43] Ebu's-su'ud Efendi is listed as one of the earliest flower breeders.[44] He created three varieties of narcissus flowers that he named after himself.

ʿAbdullāh Çelebi's writings highlight examples of trials and experiments with soil and ascribe these techniques to contemporary individuals. What motivated ʿAbdullāh Çelebi to put practical knowledge about soil, and methods

---

Millî Tetebbular Mecmuası, vol. 1.1 (Istanbul: Âsâr-i İslâmiye ve Milliye Tedkik Encümeni, 1331 AH/1915), 51.

39 Shopov, "When Istanbul Was a City." A similar development is noted for Ottoman Egypt, where, in 1552, the jurist Ibn Nujaym completed a treatise defending the rights of the rentier and landowning class, arguing for their tax exemption. See Baber Johansen, *The Islamic Law on Land Tax and Rent: The Peasants' Loss of Property Rights as Interpreted in the Hanafite Legal Literature of the Mamluk and Ottoman Periods* (London: Croom Helm, 1988).

40 Halil Inalcik, "The Ottoman State and Society: Economy and Society, 1300–1600," in *An Economic and Social History of the Ottoman Empire, 1300–1914*, ed. Halil Inalcik and Donald Quataert (Cambridge: Cambridge University Press, 1994), 11–411, on 113. The sale of land in the seventeenth century was frequently disguised as a sale of vineyards and gardens *on* the land. See Ömer Lütfi Barkan, "Edirne Askerî Kassam'ına Âit Tereke Defterleri (1545–1659)," *Belgeler* 3, nos. 5–6 (1966): 1–479, on 55; Shopov, "Balsam Orchard," 273–275.

41 Shopov, 274–275.

42 For example, pomegranate trees brought to Istanbul from Aleppo and Diyarbekir following the conquest of Iraq in the 1530s were planted on the lower slopes of the royal garden of Topkapi Palace. See Shopov, "Balsam Orchard," 273–275.

43 Kahraman, *Şükûfenâme*, 66.

44 MS Nuruosmaniye no. 3704, fol. 8b, Süleymaniye Kütüphanesi, Istanbul; Kahraman, *Şükûfenâme*, 66.

of improving it, into writing? In the following section, we demonstrate how the new expertise and authority on matters related to soil and flower breeding was predicated on earlier intervention by the Ottoman state and contemporary debates in matters related to flower breeding.

2 Establishing Authority among Communities of Experts

Establishing his credentials, ʿAbdullāh Çelebi describes himself as the nephew of an Ottoman bureaucrat and as a well-known flower breeder. He also lists several other flower breeders, most of them scholars and learned bureaucrats, all of whom, he states, "have the capacity to reject and accept expert opinions [on matters related to flowers]" (*redd ve kabul sahibleri*). He says that he had the desire to collect "all the pearl-like precious opinions" of these people as well as the useful knowledge that he himself had obtained through experience (*tecrübe*). These "opinions," he emphasizes, were circulated and "exchanged" (*mutedavil*) among these flower breeders. Furthermore, ʿAbdullāh Çelebi's writing was strongly supported by the grand vizier Mustafa Paşa, to whom his treatise was dedicated. According to an inventory of the grand vizier's private possessions, drafted following his execution, he owned agriculturally productive properties, including a farm estate near Edirne and several gardens with pavilions around Istanbul.[45] The involvement of this type of patron, a landowner interested in agricultural techniques, was related to Ottoman thought on statecraft during this period, which considered agriculture one of the major means for high-ranking Ottoman officials to gain wealth.[46] This practice of patronage offered by such government officials to the producers of works on flowers continued until the early eighteenth century. One late example is a treatise on tulips dedicated to the grand vizier Ibrahim Paşa (1666–1730).[47]

45 Here, ʿAbdullāh Çelebi is most likely referring to Merzifonlu Kara Mustafa Paşa, who led the unsuccessful second siege of Vienna in 1683. On Merzifonlu Kara Mustafa Paşa's agricultural properties, see Hedda Reindl-Kiel's article, "The Must-Haves of a Grand Vizier: Merzifonlu Kara Mustafa Pasha's Luxury Assets," *Wiener Zeitschrift für die Kunde des Morgenlandes* 106 (2016): 179–221.

46 The grand vizier Derviş Mehmed Paşa (in office, 1653–1654), made agricultural investments and profits in Iraq while serving as governor, a fact which was noted and commented on by Mustafa Naima, an Ottoman bureaucrat and historian (1655–1716). See İbrahim Metin Kunt, "Derviş Mehmed Paşa, *Vezir* and Entrepreneur: A Study in Ottoman Political-Economic Theory and Practice," *Turcica* 9 (1977): 197–214, on 203–207.

47 Lalezari Mehmed, *Lale Risalesi*, MS Pertsch Türkisch 292, fols. 3a–4a, Staatsbibliothek zu Berlin.

The introduction of new leasing practices for waqf (endowed) land in the seventeenth century had allowed Istanbulites like ʿAbdullāh Çelebi to appropriate agricultural land in and around the city.[48] Such urbanites included a number of women who, according to rules governing the lease contracts known as *icareteyn* (double lease), were allowed to inherit such leases.[49] This also explains the presence of women among the lists of flower breeders in this period. ʿAbdullāh Çelebi's reporting of his father-in-law Hasan Kapudanzade's techniques of soil fertilizing indicates that marriage may have played an important role in the transmission of flower breeding techniques between different households.[50]

ʿAbdullāh Çelebi's text shows an awareness of being part of the newly emerging science of flower breeding, which would continue to engage Ottoman scholars until the nineteenth century. Noting that no one person is capable of knowing everything, his conclusion recommends reading his treatise critically, encouraging readers to correct any mistakes he might have made, and even to insert their knowledge on the matter into the margins of the page.[51] The treatise is presented as an open text to be altered, amended, and corrected, as a space for debate and the exchange of ideas, an invitation that is not to be found in earlier Islamic agronomic literature.

Despite his generous invitation to readers to comment on and correct their copy of the manuscript, ʿAbdullāh Çelebi was not indifferent to the disagreements and debate that arose in the seventeenth century between different groups of flower breeders in Istanbul. In the conclusion, he states: "If some people attack my writings and argue that they have created desirable flowers without following these proper methods, this should not be taken seriously or listened to."[52] He then discusses examples of how certain varieties of flowers and their seeds had been created by his contemporaries, criticizing those who had misattributed these flowers to other individuals. For example, he describes a dispute over the identity of the creator of the seeds of a narcissus variety called "world adorning" (*ʿālem-ārā*): some attribute the creation of these seeds to the calligrapher Mahmud Çelebi, while others believe that Mahmud Çelebi obtained the seeds from the chief physician upon his death.[53] Another example is the *süleymani* variety of narcissus that, according to ʿAbdullāh Çelebi, was

---

48 Shopov, "When Istanbul Was a City," 287–292.
49 Shopov, 288–289.
50 Shopov, "Flower Breeding," 591; Kahraman, *Şükûfenâme*, 120.
51 Kahraman, *Şükûfenâme*, 129.
52 Kahraman, 129.
53 Kahraman, 129.

created by Ahmed Dede—whom the text identifies elsewhere as the initiator of the science of flower breeding—but which had been wrongly attributed to Salih Efendi. According to ʿAbdullāh Çelebi, Salih Efendi had purchased the seeds for the price of just enough woolen cloth to clothe one single person.

Alternative attributions for the *alem-ara* and *süleymani* varieties of narcissus to which ʿAbdullāh Çelebi alludes are found in the 1667 *Şükūfe-nāme* by ʿAli Çelebi, who attributes *ʿālem-ārā* to the calligrapher Mahmud Çelebi and the *süleymani* variety to Salih Çelebi.[54] The texts of ʿAbdullāh Çelebi and ʿAli Çelebi differ not only in terms of their attributions of varieties to different breeders but also in terms of their content. ʿAli Çelebi's *Şükūfe-nāme* treatise contains four chapters, each devoted to a particular kind of flower—narcissus, tulip, hyacinth, rose—and provides information about the color, form, and name of the newly bred varieties. He also lists both male and female Ottoman flower breeders, many of them artisans living in or near Istanbul. ʿAbdullāh Çelebi, in contrast, divides the chapters according to various methods and techniques for creating these breeds, including the treatment and improvement of soil. The Şükūfe-nāme of ʿAli Çelebi and ʿAbdullāh Çelebi provide insights into debates about the question of manure and soil among the Istanbul flower breeding community. One instance is found in ʿAli Çelebi's introduction, where he interprets the ability of some gardeners to develop new varieties within a span of five to ten years as a sign of technical expertise. However, ʿAbdullāh Çelebi objects to this view in his treatise on the techniques used to create such varieties.[55] In the chapter on soil, he writes that some flower breeders grew flowers from seeds by applying manure to the soil in which the flower seeds were sown. This accelerating method, he noted, produced bulbs in a time period as brief as five years, albeit with a reduction in the quality of the bulbs.[56] In essence, ʿAbdullāh Çelebi's chapter on soil is a description of alternative soil treatment methods aimed at cultivating superb flowers over a longer period of time, in some cases up to seventeen years. As we saw earlier, he peppers the chapter with information on individual flower breeders and places where these techniques were invented. One of ʿAbdullāh Çelebi's final chapters comprises a list of the twenty-three traits that he claimed made a "distinct and indisputable flower" which, when achieved, could increase the price of the bulb, in his words, to a large sum (seventy-five ġurūş).

While ʿAbdullāh Çelebi criticizes many of the attributions found in ʿAli Çelebi's work, ʿAli Çelebi's treatise praises ʿAbdullāh Çelebi for the creation of

54 Kahraman, 34.
55 Kahraman, 30.
56 Kahraman, 119.

the *nevrūzīyye* variety of narcissus. He also notes that ʿAbdullāh Çelebi is still among the living and that he lives in Galata, just across the bay known as the Haliç (Golden Horn) overlooking Istanbul. He describes the new variety created by ʿAbdullāh Çelebi as a narcissus with "almond-like petals and some of them truncated in places toward the edges ... and the seed houses are short and thick and a perfect yellow." He also details its growth and development from bulb to flower.[57] And illustration of this variety is found in ʿAli Çelebi's manuscripts. Painted in full color, the image fills an entire page, illustrating his description of the flower's development, showing multiple stages simultaneously as if to simulate the flower's life cycle as an animation.[58] According to ʿAli Çelebi, ʿAbdullāh Çelebi "discovered" (*bulunmuş*) this flower, growing it new from seed on Nevruz (the first day of spring), hence its name, *nevrūzīyye*.[59]

Almost a century after the recognition of flower breeding as both a science and a guild by the Ottoman government, biweekly learned gatherings in grand vizierial households were still a common occurrence.[60] Flower breeders also met regularly in dervish lodges as well as in private households and gardens.[61] An Imperial edict from 1725 begins by highlighting that, for a long time, the residents of Istanbul had a deep inclination toward flower cultivation (*şükūfe perverliğe meyl u rağbet ediyorlar*). It also notes the existing flower market in the city, underscoring a notable surge in flower prices attributed to heightened demand and speculation.[62] In an attempt to regulate the issue of flower breeders and pricing, an Imperial decree established directives for pricing and mandated the presence of guild members, alongside the head of the flower breeders' guild, who were expected to create a list of flowers and their prices.

The same Istanbul court recorded an entry on December 23, 1725, with the prices of 224 varieties of *rumi* tulip.[63] This list shows the sheer amount of new

57 Kahraman, 38–39.

58 Kahraman, 38.

59 Shopov, "Flower Breeding," 593.

60 Halil Inalcik, *Şair ve Patron: Patrimonyal Devlet ve Sanat Üzerine Sosyolojik Bir Inceleme* (Ankara: Doğü Batı Yaınları, 2003), 73, quoted in M. Fatih Çalışır, "Vişnezâde İzzetî Mehmed Efendi (ö. 1092/1681): Kazasker, Şâir ve Hâmî," *Hikmet: Akademik Edebiyat Dergisi* (2021): 179–192, on 184–185.

61 ʿAli Çelebi writes that the Sufi Şeyh Ḥāsan Efendi organized learned gatherings every Friday in the dervish lodge of Koca Mustafa Paşa. See Kahraman, *Şükûfenâme*, 29; Shopov, "Flower Breeding," 593.

62 Fuat Recep, Mehmed Akan, and Fikret Sarıcaoğlu, *İstanbul Kadı Sicilleri Istanbul Mahkemesi 24 Numaralı Sicil* (*H. 1138–1151 / M. 1726–1738*) (Istanbul: ISAM, 2011), 120.

63 Fuat Recep, Mehmed Akan, and Fikret Sarıcaoğlu, *İstanbul Kadı Sicilleri Istanbul Mahkemesi 24 Numaralı Sicil* (*H. 1138–1151 / M. 1726–1738*) (Istanbul: ISAM, 2011), 165.

varieties that had been created throughout the seventeenth and early eighteenth centuries, many of them of high value.[64] Two years later, the prices of some of the flower varieties increased, which could be an indication that the Imperial Council was not able to control the high demand and speculation on Istanbul's flower market.[65] Soil, too, became an object of trade, as it was transported and sold to people constructing gardens within and around Istanbul. For example, the steward (*kethüda*) of the grand vizier Ibrahim Paşa purchased soil and pebbles for the price of 134 ġurūş, including transportation.[66]

3 Conclusion

This chapter has tried to show how and why practices surrounding flower breeding, particularly the improvement of soil, moved into the realm of textual production in mid-seventeenth-century Ottoman Istanbul. In ʿAli Çelebi's work, flower varieties are attributed to specific breeders. ʿAbdullāh Çelebi's work focuses on techniques for cultivating flowers, including improving the soil. It also identifies techniques with contemporary individuals who practiced and advocated for them, even elevating practitioners to the level of founders of "schools" in matters related to soil—a departure from earlier agricultural writings. The texts reflect an experimental culture in which flower breeding techniques, and flower genealogies, were debated within a community of practitioners. The two works were in conversation with each other. While ʿAli Çelebi praises the speed with which new varieties were being created, ʿAbdullāh Çelebi criticizes techniques for treating the soil that shortened the time period between the planting of the seed and the development of bulbs and flowers. These early examples of Şükūfe-nāme writings followed the recognition of flower breeding as a distinct science by the Ottoman government and the establishment of a flower guild with an appointed leader, a development with no precedent in either Ottoman or Islamic history.

While Ottoman flower breeding may be considered one of many transcultural global exchanges in the early modern period, the discussions of soil in the

64 The breeding of tulips also found expression in Ottoman visual culture. See Demiriz, "Tulips."

65 Munir Aktepe, "Damat Ibrahim Paşa Devrinde Laleye Dair bi Vesika," *Turkiyat Mecmuası* 11 (1954): 117–130, on 117.

66 Abdullah Sivridağ and Ali Coşkun, *Istanbul Kadı Sicilleri 66: Bab Mahkemesi 151 Numaralı Sicil (H. 1143–1144 / M. 1731)* (Istanbul: Kültür AŞ, 2019), 204.

Şükūfe-nāme of ʿAbdullāh Çelebi reflect the birth of a community of practitioners who were putting their experimental and innovation-oriented culture into writing. Far from constraining technical developments, the establishment of a guild coincided with the production of texts in which the nitty-gritty technical aspects of flower breeding were hotly debated. ʿAbdullāh Çelebi constructed an authoritative language based on *experience* and *experimentation* by applying terminology borrowed from Islamic legal discourse and by invoking contemporary practitioners and their experiments in the treatment of soil. The case of flower breeding in seventeenth-century Istanbul together with the generation of textual knowledge and authority spurred on by this emerging science—in particular on the matter of soil improvement—exemplifies the transformation of soil into a new object of study within Ottoman learned discourse.

CHAPTER 11

Taste and the Quality of Soil in Early Modern South Asia

Flavors of Fertility

Nicolas Roth

1 Introduction

Sometime around 1577, Cakrapāṇi Miśra, a scholar at the court of Mahārāṇā Pratāp (1540–1597) in Mewar, northwestern India, wrote a gardening treatise in Sanskrit. In doing so, he was continuing a tradition of horticultural writing that was by then about a millennium old and generally known as *vṛkṣāyurveda*, or "plant medicine."[1] At first glance, Cakrapāṇi Miśra's work, entitled *Viśvavallabha* (Dear to all) seems to hew closely to the conventions and standard inherited content of the generally conservative genre. Upon close reading, however, it soon reveals itself to be subtly innovative, full of attempts to update received tradition and bring it more in line with contemporary horticultural realities and the importance of practical experience. This is made quite explicit in the opening verses of the third chapter of the *Viśvavallabha*, which discusses the characteristics of soils:

1 *Vṛkṣāyurveda* is a compound of *vṛkṣa*, "tree, plant," and *āyurveda*, the name of one of India's traditional systems of medicine, which translates roughly to "science of life spans." The term occurs as early as the *Arthaśāstra* of Kauṭilya and the *Kāmasūtra* of Vatsyayana, dating roughly to the fourth century CE, while the first surviving full, albeit brief, chapter on *vṛkṣāyurveda* appears in the sixth century CE in the encyclopedic *Bṛhatsaṁhitā* of Varāhamihira. Surapāla's *Vṛkṣāyurveda*, produced in Bengal in the late eleventh or early twelfth century, is the oldest extant independent treatise in the genre. See Kauṭilya, *King, Governance, and Law in Ancient India: Kauṭilya's Arthaśāstra*, trans. Patrick Olivelle (New York: Oxford University Press, 2013), 29–31; Daud Ali, "Botanical Technology and Garden Culture in Someśvara's Mānasollāsa," in *Garden and Landscape Practices in Precolonial India: Histories from the Deccan*, ed. Daud Ali and Emma Flatt (Delhi: Routledge, 2011), 39–53, on 40; Vatsyayana, *Kamasutra*, trans. Wendy Doniger and Sudhir Kakar (Oxford: Oxford University Press, 2002), xi–xii, 14–15; Varāhamihira, *Bṛhat-Saṁhitā of Varāha-Mihira*, ed. Johan Hendrik Caspar Kern and Shrikrishna "Jugnu," trans. N. Chidambaram Iyer (Delhi: Parimal Publications, 2013), 1:199–204, 242–244, 2:64–101; Rahul Peter Das, introduction to *Das Wissen von der Lebensspanne der Bäume: Surapālas Vṛkṣāyurveda*, by Surapāla, ed. and trans. Rahul Peter Das (Stuttgart: Franz Steiner Verlag, 1988), 1–56, on 17–19.

 | DOI:10.1163/9789004748484_013

jāṅgalānūpasāmānyasvabhāvāt trividhā dharā |
rasaiśca ṣaḍbhiḥ sābhinnā jñeyāstadvarṇato rasāḥ ||
malinā pāṇḍurā bhūmi śyāmalādhavalārūṇā |
pītā miṣṭāmlalavaṇakaṭutiktakaṣāyikā ||
krameṇa ca rasāstvatra jñeyā jīrṇamataṃ tvidam |
mṛttikāsvādanātte ca jñāyante me mati dhruvam ||[2]

Arid, marshy, or of average nature—soil is of [these] three kinds,
And it is distinguished according to six tastes, [and] these tastes are to be known from its color.
Soil is gray, pale, dark, white, reddish, and yellow,
Sweet, sour, salty, bitter, pungent, and astringent.
And according to this system [of correspondence to color], the tastes are to be known
—that, however, is an old opinion,
And it is my firm view that they are to be ascertained by tasting the soil.[3]

Here, Cakrapāṇi Miśra was explicitly contrasting received wisdom with his own belief in the importance of practical, hands-on experience. He was making a claim for modernization, though paradoxically the substance of his intervention—that one should literally taste soil to find out its characteristics—might appear rather alien to a modern audience. Yet while eating what in colloquial English is often rather pejoratively referred to as "dirt" may seem strange today, in Cakrapāṇi Miśra's time it would have arguably been a logical extension of "flavor" as one of the established characteristics through which to assess, categorize, and understand soil. What is at times less clear, independently of whether the two sets of characteristics correspond to each other in a predictable fashion or not, is how these distinctions of taste or color relate to the fertility of the soil or its suitability for the cultivation of particular plants. To an extent, they may simply reflect the obsession with comprehensive classifications evident throughout much Sanskrit *śāstra*, or technical literature, rather than directly actionable information. However, by insisting that gardeners actually taste the soil in order to assess its characteristics, Cakrapāṇi Miśra pushed back against mere theoretical classification, demanding instead that knowledge be implemented through practical, material experience. In the verse that follows the ones cited above, he also gave some indication of the taste that matters the most when trying to ascertain the quality of soil, though

2 Cakrapāṇi Miśra, *Viśvavallabha*, trans. Nalini Sadhale, commentaries by Nalini Sadhale and Y. L. Nene (Secunderabad: Asian Agri-History Foundation, 2004), 34–35.

3 Translations from Sanskrit and Persian in this article are the author's own.

the flavor in question appears interspersed with various aspects of texture and location:

valmīkagartapāṣāṇabahulādūṣarāshubhā |
dūrodakā śārkarilā saviṣā taruropane ||[4]

[Soil] that is afflicted with anthills, holes, or salt, and is foul,
Far from water, or gravelly, is poisonous to the cultivation of plants.

Tasting the soil, then, is primarily about detecting potential salinity, as well as, perhaps, any broader sign of "foulness" or contamination. Flavor is thus mainly a means of figuring out which soils are *not* fit for cultivation, and as will be discussed in detail below, the most important taste dichotomy in making this determination is between salty and "bad" flavors on the one hand and (metaphorical) sweetness on the other. In the broader context of gardening and farming treatises available in northern India in the sixteenth century, both Cakrapāṇi Miśra's insistence on tasting the soil and his specific definition of brackish and contaminated soil (the term he uses is *aśubha*, which has a broad semantic range including "inauspicious," "impure," and "foul") as soil incapable of nurturing plant growth underscore the practical modernizing impulse that he himself explicitly claimed. His instructions also point to an ongoing encounter between the indigenous Sanskrit tradition of *vṛkṣāyurveda* horticultural writings and a Persianate gardening culture that was increasingly taking root in northern India, bringing with it distinctive garden layouts, new temperate-climate plants, and Persian agronomic texts.

2 Assessing Soil in the Sanskrit Tradition of Agricultural Writing

The basic set of possible soil colors and flavors itself long predates the *Viśvavallabha*. It occurs in almost identical form in verses contained in the *Vṛkṣāyurveda* of Surapāla composed in Bengal on the opposite side of northern India in the late eleventh or early twelfth century, as well as in the more popular abridgment of that text contained in a chapter of the encyclopedic *Śārṅgadharapaddhati* compiled in 1363, most likely just to the northeast of Mewar:[5]

4 Miśra, *Viśvavallabha*, 34.

5 Lallanji Gopal, *Vṛkṣāyurveda in Ancient India* (New Delhi: Sundeep Prakashan, 2000), 105; Śivaprasāda Dvivedī, ed., *Agnipurāṇam: Hindīvyākhyopetam* (Delhi: Caukhambā Saṃskṛta Pratiṣṭhāna, 2004), 76–77.

jāṅgalānūpasāmānyabhedaiḥ sā bhidyate mahī |
punaḥ sā bhidyate ṣoṛhā varṇato rasatastathā ||
asitavipāṇḍuśyāmalalohitāsitapītarociṣaḥ kramaśaḥ |
madhurāmlalavaṇatiktakaṭukaṣāyā bhuvo rasataḥ ||[6]

Arid, marshy, and average—the earth is divided into these kinds;
Then it is divided into six kinds according to color and flavor.
Dark [gray], pale, dark [black], red, deep yellow, and bright [white], in order;
Soil is sweet, sour, salty, bitter, pungent, or astringent [in terms] of taste.

Cakrapāṇi Miśra is likely to have known these verses or very similar ones; the "in order" at the end of the penultimate line stresses a rigid correspondence of color and flavor that he decries as a *jīrṇamata*, or old/outdated opinion. Note that the order in which the colors are listed here is slightly different from that in the *Viśvavallabha*. In the latter, the fourth color is white, the fifth is red, and the sixth yellow, while in the earlier Surapāla/*Śārṅgadharapaddhati* text, the fourth is red, the fifth yellow, and the sixth white. Since the order of tastes does not change, any supposed correspondence between color and taste would not be consistent between the two sets of verses. It is conceivable that Cakrapāṇi Miśra had multiple older *vṛkṣāyurveda* texts at his disposal and noticed such discrepancies between them, which caused him to distrust the traditional—or at least textually prescribed—method of deducing the taste of soil from its visual characteristics. However, as will be discussed at greater length below, the ongoing encounter with Persianate garden culture may well have played a greater role.

3 Theorizing Taste in Sanskrit

The broader cultural theorization of taste that informs the use of this sense in assessing and classifying soil, meanwhile, goes back significantly further than even the *Vṛkṣāyurveda* of Surapāla. A discussion of the senses in the epic *Mahābhārata*, dated to around the beginning of the Common Era, states: "sweet, salty, bitter, astringent, sour, also pungent. This sixfold enumeration of

6 *Upavanavinodah, Kṛṣṇānandajhā-viracitayā Śyamalayā Hindīvyākhyayopetaḥ* (Darbhanga: Kāmeśvarasiṃha Darabhaṅgā Saṃskṛta Viśvavidyālayaḥ, 1984), 10.

taste, made of water, is taught."[7] Also important to note is the multivalence of the term *rasa*, which is used for "taste" or "flavor" in all of these texts. Beyond the aforementioned meanings, its semantic range encompasses "juice" and "essence." Most prominently, however, it serves as the central technical concept in traditional Indian literary aesthetics. Beginning with the *Nāṭyaśāstra* (Treatise on drama) of Bharata, tentatively dated to the early first millennium of the Common Era, scholars of Sanskrit as well as later vernacular literatures assessed the aesthetic response to a work of art in terms of eight *rasa*s, or generalized emotional categories, namely the erotic, the comedic, the heroic, disgust, horror, fury, compassion, and wonder. Later, a ninth *rasa*, the peaceful or quiescent, was added and held by some to be the supreme *rasa*, as it represents the ultimate peace and release from the cycle of rebirth that is the highest goal of human existence in most indigenous Indic religious systems.[8] There appears to be no explicit link between *rasa* as flavor and *rasa* as aestheticized emotion, and the set of aesthetic "flavors" notably includes two or even three *rasa*s more than the somewhat more stable, codified inventory of tastes. It is intriguing, however, that the erotic, or *śr̥ṅgāra*, *rasa*—often dominant in poetic works, even if the *śānta*, or peaceful, *rasa* is theoretically more important—evolves into the *mādhurya rasa*, or the *rasa* of sweetness, in the context of prominent North Indian devotional traditions centering on the god Kr̥ṣṇa and his consort Rādhā. At least figuratively, then, the sense of *rasa* as flavor returns in the devotional ideal in which erotic love-in-union converges with the experience of union with the divine. Moreover, just as the erotic is generally the most prominent *rasa* in a literary context, de facto if not de jure, sweetness mostly comes first when it comes to flavor—quite literally in the case of all of the enumerations of the six tastes already cited here.

4 Horticultural Change in Early Modern South Asia

The world in which Cakrapāṇi Miśra wrote about gardening, however, was no longer purely that of Sanskrit literature and its mostly tropical South Asian vegetation, cultivated in informal woodland gardens. Muslim dynasties had ruled large parts of the north and middle of the subcontinent since the thirteenth century and most of its northwestern regions for even longer. Cakrapāṇi Miśra's

7 James McHugh, *Sandalwood and Carrion: Smell in Indian Religion and Culture* (Oxford: Oxford University Press, 2012), 30–31.

8 Sheldon Pollock, ed. and trans., *A Rasa Reader: Classical Indian Aesthetics* (New York: Columbia University Press, 2016), 6–7, 187.

home region of Mewar had sat between the Gujarat Sultanate to the south and the Delhi Sultanate to the northwest for well over a century; by the time of the writing of the *Viśvavallabha*, both had been replaced by the newly ascendant Mughal Empire 1526–1857. All of these state formations fostered Iranian and Central Asian cultural influences as well as the use of the Persian language in literature, science, and courtly correspondence, and they prompted an ongoing influx of people from Iran and Central Asia seeking career opportunities. Their Hindu-ruled neighbors, vassals, and rivals, like Mewar, came to share in much of the resultant Indo-Persian elite culture.[9] This included new horticultural practices and ideals of garden design, most notably a shift to formal walled gardens with symmetrical layouts, with parterres of herbaceous and annual flowers and, when possible, temperate-climate fruit and nut trees like peaches and almonds alongside the (sub)tropical mangoes and bananas. The impact of this change is discernible if one compares what little the *Viśvavallabha* has to say about garden design with the visions of what a garden should look like found in earlier *vṛkṣāyurveda* texts. In the *Vṛkṣāyurveda* of Surapāla, the following verse summarizes an older garden typology typical of Sanskrit textual tradition:

> *madhye tasmiñchaśiraśikhari spardhi veśma pravātaṃ guḍhopāntaṃ surabhikusumaiḥ śākhibhirnamraśākhaiḥ* |
> *sthāne sthāne sphaṭikadhavalaṃ maṇḍapaṃ maṇḍanārddhaṃ kuryāt tasminnapi ca kadalī maṇḍitaṃ mandavāyu* ||[10]

> In its middle should be an airy building vying with lofty peaks, its immediate surroundings hidden by branches bearing fragrant flowers and weeping branches;
> Here and there, crystal-white and partially ornamented pavilions are to be built, adorned with banana plants and with a gentle breeze.

This, then, is a vision of a sprawling woodland garden of irregular layout and structures picturesquely dotted among the vegetation. The contrast in the *Viśvavallabha* is stark:

9 On the history and cultural developments of Persianate South Asia, see, e.g., Richard Eaton, *India in the Persianate Age, 1000–1765* (Oakland: University of California Press, 2019); Emma Flatt, *The Courts of the Deccan Sultanates: Living Well in the Persian Cosmopolis* (Cambridge: Cambridge University Press, 2019).

10 Gopal, *Vṛkṣāyurveda*, 220.

durge yathāmānayathāvakāśaṃ kuryācca sādhāraṇavāṭikā ca |[11]

Inside the fort, a regular garden is to be laid out according to particular dimensions and the space available.

The emphasis here is on structure and regularity; rather than the garden containing sundry architectural follies, it is itself embedded in a larger built context. This certainly reflects the realities of the palace-fort architecture of Mewar's nobility in the early modern period, but it also coheres with the Persianate vision of a "proper"—regular, symmetrical, walled—garden. Also suggestive is that the subsequent lengthy and partially obscure list of flowers to be planted in this space, immediately following the passage cited above, contains at least two types of roses that, while not unknown in Sanskrit texts, are much more central to Persianate horticulture than to older South Asian practices.[12] Cakrapāṇi Miśra, in other words, was beginning to reflect real-world changes in his gardening manual, even as he wrote in Sanskrit within the largely conservative *vṛkṣāyurveda* tradition.

Further evidence for the text's engagement with Persianate horticulture can be found in the garden plants it mentions and the specific names it uses for them. It lists species not previously discussed in *vṛkṣāyurveda* works and generally foreign to the world of Sanskrit literature and scholarship. These include the corn poppy (*Papaver rhoeas*), which appears as *gulāla*, from Persian *gul-i lālah*,[13] *fālsah* (*Grewia asiatica*), which here becomes *phirasā*, and walnut

11 Miśra, *Viśvavallabha*, 36.

12 The terms listed are *sevatī* and *kubja*, which generally refer variously to the musk rose (*Rosa moschata*) or the very similar Himalayan musk rose (*R. brunonii*), though the fact that both are listed here separately raises the question of whether one is meant to refer to a different kind of rose, such as the pink Damask cultivars (*R. x damascena*) common in Persianate cultural contexts and widely cultivated for the production of rose water. In contemporary usage in some Indian languages, especially those in the southwest of the subcontinent such as Marathi, Kannada, and Konkani, *sevatī* and its variations are also used to refer to chrysanthemums, especially those that are white and yellow like the musk rose. It is not entirely clear how early or widespread this usage is historically. However, a depiction of a double-flowered chrysanthemum cultivar appears labeled in Konkani as *seva(n)tī* in the *Hortus Malabaricus*, the large illustrated flora of Kerala published in Amsterdam by Hendrik Adriaan van Rheede tot Drakenstein (1636–1691) between 1678 and 1693. Even if *sevatī* in the *Viśvavallabha* were to refer to chrysanthemums, though, it would still indicate a shift to herbaceous bedding typical of more Persianate garden models. See Hendrik Adriaan van Rheede tot Drakenstein, *Hortus Indicus Malabaricus* (Amsterdam, 1690), 10:86, https://www.biodiversitylibrary.org/item/14379#page/303/mode/1up.

13 In Iranian Persian usage, the term *lālah* usually refers to tulips; however, in early modern South Asia, its meaning was extended to include the corn poppy, which like many species

and pistachio, which appear respectively as *āṣoṭa* and *ākhoṭa*—from Urdu/Hindi *akhroṭ*—and *pista*, from the same name current in both Persian and North Indian vernaculars.[14] Similarly, the apple appears as *seva*, from Persian *sīb*, and the almond (*Prunus dulcis*) appears as both *vidāma* and *badāma*, both from Persian *bādām*.[15] Given these obvious Persianate influences in the *Viśvavallabha*, it appears more than reasonable that the encounter with Persianate garden culture would have also colored its treatment of soil "science." Indeed, Cakrapāṇi Miśra's forceful call to move beyond the *jīrṇamata* of color correspondences by actually tasting the soil and his specific identification of salinity and "impurity" or contamination as undesirable characteristics that could be avoided in this manner becomes less surprising when seen in the light of the Persianate gardening tradition's own body of agronomic technical literature.

5 The Tradition of Agronomic Writing in Persian

In Persian-language horticultural manuals that circulated or were compiled in South Asia in the sixteenth and seventeenth centuries, the relationship between knowing and tasting soil is discussed quite differently than in the older *vṛkṣāyurveda* texts; in the advice given in the *Viśvavallabha*, however, there are certain similarities. The *Irshād al-zirāʿah* of Qāsim ibn Yūsuf Abū Naṣrī Haravī, completed in Herat in 1515, prescribes the following method to test the quality of soil:

> *Zamīn rā bikanand dū gaz yā sih gaz pas qadrī kulūkh az maghz-i ān bigīrand va dar hāvun narm bisāyand va dar āb-i bārān āghashtah kunand chandānkih āb az sar-i ān biguẕarad va hamchunān biguẕārand tā āb ṣāfī shavad agar ṭaʿam-i āb bahāl-i khvud bāshad zamīn nīk buvad va agar āb shūr shavad ān zamīn shūrahzār buvad va har zamīn kih khāk-i ū rā bū-yi*

of tulip has bright red flowers and a brown-black basal blotch at the center of the flower. Within the Indian subcontinent, tulips were limited to the Himalayan regions in the far northwest, while poppies were easily cultivated throughout much of South Asia as winter annuals. In addition to the visual parallel, this means that, like tulips, they also bloom primarily in spring. Moreover, it made them one of the most prominent flowers of the herbaceous, seasonally changing parterres that constituted one of the most significant departures of the new Persianate garden typology from older Indic practices.

14 Miśra, *Viśvavallabha*, 36–38.

15 Miśra, 35, 37–38.

> *bad buvad ṣāliḥ-i zirāʿat nabuvad va zamīnī kih shūr buvad darān hīch dirakht nayāyad magar dirakht-i khurmā va nay va ghubayrā.*[16]
>
> The ground is dug up two *gaz* or three *gaz*[17] deep, then a little clod is taken from the core [of the excavated soil] and ground soft in a mortar and mixed with rain water so that the water level is above the soil. And it is left [to stand] such that the water becomes clear. If the water retains its own taste, the soil is good, and if the water is salty, that ground is brackish. And any ground whose dust has a bad smell is not fit for cultivation, and in ground that is salty, no plant will grow except date palms,[18] reeds, and Russian olive.[19]

These instructions bear a clear resemblance to the "technique" outlined in the 1638 Ottoman verse treatise by Kemānī discussed in this volume by Aleksandar Shopov and Himmet Taşkömür. Unlike in Sanskrit, there is no list here of every possible flavor that has been theorized, nor any suggestion that each of those flavors may be found in different kinds of soil. Nor is there any proposed relationship to an equally extensive range of potential soil colors. Instead, it is the physical act of tasting and smelling the soil—in the former case, via the intermediary of water in which it has been soaked—that is at the heart of the assessment of its suitability for cultivation. Rather than the presence of particular aromas, it is mostly their absence that marks good, "clean" soil that is *ṣāliḥ-i zirāʿat*, or fit for cultivation. The only flavor singled out for discussion is salty, perhaps in part because saltiness is particularly distinctive and clearly detectable by tasting. More importantly, however, in the context of gardening and farming under intensive irrigation in an arid climate such as that of Herat or large parts of northwestern India, brackish soils and creeping salinization are a frequent issue due to intense surface evaporation. Tellingly, the saline-tolerant plants Qāsim ibn Yūsuf mentions—date palm, reed, and Russian olive—are typical oasis plants, adapted to grow in and around brackish bodies of water. A later passage discusses a procedure of repeated thorough watering, ideally at the time of winter frost, to wash out the salt from saline soil and turn it "sweet." The text acknowledges that this is likely to be of limited success, but

16 Qāsim ibn Yūsuf Abū Naṣrī Haravī, *Irshād al-zirāʿah*, ed. Muḥammad Mushīrī (Tehran: Dānishgāh-i Tihrān, 1967), 54.

17 A variable measure of length in early modern South Asia, though generally roughly equivalent to a yard.

18 *Phoenix dactylifera*; in South Asia, also the wild date palm, *P. sylvestris*.

19 *Elaeagnus angustifolia*, a drought-tolerant tree native to Central and Southwestern Asia with fragrant flowers and edible fruit, widely naturalized in many parts of the world.

the conceptual juxtaposition of literally salty and figuratively sweet—that is, fertile—soil is nonetheless significant if we recall the primacy of sweetness in the Indic classification of flavors.

The presence of the *Irshād al-zirāʿah* in the Indian subcontinent is attested by multiple Mughal-period Indian manuscript copies of the text, including one that bears the seal of the emperor Aurangzeb (r. 1658–1707).[20] In addition, Maria Subtelny has traced the subsequent Indian careers of several people involved with the production of the text. Qāsim ibn Yūsuf identifies the prominent gardener and landscape architect Sayyid Niẓām al-Dīn Amīr Sulṭān Maḥmūd, known as Mīrak-i Sayyid Ghiyās̱, as the primary informant for his work.[21] Part of a long line of agronomists and garden designers, this Mīrak-i Sayyid Ghiyās̱ began his career in Herat under Sulṭān Ḥusain-i Bāyqarā (r. 1469–1506). In 1529, three years after the founding of his South Asian empire, the first Mughal ruler, Ẓahīr al-Dīn Muḥammad Bābur (1483–1530), employed him in construction projects in Agra and Dholpur.[22] Later, Mīrak-i Sayyid Ghiyās̱ was active in Bukhara, then the capital of the Uzbeks, modern-day Uzbekistan, where he appears to have died sometime in the 1550s.[23] However, a son of his known as Sayyid Muḥammad-i Mīrak appears to have remained in India or later returned there. Following the family tradition, he is recorded as having been in charge of the construction of the monumental garden tomb complex of the emperor Humāyūn (r. 1530–1540 and 1555–1556) in Delhi, completed in 1571 and an important precursor to later grand garden tombs of Mughal elites, including most famously the Taj Mahal.[24] There is also the

20 Maria E. Subtelny, "Mīrak-i Sayyid Ghiyās and the Timurid Tradition of Landscape Architecture," *Studia Iranica* 24, no. 1 (1995): 19–60, on 49. The seventeenth-century manuscript in question is bound up with two other treatises on gardening and agriculture and has marginalia giving the equivalent in Indian months of operational timings stated using Perso-Arabic zodiac names, translations into simpler Persian of various Arabisms, general explanations such as "oak is a type of tree," and, in at least one instance, the local Hindustani name of a plant (*methī*, written next to the Arabic *ḥilbah*, for fenugreek). See Qāsim ibn Yūsuf Abū Naṣrī Haravī, *Irshād al-zirāʿah*, 1515, MS BL Or. 7557, British Library, London. Further South Asian manuscript copies are found in *Majmūʿah-i Rasāʾil*, MS C-391/ac. nos. 672–674, Dr. Zakir Husain Central Library, Jamia Millia Islamia, Delhi; as well as *Irshād al-zirāʿah*, MS 1614, Society Collection, Asiatic Society, Kolkata; and *Irshād al-zirāʿah*, MS 628, Curzon Collection, Asiatic Society, Kolkata.

21 Subtelny, "Mīrak," 21–24.

22 Subtelny, 27–28.

23 Subtelny, 32, 34–35.

24 Subtelny, 31, 48–49. See also, among various others, Lisa Golombek, "From Tamerlane to the Taj Mahal," in *Islamic Art and Architecture: Essays in Honor of Katharina Otto-Dorn*, ed. Abbas Daneshvari (Malibu: Undena Publications, 1981), 43–50; Glenn D. Lowry, "Humayun's Tomb: Form, Function, and Meaning in Early Mughal Architecture,"

possibility that the author of the *Irshād al-zirāʿah* himself may have migrated to India, since a "Mullā Qāsim" is mentioned as having been employed at Agra and Dholpur alongside Mīrak-i Sayyid Ghiyāṣ, and Humāyūn's sister Gulbadan Begam, in her *Humāyūnnāmah*, mentions "the architect Khvājah Qāsim," who was instructed by Bābur to carry out in Agra whatever building projects the emperor's aunts ordered.[25]

Virtually the same passage on testing soil by taste and smell found in the *Irshād al-zirāʿah* recurs in at least two later Persian gardening manuals that appear to have circulated widely in early modern South Asia. The first is an anonymous treatise variously called the *Maʿrifat-i falāḥat* or *Davāzdah bāb-i kishāvarzī*. Its most recent Iranian editor, Īraj Afshār, dates this work to 1523 and attributes it to the astronomer and mathematician ʿAbd al-ʿAlī bin Muḥammad bin Ḥusayn Bīrjandī (d. 1528).[26] Judging by surviving manuscripts, this text was particularly popular in both Iran and India.[27] The second text in question is the lengthy gardening chapter of the early seventeenth-century *Ganj-i bādāvard*, a gargantuan pharmacopoeia and medical encyclopedia composed by the physician and courtier Amān Allāh Khān Zamān Ḥusaynī (d. 1636), eldest son of the prominent Mughal general and governor Mahābat Khān (d. 1634). Under the title "ʿAmal-i yazdahum" (Eleventh chapter), this piece came to be copied widely as an independent treatise.[28] The only divergence in

Muqarnas 4 (1987): 133–148 (although he erroneously conflates Sayyid Muḥammad-i Mīrak and his father Mīrak-i Sayyid Ghiyāṣ); Laura Parodi, "'The Distilled Essence of the Timurid Spirit': Some Observations on the Taj Mahal," *East and West* 50, nos. 1–4 (2000): 535–542.

25 Subtelny, "Mīrak," 37; Gulbadan Begam, *Humāyūnnāmah*, ed. Annette S. Beveridge (London: Royal Asiatic Society, 1902), 14.

26 Īraj Afshār, preface to ʿAbd al-ʿAlī bin Muḥammad bin Ḥusayn Bīrjandī, *Maʿrifat-i falāḥat (Davāzdah bāb-i kishāvarzī)*, ed. Īraj Afshār (Tehran: Markaz-i Pizhūhishī-i Mīrās-i Maktūb, 2008), 17–19.

27 Afshār identifies twenty-seven copies produced between 1550 and the early twentieth century, at least nine of which are almost certainly South Asian. Further research so far has turned up an additional Iranian manuscript dated to 1697 and three more nineteenth-century Indian copies. The Iranian manuscript appeared at auction at Bonhams, London, on October 23, 2017. See "Islamic and Indian Art Including Modern and Contemporary South Asian Art," Bonhams, October 23, 2017, lot 27, https://www.bonhams.com/auctions/24198/lot/27/?category=list&length=12&page=3. The Indian copies in question are *Majmūʿah-i rasāʾil*, MS C-391/ac. nos. 672–674, Dr. Zakir Husain Central Library, Jamia Millia Islamia, New Delhi; *Falāḥnāmah/Hamīshah bahār*, 1835, MS Codrington/Reade no. 212, box 82, Royal Asiatic Society, London; and *Kitāb-i khizān va bahār*, MS Mutafarriqāt 686, Government Oriental Manuscripts Library and Research Centre, Hyderabad (India).

28 Nabi Hadi, *Dictionary of Indo-Persian Literature* (New Delhi: Indira Gandhi National Centre for the Arts, 1995), 74–75. The supposed tomb of Amān Allāh Ḥusaynī is apparently still kept in the town of Samana in the Indian Punjab by Shia *sayyid* families that trace

content between the *Irshād al-zirāʿah* and the versions of the passage found in the two latter texts is that, in the list of salt-tolerant plants, Russian olive is replaced by tamarisk (*Tamarix* sp.), another plant well adapted to desert and oasis environments.[29]

6 Ancient Mediterranean Sources

The Persian tradition of horticultural and farming treatises and encyclopedia entries drew heavily on ancient Greek and Roman agronomic knowledge by way of Byzantine works. Writing on agriculture and gardening in New Persian appears to begin at the western end of the Persophone world with the eleventh- or twelfth-century *Varznāmah*, believed to be a translation of a Middle Persian version of the now-lost sixth-century Greek compilation *Peri georgias eklogai* by Cassianus Bassus Scholasticus, which was also a source of the tenth-century Byzantine *Geoponika* and itself based on the fourth-century compilation by Vindonius Anatolius of Beirut.[30] It was also translated into Arabic, apparently both from the Greek original and from the Middle Persian *Varznāmak*, resulting in at least two often-confused and debated lines of transmission under various versions of the titles *Filāḥa al-rūmiyya*, *Filāḥa al-yūnāniyyā*, and *Filāḥa al-fārisiyya*, translating respectively to "Roman/Byzantine agriculture," "Greek agriculture," and "Persian agriculture."[31] Through these various channels of transmission, Cassianus's work was also an important source for medieval Arabic agricultural texts, which in turn informed early modern Ottoman gardening manuals. Cassianus is cited frequently, for example, throughout the *Kitāb al-filāḥa* of the Andalusian agronomist Ibn al-ʿAwwām, completed in Seville in 1180, and also discussed in this volume by Shopov and Taşkömür. He appears as both Qusṭūs and Kasiyūs, with Ibn

---

their ancestry through him. See Aanchal Malhotra, *Remnants of a Separation: A History of the Partition through Material Memory* (Noida: HarperCollins India, 2017), 174–175.

29 Bīrjandī, *Maʿrifat-i falāḥat*, 16–17; *ʿAmal-i yazdahum: Fragment einer Abhandlung zur Landwirtschaft*, Ms. or. fol. 310, Staatsbibliothek zu Berlin, fol. 1r.

30 Ḥasan ʿĀtifī, introduction to *Varznāmah*, by Fastiyūs bin Askūrāsīkah (Tehran: Markaz-i nashr-i dānishgāhī, 2009–2010), 7–9; Robert Rodgers, "Garden Making and Garden Culture in the *Geoponika*," in *Byzantine Garden Culture*, ed. Antony Littlewood, Henry Maguire, and Joachim Wolschke-Bulmahn (Washington, DC: Dumbarton Oaks, 2002), 161–162.

31 See, e.g., Angelo Alves Carrara, "Geoponica and Nabatean Agriculture: A New Approach into Their Sources and Authorship," *Arabic Sciences and Philosophy* 16, no. 1 (2006): 103–132; Francisco Javier Mariscal Linares, "Comentarios a la edición jordana de la *Filāḥa al-Rūmiyya*," *Boletín de la Asociación Española de Orientalistas* 39 (2003): 67–77.

al-ʿAwwām clearly distinguishing between the two, which suggests that the Andalusian author had access to two rather different recensions of the text.[32] Aleksandar Shopov has also previously highlighted Cassianus's importance as one of the sources of the Ottoman *Revnaḳ-ı Būstān*, a work that notably appeared at almost exactly the same time as the *Viśvavallabha* at the opposite end of the Persianate world.[33]

Descriptions of soil-testing methods turn out to be one of the most persistent elements of this textual heritage, reproduced with little change across centuries and huge swaths of geography. Thus, the passage cited above that appears in the Persian *Irshād al-zirāʿah, Maʿrifat-i falāḥat*, and *Ganj-i bādāvard* has clear echoes of the following lines from the Latin *Georgics* of Virgil, completed around 29 BCE:

> But brackish ground, called bitter, bane to germination,
> unmellowed by tillage, conserving not the lineage of vines,
> the names of apples—it will give this proof:
> fetch down from smoky rafters your tight-wickered
> baskets and wine-strainers. Fill them with that worthless soil
> tamped down, and spring-fresh sweetwater.
> You'll see the water work through, great drops between
> the willows, but its taste will testify, its bitter tang
> will twist the taster's bittered tongue.[34]

Not surprisingly, the correspondence is even more striking in the Byzantine Greek *Geoponika*, in a section attributed specifically to Vindonius Anatolius, and hence likely also contained in some form in the lost work of Cassianus:

> Others, not satisfied with testing by observation, devised a method using other senses. Having dug to a certain depth at the relevant place they extract some soil, and first of all evaluate it by smell. Not yet satisfied, they put it in a vessel, pour on drinking water and carry out an examination by taste; the taste of the water, after this mixing, shows how the soil will be. For seed crops the soil should be taken at one foot depth; for vines

32 Yaḥya ibn Muḥammad ibn al-ʿAwwām, *Libro de agricultura*, ed. and trans. Josef Antonio Banqueri (Madrid: Ministerio de Agricultura, Pesca y Alimentación, 1988), 9.

33 Aleksandar Shopov, "Between the Pen and the Fields: Books on Farming, Changing Land Regimes, and Urban Agriculture in the Ottoman Eastern Mediterranean ca. 1500–1700" (PhD diss., Harvard University, 2016), 42.

34 Virgil [Publius Vergilius Maro], *Georgics*, trans. Kimberly Johnson (London: Penguin Classics, 2009), 52–53.

three feet; for fruit trees, four feet. Some assume that the soil is sweet on the evidence of club-rush, reed, clover, or bramble growing in it—the same plants that act as evidence for water-finders—but better evidence is the actual taste.

Salty earth must be avoided, so the ancients thought. Since we keep salt out of compost, and experts tell us to apply only *amorge*[35] from unsalted olives to the roots of trees, and we water compost-heaps with sweet and not salty water, it is no surprise that they condemn salty soil, which is indeed unsuitable for anything except date palms: these, however, will flourish and fruit heavily. This is why in our own country, where the soil is entirely salty, the date palm is the only tree that gives a good crop. In salty soil, then, plant only date palms, or give it up, or improve it so far as possible, mixing sweet soil with it in the same way that compost is added.

Bad-smelling soil is wholly useless and must be avoided altogether.[36]

Here are all the elements of the later Persian tradition's evaluation of soil by taste: from the exact method of tasting by retrieving a sample from a certain depth and dissolving it in clean fresh water, to salinity as the key problem to be identified, as well as the suitability of brackish soils for date palm cultivation, and the vaguer admonition that bad-smelling or contaminated soil is to be avoided entirely. Significantly, these are also the key aspects of Cakrapāṇi Miśra's—admittedly much terser—instructions: the true, reliable method of ascertaining the quality of the soil is to taste it instead of relying on visual examination; a salty flavor indicating a possibly harmful level of salinity is the most important characteristic to be found out this way; and in general, soil that is "impure" is best avoided. In the *Geoponika* and the Persian texts, this latter characteristic is specifically linked to bad smell, while the *Viśvavallabha* refers to a more general quality of being inauspicious or polluted. Even so, the parallel appears more than coincidental. It seems likely that Cakrapāṇi Miśra was not just attempting to respond to the increasingly Persianate garden culture of his time and place, tentatively incorporating its ideals of garden design and some temperate-climate Central Asian and Mediterranean plants not previously featured in Sanskrit-language accounts of gardens and gardening. Rather, he most likely had some exposure to the Persian textual tradition that accompanied the new gardening practices, whether directly or through some interlocutor,

35 The remnants of olive oil pressing.

36 Andrew Dalby, trans., *Geoponika: Farm Work; A Modern Translation of the Roman and Byzantine Farming Handbook* (London: Prospect Books, 2011), 80.

with the *Viśvavallabha*'s instructions on tasting soil ultimately deriving from the Greco-Roman and Persian tradition.

7 Conclusion

Of course, as has been discussed, flavor was recognized as a characteristic of soil in Cakrapāṇi Miśra's own Sanskrit tradition long before this encounter. Yet this preexisting indigenous South Asian model of soil assessment was concerned primarily with attributing to the earth a full—if largely theoretical—complement of all of the features that material substances were considered able to possess, specifically a culturally fixed range of colors and potential flavors. The author of the *Viśvavallabha* simultaneously acknowledged this received wisdom and partly rejected it, thus aligning his writing with an alternate tradition that had reached South Asia from Iran and Central Asia, although much of its content ultimately reached back to the Byzantine Mediterranean and even further. That both traditions singled out taste as an important aspect of soil and a way to understand and evaluate it, thereby allowing for this to become a point of convergence and negotiation between the two bodies of knowledge, speaks to how salient this now largely ignored feature once was. Miśra's exhortation to the reader to engage their own sense of taste, although a small detail, reveals shifts in horticultural theory and practice as well as in ways of knowing, both at the time of its writing and over the centuries since then.

Index

Page numbers in *italics* denote illustrations. The prefix *n* refers to the footnotes.

www.ingramcontent.com/pod-product-compliance
Lightning Source LLC
Chambersburg PA
CBHW070216190526
45161CB00002B/95
9789004727489